T0282546

CAMBRIDGE LIBRARY COLLECTION

Books of enduring scholarly value

Darwin

Two hundred years after his birth and 150 years after the publication of 'On the Origin of Species', Charles Darwin and his theories are still the focus of worldwide attention. This series offers not only works by Darwin, but also the writings of his mentors in Cambridge and elsewhere, and a survey of the impassioned scientific, philosophical and theological debates sparked by his 'dangerous idea'.

The Variation of Animals and Plants under Domestication

Charles Darwin (1809–82) first published this work in 1868 in two volumes. The book began as an expansion of the first two chapters of *On the Origin of Species:* 'Variation under Domestication' and 'Variation under Nature',and it developed into one of his largest works; Darwin referred to it as his 'big book'. Volume 1 deals with the variations introduced into species as a result of domestication, through changes in climate, diet, breeding and an absence of predators. He began with an examination of dogs and cats, comparing them with their wild counterparts, and moved on to investigate horses and asses; pigs, cattle, sheep, and goats; domestic rabbits; domestic pigeons; fowl; and finally cultivated plants. The work is a masterpiece of nineteenth-century scientific investigation; it is a key text in the development of Darwin's own thought and of the wider discipline of evolutionary biology.

Cambridge University Press has long been a pioneer in the reissuing of out-of-print titles from its own backlist, producing digital reprints of books that are still sought after by scholars and students but could not be reprinted economically using traditional technology. The Cambridge Library Collection extends this activity to a wider range of books which are still of importance to researchers and professionals, either for the source material they contain, or as landmarks in the history of their academic discipline.

Drawing from the world-renowned collections in the Cambridge University Library, and guided by the advice of experts in each subject area, Cambridge University Press is using state-of-the-art scanning machines in its own Printing House to capture the content of each book selected for inclusion. The files are processed to give a consistently clear, crisp image, and the books finished to the high quality standard for which the Press is recognised around the world. The latest print-on-demand technology ensures that the books will remain available indefinitely, and that orders for single or multiple copies can quickly be supplied.

The Cambridge Library Collection will bring back to life books of enduring scholarly value (including out-of-copyright works originally issued by other publishers) across a wide range of disciplines in the humanities and social sciences and in science and technology.

The Variation of Animals and Plants under Domestication

VOLUME 1

CHARLES DARWIN

CAMBRIDGE UNIVERSITY PRESS

Cambridge, New York, Melbourne, Madrid, Cape Town, Singapore,
São Paolo, Delhi, Dubai, Tokyo

Published in the United States of America by Cambridge University Press, New York

www.cambridge.org
Information on this title: www.cambridge.org/9781108014229

This edition first published 1868
This digitally printed version 2010

ISBN 978-1-108-01422-9 Paperback

THE VARIATION

OF

ANIMALS AND PLANTS

UNDER DOMESTICATION.

BY CHARLES DARWIN, M.A., F.R.S., &c.

IN TWO VOLUMES.—VOL. I.

WITH ILLUSTRATIONS.

LONDON:
JOHN MURRAY, ALBEMARLE STREET.
1868.

BY THE SAME AUTHOR.

ON THE ORIGIN OF SPECIES BY MEANS OF NATURAL SELECTION; or The Preservation of Favoured Races in the Struggle for Life. Fourth Edition (*Eighth Thousand*), with Additions and Corrections. 1866. Murray.

A NATURALIST'S VOYAGE ROUND THE WORLD; or, A Journal of Researches into the Natural History and Geology of the Countries visited during the Voyage of H.M.S. Beagle, under the Command of Capt. Fitz-Roy, R.N. *Tenth Thousand.* Murray.

ON THE STRUCTURE AND DISTRIBUTION OF CORAL REEFS. Smith, Elder, & Co.

GEOLOGICAL OBSERVATIONS ON VOLCANIC ISLANDS. Smith, Elder, & Co.

GEOLOGICAL OBSERVATIONS ON SOUTH AMERICA. Smith, Elder, & Co.

A MONOGRAPH OF THE CIRRIPEDIA. With numerous Illustrations. 2 vols. 8vo. Hardwicke.

ON THE VARIOUS CONTRIVANCES BY WHICH BRITISH AND FOREIGN ORCHIDS ARE FERTILISED BY INSECTS; and on the Good Effects of Crossing. With numerous Woodcuts. Murray.

ON THE MOVEMENTS and HABITS of CLIMBING PLANTS. With Woodcuts. Williams & Norgate.

CONTENTS OF VOLUME I.

INTRODUCTION .. .. Page 1

CHAPTER I.

DOMESTIC DOGS AND CATS.

ANCIENT VARIETIES OF THE DOG — RESEMBLANCE OF DOMESTIC DOGS IN VARIOUS COUNTRIES TO NATIVE CANINE SPECIES — ANIMALS NOT ACQUAINTED WITH MAN AT FIRST FEARLESS — DOGS RESEMBLING WOLVES AND JACKALS — HABIT OF BARKING ACQUIRED AND LOST — FERAL DOGS — TAN-COLOURED EYE-SPOTS — PERIOD OF GESTATION — OFFENSIVE ODOUR — FERTILITY OF THE RACES WHEN CROSSED — DIFFERENCES IN THE SEVERAL RACES IN PART DUE TO DESCENT FROM DISTINCT SPECIES — DIFFERENCES IN THE SKULL AND TEETH — DIFFERENCES IN THE BODY, IN CONSTITUTION — FEW IMPORTANT DIFFERENCES HAVE BEEN FIXED BY SELECTION — DIRECT ACTION OF CLIMATE — WATER-DOGS WITH PALMATED FEET — HISTORY OF THE CHANGES WHICH CERTAIN ENGLISH RACES OF THE DOG HAVE GRADUALLY UNDERGONE THROUGH SELECTION — EXTINCTION OF THE LESS IMPROVED SUB-BREEDS.

CATS, CROSSED WITH SEVERAL SPECIES — DIFFERENT BREEDS FOUND ONLY IN SEPARATED COUNTRIES — DIRECT EFFECTS OF THE CONDITIONS OF LIFE — FERAL CATS — INDIVIDUAL VARIABILITY 15

CHAPTER II.

HORSES AND ASSES.

HORSE. — DIFFERENCES IN THE BREEDS — INDIVIDUAL VARIABILITY OF — DIRECT EFFECTS OF THE CONDITIONS OF LIFE — CAN WITHSTAND MUCH COLD — BREEDS MUCH MODIFIED BY SELECTION — COLOURS OF THE HORSE — DAPPLING — DARK STRIPES ON THE SPINE, LEGS, SHOULDERS, AND FOREHEAD — DUN-COLOURED HORSES MOST FREQUENTLY STRIPED — STRIPES PROBABLY DUE TO REVERSION TO THE PRIMITIVE STATE OF THE HORSE.

ASSES. — BREEDS OF — COLOUR OF — LEG- AND SHOULDER- STRIPES — SHOULDER-STRIPES SOMETIMES ABSENT, SOMETIMES FORKED 49

CHAPTER III.

PIGS — CATTLE — SHEEP — GOATS.

PIGS BELONG TO TWO DISTINCT TYPES, SUS SCROFA AND INDICA — TORF-SCHWEIN — JAPAN PIG — FERTILITY OF CROSSED PIGS — CHANGES IN THE SKULL OF THE HIGHLY CULTIVATED RACES — CONVERGENCE OF CHARACTER — GESTATION — SOLID-HOOFED SWINE — CURIOUS APPENDAGES TO THE JAWS — DECREASE IN SIZE OF THE TUSKS — YOUNG PIGS LONGITUDINALLY STRIPED — FERAL PIGS — CROSSED BREEDS.

CATTLE. — ZEBU A DISTINCT SPECIES — EUROPEAN CATTLE PROBABLY DESCENDED FROM THREE WILD FORMS — ALL THE RACES NOW FERTILE TOGETHER — BRITISH

PARK CATTLE — ON THE COLOUR OF THE ABORIGINAL SPECIES — CONSTITUTIONAL DIFFERENCES — SOUTH AFRICAN RACES — SOUTH AMERICAN RACES — NIATA CATTLE — ORIGIN OF THE VARIOUS RACES OF CATTLE.

SHEEP. — REMARKABLE RACES OF — VARIATIONS ATTACHED TO THE MALE SEX — ADAPTATIONS TO VARIOUS CONDITIONS — GESTATION OF — CHANGES IN THE WOOL —SEMI-MONSTROUS BREEDS.

GOATS. — REMARKABLE VARIATIONS OF .. .. Page 65

CHAPTER IV.

DOMESTIC RABBITS.

DOMESTIC RABBITS DESCENDED FROM THE COMMON WILD RABBIT — ANCIENT DOMESTICATION — ANCIENT SELECTION — LARGE LOP-EARED RABBITS — VARIOUS BREEDS — FLUCTUATING CHARACTERS — ORIGIN OF THE HIMALAYAN BREED — CURIOUS CASE OF INHERITANCE — FERAL RABBITS IN JAMAICA AND THE FALKLAND ISLANDS — PORTO SANTO FERAL RABBITS — OSTEOLOGICAL CHARACTERS — SKULL — SKULL OF HALF-LOP RABBITS — VARIATIONS IN THE SKULL ANALOGOUS TO DIFFERENCES IN DIFFERENT SPECIES OF HARES — VERTEBRÆ — STERNUM — SCAPULA — EFFECTS OF USE AND DISUSE ON THE PROPORTIONS OF THE LIMBS AND BODY — CAPACITY OF THE SKULL AND REDUCED SIZE OF THE BRAIN — SUMMARY ON THE MODIFICATIONS OF DOMESTICATED RABBITS 103

CHAPTER V.

DOMESTIC PIGEONS.

ENUMERATION AND DESCRIPTION OF THE SEVERAL BREEDS — INDIVIDUAL VARIABILITY — VARIATIONS OF A REMARKABLE NATURE — OSTEOLOGICAL CHARACTERS: SKULL, LOWER JAW, NUMBER OF VERTEBRÆ — CORRELATION OF GROWTH: TONGUE WITH BEAK; EYELIDS AND NOSTRILS WITH WATTLED SKIN — NUMBER OF WING-FEATHERS, AND LENGTH OF WING — COLOUR AND DOWN — WEBBED AND FEATHERED FEET — ON THE EFFECTS OF DISUSE — LENGTH OF FEET IN CORRELATION WITH LENGTH OF BEAK — LENGTH OF STERNUM, SCAPULA, AND FURCULA — LENGTH OF WINGS — SUMMARY ON THE POINTS OF DIFFERENCE IN THE SEVERAL BREEDS 131

CHAPTER VI.

PIGEONS—*continued.*

ON THE ABORIGINAL PARENT-STOCK OF THE SEVERAL DOMESTIC RACES — HABITS OF LIFE — WILD RACES OF THE ROCK-PIGEON — DOVECOT-PIGEONS — PROOFS OF THE DESCENT OF THE SEVERAL RACES FROM COLUMBA LIVIA — FERTILITY OF THE RACES WHEN CROSSED — REVERSION TO THE PLUMAGE OF THE WILD ROCK-PIGEON — CIRCUMSTANCES FAVOURABLE TO THE FORMATION OF THE RACES — ANTIQUITY AND HISTORY OF THE PRINCIPAL RACES — MANNER OF THEIR FORMATION — SELECTION — UNCONSCIOUS SELECTION — CARE TAKEN BY FANCIERS IN SELECTING THEIR BIRDS — SLIGHTLY DIFFERENT STRAINS GRADUALLY CHANGE INTO WELL-MARKED BREEDS — EXTINCTION OF INTERMEDIATE FORMS — CERTAIN BREEDS REMAIN PERMANENT, WHILST OTHERS CHANGE — SUMMARY 180

CHAPTER VII.

FOWLS.

BRIEF DESCRIPTIONS OF THE CHIEF BREEDS — ARGUMENTS IN FAVOUR OF THEIR DESCENT FROM SEVERAL SPECIES — ARGUMENTS IN FAVOUR OF ALL THE BREEDS HAVING DESCENDED FROM GALLUS BANKIVA — REVERSION TO THE PARENT-STOCK IN COLOUR — ANALOGOUS VARIATIONS — ANCIENT HISTORY OF THE FOWL — EXTERNAL DIFFERENCES BETWEEN THE SEVERAL BREEDS — EGGS — CHICKENS — SECONDARY SEXUAL CHARACTERS — WING- AND TAIL- FEATHERS, VOICE, DISPOSITION, ETC. — OSTEOLOGICAL DIFFERENCES IN THE SKULL, VERTEBRÆ, ETC. — EFFECTS OF USE AND DISUSE ON CERTAIN PARTS — CORRELATION OF GROWTH Page 225

CHAPTER VIII.

DUCKS — GOOSE — PEACOCK — TURKEY — GUINEA-FOWL — CANARY-BIRD — GOLD-FISH — HIVE-BEES — SILK-MOTHS.

DUCKS, SEVERAL BREEDS OF — PROGRESS OF DOMESTICATION — ORIGIN OF, FROM THE COMMON WILD-DUCK — DIFFERENCES IN THE DIFFERENT BREEDS — OSTEOLOGICAL DIFFERENCES — EFFECTS OF USE AND DISUSE ON THE LIMB-BONES.

GOOSE, ANCIENTLY DOMESTICATED — LITTLE VARIATION OF — SEBASTOPOL BREED.

PEACOCK, ORIGIN OF BLACK-SHOULDERED BREED.

TURKEY, BREEDS OF — CROSSED WITH THE UNITED STATES SPECIES — EFFECTS OF CLIMATE ON.

GUINEA-FOWL, CANARY-BIRD, GOLD-FISH, HIVE-BEES.

SILK-MOTHS, SPECIES AND BREEDS OF — ANCIENTLY DOMESTICATED — CARE IN THEIR SELECTION—DIFFERENCES IN THE DIFFERENT RACES—IN THE EGG, CATERPILLAR, AND COCOON STATES — INHERITANCE OF CHARACTERS — IMPERFECT WINGS — LOST INSTINCTS — CORRELATED CHARACTERS 276

CHAPTER IX.

CULTIVATED PLANTS: CEREAL AND CULINARY PLANTS.

PRELIMINARY REMARKS ON THE NUMBER AND PARENTAGE OF CULTIVATED PLANTS — FIRST STEPS IN CULTIVATION — GEOGRAPHICAL DISTRIBUTION OF CULTIVATED PLANTS.

CEREALIA. — DOUBTS ON THE NUMBER OF SPECIES. —— WHEAT: VARIETIES OF — INDIVIDUAL VARIABILITY — CHANGED HABITS — SELECTION — ANCIENT HISTORY OF THE VARIETIES. —— MAIZE: GREAT VARIATION OF — DIRECT ACTION OF CLIMATE ON.

CULINARY PLANTS. — CABBAGES: VARIETIES OF, IN FOLIAGE AND STEMS, BUT NOT IN OTHER PARTS — PARENTAGE OF — OTHER SPECIES OF BRASSICA. —— PEAS: AMOUNT OF DIFFERENCE IN THE SEVERAL KINDS, CHIEFLY IN THE PODS AND SEED — SOME VARIETIES CONSTANT, SOME HIGHLY VARIABLE — DO NOT INTERCROSS. —— BEANS. —— POTATOES: NUMEROUS VARIETIES OF — DIFFERING LITTLE, EXCEPT IN THE TUBERS — CHARACTERS INHERITED 305

CHAPTER X.

PLANTS *continued* — FRUITS — ORNAMENTAL TREES — FLOWERS.

FRUITS. — GRAPES — VARY IN ODD AND TRIFLING PARTICULARS. —— MULBERRY. —— THE ORANGE GROUP — SINGULAR RESULTS FROM CROSSING. —— PEACH AND NECTARINE — BUD-VARIATION — ANALOGOUS VARIATION — RELATION TO THE ALMOND. —— APRICOT. —— PLUMS — VARIATION IN THEIR STONES. —— CHERRIES — SINGULAR VARIETIES OF. —— APPLE. —— PEAR. —— STRAWBERRY — INTERBLENDING OF THE ORIGINAL FORMS. —— GOOSEBERRY — STEADY INCREASE IN SIZE OF THE FRUIT — VARIETIES OF. —— WALNUT. —— NUT. —— CUCURBITACEOUS PLANTS — WONDERFUL VARIATION OF.

ORNAMENTAL TREES — THEIR VARIATION IN DEGREE AND KIND — ASH-TREE — SCOTCH-FIR — HAWTHORN.

FLOWERS — MULTIPLE ORIGIN OF MANY KINDS — VARIATION IN CONSTITUTIONAL PECULIARITIES — KIND OF VARIATION. —— ROSES — SEVERAL SPECIES CULTIVATED. —— PANSY. —— DAHLIA. —— HYACINTH, HISTORY AND VARIATION OF

Page 332

CHAPTER XI.

ON BUD-VARIATION, AND ON CERTAIN ANOMALOUS MODES OF REPRODUCTION AND VARIATION.

BUD-VARIATIONS IN THE PEACH, PLUM, CHERRY, VINE, GOOSEBERRY, CURRANT, AND BANANA, AS SHOWN BY THE MODIFIED FRUIT — IN FLOWERS: CAMELLIAS, AZALEAS, CHRYSANTHEMUMS, ROSES, ETC. — ON THE RUNNING OF THE COLOUR IN CARNATIONS — BUD-VARIATIONS IN LEAVES — VARIATIONS BY SUCKERS, TUBERS, AND BULBS — ON THE BREAKING OF TULIPS — BUD-VARIATIONS GRADUATE INTO CHANGES CONSEQUENT ON CHANGED CONDITIONS OF LIFE — CYTISUS ADAMI, ITS ORIGIN AND TRANSFORMATION — ON THE UNION OF TWO DIFFERENT EMBRYOS IN ONE SEED — THE TRIFACIAL ORANGE — ON REVERSION BY BUDS IN HYBRIDS AND MONGRELS — ON THE PRODUCTION OF MODIFIED BUDS BY THE GRAFTING OF ONE VARIETY OR SPECIES ON ANOTHER — ON THE DIRECT OR IMMEDIATE ACTION OF FOREIGN POLLEN ON THE MOTHER-PLANT — ON THE EFFECTS IN FEMALE ANIMALS OF A FIRST IMPREGNATION ON THE SUBSEQUENT OFFSPRING — CONCLUSION AND SUMMARY 373

ERRATA.

Vol. I., p. 76, 13 lines from top, *for* Phascochœrus *read* Phacochœrus.

" " 104, transpose footnotes 4 and 5.

" " 275, 7 lines from bottom, *for* Amherstii *read* Amherstiæ.

" " 282, 14 lines from top, *for* Tadorna Ægyptiaca *read* Anser Ægyptiacus.

" " 286, in bottom line in the Table, *for* 713 *read* 717.

LIST OF ILLUSTRATIONS.

PAGE

1. Dun Devonshire Pony, with shoulder, spinal, and leg stripes .. 56
2. Head of Japan or Masked Pig .. 69
3. Head of Wild Boar, and of "Golden Days," a pig of the Yorkshire large breed .. 72
4. Old Irish Pig, with jaw-appendages .. 75
5. Half-lop Rabbit .. 108
6. Skull of Wild Rabbit .. 117
7. Skull of large Lop-eared Rabbit .. 117
8. Part of Zygomatic Arch, showing the projecting end of the malar-bone, and the auditory meatus, of Rabbits .. 118
9. Posterior end of Skull, showing the inter-parietal bone, of Rabbits .. 118
10. Occipital Foramen of Rabbits .. 118
11. Skull of Half-lop Rabbit .. 119
12. Atlas Vertebræ of Rabbits .. 121
13. Third Cervical Vertebræ of Rabbits .. 121
14. Dorsal Vertebræ, from sixth to tenth inclusive, of Rabbits .. 122
15. Terminal Bone of Sternum of Rabbits .. 123
16. Acromion of Scapula of Rabbits .. 123
17. The Rock-Pigeon, or Columbia Livia .. 135
18. English Pouter .. 137
19. English Carrier .. 140
20. English Barb .. 145
21. English Fantail .. 147
22. African Owl .. 149
23. Short-faced English Tumbler .. 152
24. Skulls of Pigeons, viewed laterally .. 163
25. Lower Jaws of Pigeons, seen from above .. 164
26. Skull of Runt, seen from above .. 165
27. Lateral view of Jaws of Pigeons .. 165
28. Scapulæ of Pigeons .. 167
29. Furculæ of Pigeons .. 167
30. Spanish Fowl .. 226
31. Hamburgh Fowl .. 228
32. Polish Fowl .. 229
33. Occipital Foramen of the Skulls of Fowls .. 261

PAGE

34. Skulls of Fowls, viewed from above, a little obliquely 262
35. Longitudinal sections of Skulls of Fowls, viewed laterally .. 263
36. Skull of Horned Fowl, viewed from above, a little obliquely .. 265
37. Sixth Cervical Vertebræ of Fowls, viewed laterally 267
38. Extremity of the Furcula of Fowls, viewed laterally 268
39. Skulls of Ducks, viewed laterally, reduced to two-thirds of the natural size 282
40. Cervical Vertebræ of Ducks, of natural size 283
41. Pods of the Common Pea 328
42. Peach and Almond Stones, of natural size, viewed edgeways .. 337
43. Plum Stones, of natural size, viewed laterally 345

THE

VARIATION OF ANIMALS AND PLANTS

UNDER DOMESTICATION.

INTRODUCTION.

The object of this work is not to describe all the many races of animals which have been domesticated by man, and of the plants which have been cultivated by him; even if I possessed the requisite knowledge, so gigantic an undertaking would be here superfluous. It is my intention to give under the head of each species only such facts as I have been able to collect or observe, showing the amount and nature of the changes which animals and plants have undergone whilst under man's dominion, or which bear on the general principles of variation. In one case alone, namely in that of the domestic pigeon, I will describe fully all the chief races, their history, the amount and nature of their differences, and the probable steps by which they have been formed. I have selected this case, because, as we shall hereafter see, the materials are better than in any other; and one case fully described will in fact illustrate all others. But I shall also describe domesticated rabbits, fowls, and ducks, with considerable fullness.

The subjects discussed in this volume are so connected that it is not a little difficult to decide how they can be best arranged. I have determined in the first part to give, under the heads of the various animals and plants, a large body of facts, some of which may at first appear but little related to our subject, and to devote the latter part to general discussions. Whenever I have found it necessary to give numerous details, in support of any proposition or conclusion, small type has been used. The reader will, I think, find this plan a convenience, for, if he does not doubt the conclusion or care about the details, he can easily pass them over; yet I may be permitted to say that some of the discussions thus printed deserve attention, at least from the professed naturalist.

It may be useful to those who have read nothing about Natural Selection, if I here give a brief sketch of the whole subject and of its bearing on the origin of species.[1] This is the more desirable, as it is impossible in the present work to avoid many allusions to questions which will be fully discussed in future volumes.

From a remote period, in all parts of the world, man has subjected many animals and plants to domestication or culture. Man has no power of altering the absolute conditions of life; he cannot change the climate of any country; he adds no new element to the soil; but he can remove an animal or plant from one climate or soil to another, and give it food on which it did not subsist in its natural state. It is an error to speak of man "tampering with nature" and causing variability. If organic beings had not possessed an inherent tendency to vary, man could have done nothing.[2] He unintentionally exposes his animals and plants to various conditions of life, and variability supervenes, which he cannot even prevent or check. Consider the simple case of a plant which has been cultivated during a long time in its native country, and which consequently has not been subjected to any change of climate. It has been protected to a certain extent from the competing roots of plants of other kinds; it has generally been grown in manured soil, but probably not richer than that of many an alluvial flat; and lastly, it has been exposed to changes in its conditions, being grown sometimes in one district and sometimes in another, in different soils. Under such circumstances,

[1] To any one who has attentively read my 'Origin of Species' this Introduction will be superfluous. As I stated in that work that I should soon publish the facts on which the conclusions given in it were founded, I here beg permission to remark that the great delay in publishing this first work has been caused by continued ill-health.

[2] M. Pouchet has recently ('Plurality of Races,' Eng. Translat., 1864, p. 83, &c.) insisted that variation under domestication throws no light on the natural modification of species. I cannot perceive the force of his arguments, or, to speak more accurately, of his assertions to this effect.

scarcely a plant can be named, though cultivated in the rudest manner, which has not given birth to several varieties. It can hardly be maintained that during the many changes which this earth has undergone, and during the natural migrations of plants from one land or island to another, tenanted by different species, that such plants will not often have been subjected to changes in their conditions analogous to those which almost inevitably cause cultivated plants to vary. No doubt man selects varying individuals, sows their seeds, and again selects their varying offspring. But the initial variation on which man works, and without which he can do nothing, is caused by slight changes in the conditions of life, which must often have occurred under nature. Man, therefore, may be said to have been trying an experiment on a gigantic scale; and it is an experiment which nature during the long lapse of time has incessantly tried. Hence it follows that the principles of domestication are important for us. The main result is that organic beings thus treated have varied largely, and the variations have been inherited. This has apparently been one chief cause of the belief long held by some few naturalists that species in a state of nature undergo change.

I shall in this volume treat, as fully as my materials permit, the whole subject of variation under domestication. We may thus hope to obtain some light, little though it be, on the causes of variability,—on the laws which govern it, such as the direct action of climate and food, the effects of use and disuse, and of correlation of growth,—and on the amount of change to which domesticated organisms are liable. We shall learn something on the laws of inheritance, on the effects of crossing different breeds, and on that sterility which often supervenes when organic beings are removed from their natural conditions of life, and likewise when they are too closely interbred. During this investigation we shall see that the principle of Selection is all important. Although man does not cause variability and cannot even prevent it, he can select, preserve, and accumulate the variations given to him by the hand of nature in any way which he chooses; and thus he can certainly produce a great result. Selection may be followed either methodically and intentionally, or unconsciously and unintentionally. Man may select and preserve each successive variation, with the distinct intention of improving and altering a breed, in accordance with a preconceived idea; and by thus adding up variations, often so slight as to be imperceptible by an uneducated eye, he has effected wonderful changes and improvements. It can, also, be clearly shown that man, without any intention or thought of improving the breed, by preserving in each successive generation the individuals which he prizes most, and by destroying the worthless individuals, slowly, though surely, induces great changes. As the will of man thus comes into play, we can understand how it is that domesticated breeds show adaptation to his wants and pleasures. We can further understand how it is that domestic races of animals and cultivated races of plants often exhibit an abnormal character, as compared with natural species; for they have been modified not for their own benefit, but for that of man.

In a second work I shall discuss the variability of organic beings in a state of nature; namely, the individual differences presented by animals and plants, and those slightly greater and generally inherited differences which are ranked by naturalists as varieties or geographical races. We shall see how difficult, or rather how impossible it often is, to distinguish between races and sub-species, as the less well-marked forms have sometimes been denominated; and again between sub-species and true species. I shall further attempt to show that it is the common and widely ranging, or, as they may be called, the dominant species, which most frequently vary; and that it is the large and flourishing genera which include the greatest number of varying species. Varieties, as we shall see, may justly be called incipient species.

But it may be urged, granting that organic beings in a state of nature present some varieties,—that their organization is in some slight degree plastic; granting that many animals and plants have varied greatly under domestication, and that man by his power of selection has gone on accumulating such variations until he has made strongly marked and firmly inherited races; granting all this, how, it may be asked, have species arisen in a state of nature? The differences between natural varieties are slight; whereas the differences are considerable between the species of the same genus, and great between the species of distinct genera. How do these lesser differences become augmented into the greater difference? How do varieties, or as I have called them incipient species, become converted into true and well-defined species? How has each new species been adapted to the surrounding physical conditions, and to the other forms of life on which it in any way depends? We see on every side of us innumerable adaptations and contrivances, which have justly excited in the mind of every observer the highest admiration. There is, for instance, a fly (Cecidomyia)[3] which deposits its eggs within the stamens of a Scrophularia, and secretes a poison which produces a gall, on which the larva feeds; but there is another insect (Misocampus) which deposits its eggs within the body of the larva within the gall, and is thus nourished by its living prey; so that here a hymenopterous insect depends on a dipterous insect, and this depends on its power of producing a monstrous growth in a particular organ of a particular plant. So it is, in a more or less plainly marked manner, in thousands and tens of thousands of cases, with the lowest as well as with the highest productions of nature.

This problem of the conversion of varieties into species,—that is, the augmentation of the slight differences characteristic of varieties into the greater differences characteristic of species and genera, including the admirable adaptations of each being to its complex organic and inorganic conditions of life,—will form the main subject of my second work. We shall therein see that all organic beings, without exception, tend to increase at so high a ratio, that no district, no station, not even the whole surface of the land or the whole ocean, would hold the progeny of a single pair after a certain number of generations. The inevitable result is an ever-recurrent Struggle for Existence. It has truly been said that all nature is at war; the strongest ultimately prevail, the weakest fail; and we well know that myriads of forms have disappeared from the face of the earth. If then organic beings in a state of nature vary even in a slight degree, owing to changes in the surrounding

[3] Léon Dufour in 'Annales des Scienc. Nat.' (3rd series, Zoolog.), tom. v. p. 6.

conditions, of which we have abundant geological evidence, or from any other cause; if, in the long course of ages, inheritable variations ever arise in any way advantageous to any being under its excessively complex and changing relations of life; and it would be a strange fact if beneficial variations did never arise, seeing how many have arisen which man has taken advantage of for his own profit or pleasure; if then these contingencies ever occur, and I do not see how the probability of their occurrence can be doubted, then the severe and often-recurrent struggle for existence will determine that those variations, however slight, which are favourable shall be preserved or selected, and those which are unfavourable shall be destroyed.

This preservation, during the battle for life, of varieties which possess any advantage in structure, constitution, or instinct, I have called Natural Selection; and Mr. Herbert Spencer has well expressed the same idea by the Survival of the Fittest. The term "natural selection" is in some respects a bad one, as it seems to imply conscious choice; but this will be disregarded after a little familiarity. No one objects to chemists speaking of "elective affinity;" and certainly an acid has no more choice in combining with a base, than the conditions of life have in determining whether or not a new form be selected or preserved. The term is so far a good one as it brings into connection the production of domestic races by man's power of selection, and the natural preservation of varieties and species in a state of nature. For brevity sake I sometimes speak of natural selection as an intelligent power;—in the same way as astronomers speak of the attraction of gravity as ruling the movements of the planets, or as agriculturists speak of man making domestic races by his power of selection. In the one case, as in the other, selection does nothing without variability, and this depends in some manner on the action of the surrounding circumstances on the organism. I have, also, often personified the word Nature; for I have found it difficult to avoid this ambiguity; but I mean by nature only the aggregrate action and product of many natural laws,—and by laws only the ascertained sequence of events.

In the chapter devoted to natural selection I shall show from experiment and from a multitude of facts, that the greatest amount of life can be supported on each spot by great diversification or divergence in the structure and constitution of its inhabitants. We shall, also, see that the continued production of new forms through natural selection, which implies that each new variety has some advantage over others, almost inevitably leads to the extermination of the older and less improved forms. These latter are almost necessarily intermediate in structure as well as in descent between the last-produced forms and their original parent-species. Now, if we suppose a species to produce two or more varieties, and these in the course of time to produce other varieties, the principle of good being derived from diversification of structure will generally lead to the preservation of the most divergent varieties; thus the lesser differences characteristic of varieties come to be augmented into the greater differences characteristic of species, and, by the extermination of the older intermediate forms, new species come to be distinctly defined objects. Thus, also, we shall see how it is that organic beings can be classed by what is called a natural method in distinct groups—species under genera, and genera under families.

As all the inhabitants of each country may be said, owing to their high rate of reproduction, to be striving to increase in numbers; as each form is related to many other forms in the struggle for life,—for destroy any one and its place will be seized by others; as every part of the organization occasionally varies in some slight degree, and as natural selection acts exclusively by the preservation of variations which are advantageous under the excessively complex conditions to which each being is exposed, no limit exists to the number, singularity, and perfection of the contrivances and co-adaptations which may thus be produced. An animal or a plant may thus slowly become related in its structure and habits in the most intricate manner to many other animals and plants, and to the physical conditions of its home. Variations in the organization will in some cases be aided by habit, or by the use and disuse of parts, and they will be governed by the direct action of the surrounding physical conditions and by correlation of growth.

On the principles here briefly sketched out, there is no innate or necessary tendency in each being to its own advancement in the scale of organization. We are almost compelled to look at the specialization or differentiation of parts or organs for different functions as the best or even sole standard of advancement; for by such division of labour each function of body and mind is better performed. And, as natural selection acts exclusively through the preservation of profitable modifications of structure, and as the conditions of life in each area generally become more and more complex, from the increasing number of different forms which inhabit it and from most of these forms acquiring a more and more perfect structure, we may confidently believe, that, on the whole, organization advances. Nevertheless a very simple form fitted for very simple conditions of life might remain for indefinite ages unaltered or unimproved; for what would it profit an infusorial animalcule, for instance, or an intestinal worm, to become highly organized? Members of a high group might even become, and this apparently has occurred, fitted for simpler conditions of life; and in this case natural selection would tend to simplify or degrade the organization, for complicated mechanism for simple actions would be useless or even disadvantageous.

In a second work, after treating of the Variation of organisms in a state of nature, of the Struggle for Existence and the principle of Natural Selection, I shall discuss the difficulties which are opposed to the theory. These difficulties may be classed under the following heads:—the apparent impossibility in some cases of a very simple organ graduating by small steps into a highly perfect organ; the marvellous facts of Instinct; the whole question of Hybridity; and, lastly, the absence, at the present time and in our geological formations, of innumerable links connecting all allied species. Although some of these difficulties are of great weight, we shall see that many of them are explicable on the theory of natural selection, and are otherwise inexplicable.

In scientific investigations it is permitted to invent any hypothesis, and if it explains various large and independent classes of facts it rises to the rank of a well-grounded theory. The undulations of the ether and even its existence are hypothetical, yet every one now admits the undulatory theory of light. The principle of natural selection may be looked at as a mere hypothesis, but rendered in some degree probable by what we positively know of the variability of organic beings in a state of nature,—by what we positively know of the struggle for existence, and the consequent almost inevitable preservation of favourable variations,—and from the analogical formation of domestic races. Now this hypothesis may be tested,—and this seems to me the only fair and legitimate manner of considering the whole question,—by trying whether it explains several large and independent classes of facts; such as the geological succession of organic beings, their distribution in past and present times, and their mutual affinities and homologies. If the principle of natural selection does explain these and other large bodies of facts, it ought to be received. On the ordinary view of each species having been independently created, we gain no scientific explanation of any one of these facts. We can only say that it has so pleased the Creator to command that the past and present inhabitants of the world should appear in a certain order and in certain areas; that He has impressed on them the most extraordinary resemblances, and has classed them in groups subordinate to groups. But by such statements we gain no new knowledge; we do not connect together facts and laws; we explain nothing.

In a third work I shall try the principle of natural selection by seeing how far it will give a fair explanation of the several classes of facts just alluded to. It was the consideration of these facts which first led me to take up the present subject. When I visited, during the voyage of H.M.S. *Beagle*, the Galapagos Archipelago, situated in the Pacific Ocean about 500 miles from the shore of South America, I found myself surrounded by peculiar species of birds, reptiles, and plants, existing nowhere else in the world. Yet they nearly all bore an American stamp. In the song of the mocking-thrush, in the harsh cry of the carrion-hawk, in the great candlestick-like opuntias, I clearly perceived the neighbourhood of America, though the islands were separated by so many miles of ocean from the mainland, and differed much from it in their geological constitution and climate. Still more surprising was the fact that most of the inhabitants of each separate island in this small archipelago were specifically different, though most closely related to each other. The archipelago, with its innumerable craters and bare streams of lava, appeared to be of recent origin; and thus I fancied myself brought near to the very act of creation. I often asked myself how these many peculiar animals and plants had been produced: the simplest answer seemed to be that the inhabitants of the several islands had descended from each other, undergoing modification in the course of their descent; and that all the inhabitants of the archipelago had descended from those of the nearest land, namely America, whence colonists would naturally have been derived. But it long remained to me an inexplicable problem how the necessary degree of modification could have been effected, and it would have thus remained for ever, had I not studied domestic productions, and thus acquired a just idea of the power of Selection. As soon as I had fully realized this idea, I saw, on reading Malthus on Population, that Natural Selection was the inevitable result of the rapid increase of all organic beings; for I was prepared to appreciate the struggle for existence by having long studied the habits of animals.

Before visiting the Galapagos I had collected many animals whilst travelling from north to south on both sides of America, and everywhere, under conditions of life as different as it is possible to conceive, American forms were met with—species replacing species of the same peculiar genera. Thus it was when the Cordilleras were ascended, or the thick tropical forests penetrated, or the fresh waters of America searched. Subsequently I visited other countries, which in all the conditions of life were incomparably more like to parts of South America, than the different parts of that continent were to each other; yet in these countries, as in Australia or Southern Africa, the traveller cannot fail to be struck with the entire difference of their productions. Again the reflection was forced on me that community of descent from the early inhabitants or colonists of South America would alone explain the wide prevalence of American types of structure throughout that immense area.

To exhume with one's own hands the bones of extinct and

gigantic quadrupeds brings the whole question of the succession of species vividly before one's mind; and I had found in South America great pieces of tesselated armour exactly like, but on a magnificent scale, that covering the pigmy armadillo; I had found great teeth like those of the living sloth, and bones like those of the cavy. An analogous succession of allied forms had been previously observed in Australia. Here then we see the prevalence, as if by descent, in time as in space, of the same types in the same areas; and in neither case does the similarity of the conditions by any means seem sufficient to account for the similarity of the forms of life. It is notorious that the fossil remains of closely consecutive formations are closely allied in structure, and we can at once understand the fact if they are likewise closely allied by descent. The succession of the many distinct species of the same genus throughout the long series of geological formations seems to have been unbroken or continuous. New species come in gradually one by one. Ancient and extinct forms of life often show combined or intermediate characters, like the words of a dead language with respect to its several offshoots or living tongues. All these and other such facts seemed to me to point to descent with modification as the method of production of new groups of species.

The innumerable past and present inhabitants of the world are connected together by the most singular and complex affinities, and can be classed in groups under groups, in the same manner as varieties can be classed under species and sub-varieties under varieties, but with much higher grades of difference. It will be seen in my third work that these complex affinities and the rules for classification receive a rational explanation on the principle of descent, together with modifications acquired through natural selection, entailing divergence of character and the extinction of intermediate forms. How inexplicable is the similar pattern of the hand of a man, the foot of a dog, the wing of a bat, the flipper of a seal, on the doctrine of independent acts of creation! how simply explained on the principle of the natural selection of successive slight variations in the diverging descendants from a single progenitor! So it is, if we look to the structure of an individual animal or plant, when we see the fore and hind limbs, the skull and vertebræ, the jaws and legs of a crab, the petals, stamens, and pistils of a flower, built on the same type or pattern. During the many changes to which in the course of time all organic beings have been subjected, certain organs or parts have occasionally become at first of little use and ultimately superfluous; and the retention of such parts in a rudimentary and utterly useless condition can, on the descent-theory, be simply understood. On the principle of modifications being inherited at the same age in the child, at which each successive variation first appeared in the parent, we shall see why rudimentary parts and organs are generally well developed in the individual at a very early age. On the same principle of inheritance at corresponding ages, and on the principle of variations not generally supervening at a very early period of embryonic growth (and both these principles can be shown to be probable from direct evidence), that most wonderful fact in the whole round of natural history, namely, the similarity of members of the same great class in their embryonic condition,—the embryo, for instance, of a mammal, bird, reptile, and fish being barely distinguishable,—becomes simply intelligible.

It is the consideration and explanation of such facts as these which has convinced me that the theory of descent with modification by means of natural selection is in the main true. These facts have as yet received no explanation on the theory of independent Creations; they cannot be grouped together under one point of view, but each has to be considered as an ultimate fact. As the first origin of life on this earth, as well as the continued life of each individual, is at present quite beyond the scope of science, I do not wish to lay much stress on the greater simplicity of the view of a few forms, or of only one form, having been originally created, instead of innumerable miraculous creations having been necessary at innumerable periods; though this more simple view accords well with Maupertuis's philosophical axiom "of least action."

In considering how far the theory of natural selection may be extended,—that is, in determining from how many progenitors the inhabitants of the world have descended,—we may conclude that at least all the members of the same class have descended from a single ancestor. A number of organic beings are included in the same class, because they present, independently of their habits of life, the same fundamental type of structure, and because they graduate into each other. Moreover, members of the same class can in most cases be shown to be closely alike at an early embryonic age. These facts can be explained on the belief of their descent from a common form; therefore it may be safely admitted that all the members of the same class have descended from one progenitor. But as the members of quite distinct classes have something in common in structure and much in common in constitution, analogy and the simplicity of the view would lead us one step further, and to infer as probable that all living creatures have descended from a single prototype.

I hope that the reader will pause before coming to any final and hostile conclusion on the theory of natural selection. It is the facts and views to be hereafter given which have convinced me of the truth of the theory. The reader may consult my 'Origin of Species,' for a general sketch of the whole subject; but in that work he has to take many statements on trust. In considering the theory of natural selection, he will assuredly meet with weighty difficulties, but these difficulties relate chiefly to subjects—such as the degree of perfection of the geological record, the means of distribution, the possibility of transitions in organs, &c.—on which we are confessedly ignorant; nor do we know how ignorant we are. If we are much more ignorant than is generally supposed, most of these difficulties wholly disappear. Let the reader reflect on the difficulty of looking at whole classes of facts from a new point of view. Let him observe how slowly, but surely, the noble views of Lyell on the gradual changes now in progress on the earth's surface have been accepted as sufficient to account for all that we see in its past history. The present action of natural selection may seem more or less probable; but I believe in the truth of the theory, because it collects under one point of view, and gives a rational explanation of, many apparently independent classes of facts.[4]

[4] In treating the several subjects included in the present and succeeding works I have continually been led to ask for information from many zoologists, botanists, geologists, breeders of animals, and horticulturists, and I have invariably received from them the most generous assistance. Without such aid I could have effected little. I have repeatedly applied for information and specimens to foreigners, and to British merchants and officers of the Government residing in distant lands, and, with the rarest exceptions, I have received prompt, open-handed, and valuable assistance. I cannot express too strongly my obligations to the many persons who have assisted me, and who, I am convinced, would be equally willing to assist others in any scientific investigation.

CHAPTER I.

DOMESTIC DOGS AND CATS.

ANCIENT VARIETIES OF THE DOG — RESEMBLANCE OF DOMESTIC DOGS IN VARIOUS COUNTRIES TO NATIVE CANINE SPECIES — ANIMALS NOT ACQUAINTED WITH MAN AT FIRST FEARLESS — DOGS RESEMBLING WOLVES AND JACKALS — HABIT OF BARKING ACQUIRED AND LOST — FERAL DOGS — TAN-COLOURED EYE-SPOTS — PERIOD OF GESTATION — OFFENSIVE ODOUR — FERTILITY OF THE RACES WHEN CROSSED — DIFFERENCES IN THE SEVERAL RACES IN PART DUE TO DESCENT FROM DISTINCT SPECIES — DIFFERENCES IN THE SKULL AND TEETH — DIFFERENCES IN THE BODY, IN CONSTITUTION — FEW IMPORTANT DIFFERENCES HAVE BEEN FIXED BY SELECTION — DIRECT ACTION OF CLIMATE — WATER-DOGS WITH PALMATED FEET — HISTORY OF THE CHANGES WHICH CERTAIN ENGLISH RACES OF THE DOG HAVE GRADUALLY UNDERGONE THROUGH SELECTION — EXTINCTION OF THE LESS IMPROVED SUB-BREEDS.

CATS, CROSSED WITH SEVERAL SPECIES — DIFFERENT BREEDS FOUND ONLY IN SEPARATED COUNTRIES — DIRECT EFFECTS OF THE CONDITIONS OF LIFE — FERAL CATS — INDIVIDUAL VARIABILITY.

THE first and chief point of interest in this chapter is, whether the numerous domesticated varieties of the dog have descended from a single wild species, or from several. Some authors believe that all have descended from the wolf, or from the jackal, or from an unknown and extinct species. Others again believe, and this of late has been the favourite tenet, that they have descended from several species, extinct and recent, more or less commingled together. We shall probably never be able to ascertain their origin with certainty. Palæontology[1] does not throw much light on the question, owing, on the one hand, to the close similarity of the skulls of extinct as well as living wolves and jackals, and owing on the other hand to the great dissimilarity of the skulls of the several breeds of the domestic dogs. It seems, however, that remains have been found in the

[1] Owen, 'British Fossil Mammals,' p. 123 to 133. Pictet's 'Traité de Pal.,' 1853, tom. i. p. 202. De Blainville, in his 'Ostéographie, Canidæ,' p. 142, has largely discussed the whole subject, and concludes that the extinct parent of all domesticated dogs came nearest to the wolf in organization, and to the jackal in habits.

later tertiary deposits more like those of a large dog than of a wolf, which favours the belief of De Blainville that our dogs are the descendants of a single extinct species. On the other hand, some authors go so far as to assert that every chief domestic breed must have had its wild prototype. This latter view is extremely improbable; it allows nothing for variation; it passes over the almost monstrous character of some of the breeds; and it almost necessarily assumes, that a large number of species have become extinct since man domesticated the dog; whereas we plainly see that the members of the dog-family are extirpated by human agency with much difficulty; even so recently as 1710 the wolf existed in so small an island as Ireland.

The reasons which have led various authors to infer that our dogs have descended from more than one wild species are as follows.[2] Firstly, the great difference between the several breeds; but this will appear of comparatively little weight, after we shall have seen how great are the differences between the several races of various domesticated animals which certainly have descended from a single parent-form. Secondly, the more important fact that, at the most anciently known historical periods, several breeds of the dog existed, very unlike each other, and closely resembling or identical with breeds still alive.

We will briefly run back through the historical records. The materials are remarkably deficient between the fourteenth century and the Roman classical period.[3] At this earlier period

[2] Pallas, I believe, originated this doctrine in 'Act. Acad. St. Petersburgh,' 1780, Part ii. Ehrenberg has advocated it, as may be seen in De Blainville's 'Ostéographie,' p. 79. It has been carried to an extreme extent by Col. Hamilton Smith in the 'Naturalist Library,' vol. ix. and x. Mr. W. C. Martin adopts it in his excellent 'History of the Dog,' 1845; as does Dr. Morton, as well as Nott and Gliddon, in the United States. Prof. Low, in his Domesticated Animals,' 1845, p. 666, comes to this same conclusion. No one has argued on this side with more clearness and force than the late James Wilson, of Edinburgh, in various papers read before the Highland Agricultural and Wernerian Societies. Isidore Geoffroy Saint Hilaire ('Hist. Nat. Gén.,' 1860, tom. iii. p. 107), though he believes that most dogs have descended from the jackal, yet inclines to the belief that some are descended from the wolf. Prof. Gervais ('Hist. Nat. Mamm.' 1855, tom. ii. p. 69, referring to the view that all the domestic races are the modified descendants of a single species, after a long discussion, says, "Cette opinion est, suivant nous du moins, la moins probable."

[3] Berjeau, 'The Varieties of the Dog; in old Sculptures and Pictures," 1863. 'Der Hund,' von Dr. F. L. Walther,

various breeds, namely hounds, house-dogs, lapdogs, &c., exist ed; but as Dr. Walther has remarked it is impossible to recognise the greater number with any certainty. Youatt, however, gives a drawing of a beautiful sculpture of two greyhound puppies from the Villa of Antoninus. On an Assyrian monument, about 640 B.C., an enormous mastiff[4] is figured; and according to Sir H. Rawlinson (as I was informed at the British Museum), similar dogs are still imported into this same country. I have looked through the magnificent works of Lepsius and Rosellini, and on the monuments from the fourth to the twelfth dynasties (*i.e.* from about 3400 B.C. to 2100 B.C.) several varieties of the dog are represented; most of them are allied to greyhounds; at the later of these periods a dog resembling a hound is figured, with drooping ears, but with a longer back and more pointed head than in our hounds. There is, also, a turnspit, with short and crooked legs, closely resembling the existing variety; but this kind of monstrosity is so common with various animals, as with the ancon sheep, and even, according to Rengger, with jaguars in Paraguay, that it would be rash to look at the monumental animal as the parent of all our turnspits: Colonel Sykes[5] also has described an Indian Pariah dog as presenting the same monstrous character. The most ancient dog represented on the Egyptian monuments is one of the most singular; it resembles a greyhound, but has long pointed ears and a short curled tail: a closely allied variety still exists in Northern Africa; for Mr. E. Vernon Harcourt[6] states that the Arab boar-hound is "an eccentric hieroglyphic animal, such as Cheops once hunted with, somewhat resembling the rough Scotch deer-hound; their tails are curled tight round on their backs,

s. 48, Giessen, 1817: this author seems carefully to have studied all classical works on the subject. *See* also 'Volz, Beiträge zur Kultur-geschichte,' Leipzig, 1852, s. 115. 'Youatt on the Dog,' 1845, p. 6. A very full history is given by De Blainville in his 'Ostéographie, Canidæ.'

[4] I have seen drawings of this dog from the tomb of the son of Esar Haddon, and clay models in the British Museum. Nott and Gliddon, in their 'Types of Mankind,' 1854, p. 393, give a copy of these drawings. This dog has been called a Thibetan mastiff, but Mr. H. A. Oldfield, who is familiar with the so-called Thibet mastiff, and has examined the drawings in the British Museum, informs me that he considers them different.

[5] 'Proc. Zoolog. Soc.,' July 12th, 1831.

[6] 'Sporting in Algeria,' p. 51.

and their ears stick out at right angles." With this most ancient variety a pariah-like dog coexisted.

We thus see that, at a period between four and five thousand years ago, various breeds, viz. pariah dogs, greyhounds, common hounds, mastiffs, house-dogs, lapdogs, and turnspits, existed, more or less closely resembling our present breeds. But there is not sufficient evidence that any of these ancient dogs belonged to the same identical sub-varieties with our present dogs.[7] As long as man was believed to have existed on this earth only about 6000 years, this fact of the great diversity of the breeds at so early a period was an argument of much weight that they had proceeded from several wild sources, for there would not have been sufficient time for their divergence and modification. But now that we know, from the discovery of flint tools embedded with the remains of extinct animals in districts which have since undergone great geographical changes, that man has existed for an incomparably longer period, and bearing in mind that the most barbarous nations possess domestic dogs, the argument from insufficient time falls away greatly in value.

Long before the period of any historical record the dog was domesticated in Europe. In the Danish Middens of the Neolithic or Newer Stone period, bones of a canine animal are imbedded, and Steenstrup ingeniously argues that these belonged to a domestic dog; for a very large proportion of the bones of birds preserved in the refuse, consists of long bones, which it was found on trial dogs cannot devour.[8] This ancient dog was succeeded in Denmark during the Bronze period by a larger kind, presenting certain differences, and this again during the Iron period, by a still larger kind. In Switzerland, we hear

[7] Berjeau gives fac-similes of the Egyptian drawings. Mr. C. L. Martin, in his 'History of the Dog,' 1845, copies several figures from the Egyptian monuments, and speaks with much confidence with respect to their identity with still living dogs. Messrs. Nott and Gliddon ('Types of Mankind,' 1854, p. 388) give still more numerous figures. Mr. Gliddon asserts that a curl-tailed greyhound, like that represented on the most ancient monuments, is common in Borneo; but the Rajah, Sir J. Brooke, informs me that no such dog exists there.

[8] These, and the following facts on the Danish remains, are taken from M. Morlot's most interesting memoir in 'Soc. Vaudoise des Sc. Nat.,' tom. vi., 1860, pp. 281, 299, 320.

from Prof. Rütimeyer,[9] that during the Neolithic period a domesticated dog of middle size existed, which in its skull was about equally remote from the wolf and jackal, and partook of the characters of our hounds and setters or spaniels (Jagdhund und Wachtelhund). Rütimeyer insists strongly on the constancy of form during a very long period of time of this the most ancient known dog. During the Bronze period a larger dog appeared, and this closely resembled in its jaw a dog of the same age in Denmark. Remains of two notably distinct varieties of the dog were found by Schmerling in a cave;[10] but their age cannot be positively determined.

The existence of a single race, remarkably constant in form during the whole Neolithic period, is an interesting fact in contrast with what we see of the changes which the races underwent during the period of the successive Egyptian monuments, and in contrast with our existing dogs. The character of this animal during the Neolithic period, as given by Rütimeyer, supports De Blainville's view that our varieties have descended from an unknown and extinct form. But we should not forget that we know nothing with respect to the antiquity of man in the warmer parts of the world. The succession of the different kinds of dogs in Switzerland and Denmark is thought to be due to the immigration of conquering tribes bringing with them their dogs; and this view accords with the belief that different wild canine animals were domesticated in different regions. Independently of the immigration of new races of man, we know from the wide-spread presence of bronze, composed of an alloy of tin, how much commerce there must have been throughout Europe at an extremely remote period, and dogs would then probably have been bartered. At the present time, amongst the savages of the interior of Guiana, the Taruma Indians are considered the best trainers of dogs, and possess a large breed, which they barter at a high price with other tribes.[11]

The main argument in favour of the several breeds of the

[9] 'Die Fauna der Pfahlbauten,' 1861, s. 117, 162.

[10] De Blainville, 'Ostéographie, Canidæ.'

[11] Sir R. Schomburgk has given me information on this head. *See* also 'Journal of R. Geograph. Soc.,' vol. xiii., 1843, p. 65.

dog being the descendants of distinct wild stocks, is their resemblance in various countries to distinct species still existing there. It must, however, be admitted that the comparison between the wild and domesticated animal has been made but in few cases with sufficient exactness. Before entering on details, it will be well to show that there is no a priori difficulty in the belief that several canine species have been domesticated; for there is much difficulty in this respect with some other domestic quadrupeds and birds. Members of the dog family inhabit nearly the whole world; and several species agree pretty closely in habits and structure with our several domesticated dogs. Mr. Galton has shown[12] how fond savages are of keeping and taming animals of all kinds. Social animals are the most easily subjugated by man, and several species of Canidæ hunt in packs. It deserves notice, as bearing on other animals as well as on the dog, that at an extremely ancient period, when man first entered any country, the animals living there would have felt no instinctive or inherited fear of him, and would consequently have been tamed far more easily than at present. For instance, when the Falkland Islands were first visited by man, the large wolf-like dog (*Canis antarcticus*) fearlessly came to meet Byron's sailors, who, mistaking this ignorant curiosity for ferocity, ran into the water to avoid them: even recently a man, by holding a piece of meat in one hand and a knife in the other, could sometimes stick them at night. On an island in the Sea of Aral, when first discovered by Butakoff, the saigak antelopes, which are "generally very timid and watchful, did not fly from us, but on the contrary looked at us with a sort of curiosity." So, again, on the shores of the Mauritius, the manatee was not at first in the least afraid of man, and thus it has been in several quarters of the world with seals and the morse. I have elsewhere shown[13] how slowly the native birds of several islands have acquired and inherited a salutary dread of man: at the Galapagos Archipelago I pushed with the muzzle of my gun hawks from a branch, and

[12] 'Domestication of Animals:' Ethnological Soc., Dec. 22nd, 1863.

[13] 'Journal of Researches,' &c., 1845, p. 393. With respect to *Canis antarcticus*, *see* p. 193. For the case of the antelope, *see* 'Journal Royal Geograph. Soc.,' vol. xxiii. p. 94.

held out a pitcher of water for other birds to alight on and drink. Quadrupeds and birds which have seldom been disturbed by man, dread him no more than do our English birds the cows or horses grazing in the fields.

It is a more important consideration that several canine species evince (as will be shown in a future chapter) no strong repugnance or inability to breed under confinement; and the incapacity to breed under confinement is one of the commonest bars to domestication. Lastly, savages set the highest value, as we shall see in the chapter on Selection, on dogs: even half-tamed animals are highly useful to them: the Indians of North America cross their half-wild dogs with wolves, and thus render them even wilder than before, but bolder: the savages of Guiana catch and partially tame and use the whelps of two wild species of *Canis*, as do the savages of Australia those of the wild Dingo. Mr. Philip King informs me that he once trained a wild Dingo puppy to drive cattle, and found it very useful. From these several considerations we see that there is no difficulty in believing that man might have domesticated various canine species in different countries. It would indeed have been a strange fact if one species alone had been domesticated throughout the world.

We will now enter into details. The accurate and sagacious Richardson says, "The resemblance between the Northern American wolves (*Canis lupus, var. occidentalis*) and the domestic dogs of the Indians is so great that the size and strength of the wolf seems to be the only difference. I have more than once mistaken a band of wolves for the dogs of a party of Indians; and the howl of the animals of both species is prolonged so exactly in the same key that even the practised ear of the Indian fails at times to discriminate them." He adds that the more northern Esquimaux dogs are not only extremely like the grey wolves of the Arctic circle in form and colour, but also nearly equal them in size. Dr. Kane has often seen in his teams of sledge-dogs the oblique eye (a character on which some naturalists lay great stress), the drooping tail, and scared look of the wolf. In disposition the Esquimaux dogs differ little from wolves, and, according to Dr. Hayes, they are capable of no attachment to man, and are so savage, that when hungry they will attack even their masters. According to Kane they readily become feral. Their affinity is so close with wolves that they frequently cross with them, and the Indians take the whelps of wolves "to improve the breed of their dogs." The half-bred wolves sometimes (Lamare-Picquot) cannot be tamed, "though this case is rare;" but they do not become thoroughly well broken in till the second or third generation. These facts show that there can be but little, if any, sterility between the Esquimaux dog and the wolf, for otherwise they would not be used to improve the breed. As Dr. Hayes says of these dogs, "reclaimed wolves they doubtless are."[14]

North America is inhabited by a second kind of wolf, the prairie-wolf (*Canis latrans*), which is now looked at by all naturalists as specifically distinct from the common wolf; and is, according to Mr. J. K. Lord, in some respects intermediate in habits between a wolf and a fox. Sir J. Richardson, after describing the Hare Indian dog, which differs in many respects from the Esquimaux dog, says, "It bears the same relation to the prairie wolf that the Esquimaux dog does to the great grey wolf." He could, in fact, detect no marked difference between them; and Messrs. Nott and Gliddon give additional details showing their close resemblance. The dogs derived from the above two aboriginal sources cross together and with the wild wolves, at least with the *C. occidentalis*, and with European dogs. In Florida, according to Bartram, the black wolf-dog of the Indians differs in nothing from the wolves of that country except in barking.[15]

[14] The authorities for the foregoing statements are as follow :—Richardson, in 'Fauna Boreali-Americana,' 1829, pp. 64, 75; Dr. Kane, 'Arctic Explorations,' 1856, vol. i. pp. 398, 455; Dr. Hayes, 'Arctic Boat Journey,' 1860, p. 167. Franklin's 'Narrative,' vol. i. p. 269, gives the case of three whelps of a black wolf being carried away by the Indians. Parry, Richardson, and others, give accounts of wolves and dogs naturally crossing in the eastern parts of North America. Seeman, in his 'Voyage of H.M.S. Herald,' 1853, vol. ii. p. 26, says the wolf is often caught by the Esquimaux for the purpose of crossing with their dogs, and thus adding to their size and strength. M. Lamare-Picquot, in 'Bull. de la Soc. d'Acclimat.,' tom. vii., 1860, p. 148, gives a good account of the half-bred Esquimaux dogs.

[15] 'Fauna Boreali-Americana,' 1829, pp. 73, 78, 80. Nott and Gliddon, 'Types of Mankind,' p. 383. The naturalist and traveller Bartram is quoted by Hamilton Smith, in 'Nat. Hist. Lib.,' vol. x. p. 156. A Mexican domestic dog seems also to resemble a wild dog

Turning to the southern parts of the New World, Columbus found two kinds of dogs in the West Indies; and Fernandez[16] describes three in Mexico: some of these native dogs were dumb —that is, did not bark. In Guiana it has been known since the time of Buffon that the natives cross their dogs with an aboriginal species, apparently the *Canis cancrivorus*. Sir R. Schomburgk, who has so carefully explored these regions, writes to me, "I have been repeatedly told by the Arawaak Indians, who reside near the coast, that they cross their dogs with a wild species to improve the breed, and individual dogs have been shown to me which certainly resembled the *C. cancrivorus* much more than the common breed. It is but seldom that the Indians keep the *C. cancrivorus* for domestic purposes, nor is the Ai, another species of wild dog, and which I consider to be identical with the *Dusicyon silvestris* of H. Smith, now much used by the Arecunas for the purpose of hunting. The dogs of the Taruma Indians are quite distinct, and resemble Buffon's St. Domingo greyhound." It thus appears that the natives of Guiana have partially domesticated two aboriginal species, and still cross their dogs with them; these two species belong to a quite different type from the North American and European wolves. A careful observer, Rengger,[17] gives reasons for believing that a hairless dog was domesticated when America was first visited by Europeans: some of these dogs in Paraguay are still dumb, and Tschudi[18] states that they suffer from cold in the Cordillera. This naked dog is, however, quite distinct from that found preserved in the ancient Peruvian burial-places, and described by Tschudi, under the name of *Canis Ingæ*, as withstanding cold well and as barking. It is not known whether these two distinct kinds of dog are the descendants of native species, and it might be argued that when man first migrated into America he brought with him from the Asiatic continent dogs

of the same country; but this may be the prairie-wolf. Another capable judge, Mr. J. K. Lord ('The Naturalist in Vancouver Island,' 1866, vol. ii. p. 218), says that the Indian dog of the Spokans, near the Rocky Mountains, "is beyond all question nothing more than a tamed Cayote or prairie-wolf," or *Canis latrans*.

[16] I quote this from Mr. R. Hill's excellent account of the Alco or domestic dog of Mexico, in Gosse's 'Naturalist's Sojourn in Jamaica,' 1851, p. 329.

[17] 'Naturgeschichte der Saeugethiere von Paraguay,' 1830, s. 151.

[18] Quoted in Humboldt's 'Aspects of Nature' (Eng. transl.), vol. i. p. 108.

which had not learned to bark; but this view does not seem probable, as the natives along the line of their march from the north reclaimed, as we have seen, at least two N. American species of Canidæ.

Turning to the Old World, some European dogs closely resemble the wolf; thus the shepherd dog of the plains of Hungary is white or reddish-brown, has a sharp nose, short, erect ears, shaggy coat, and bushy tail, and so much resembles a wolf that Mr. Paget, who gives this description, says he has known a Hungarian mistake a wolf for one of his own dogs. Jeitteles, also, remarks on the close similarity of the Hungarian dog and wolf. Shepherd dogs in Italy must anciently have closely resembled wolves, for Columella (vii. 12) advises that white dogs be kept, adding, "pastor album probat, ne pro lupo canem feriat." Several accounts have been given of dogs and wolves crossing naturally; and Pliny asserts that the Gauls tied their female dogs in the woods that they might cross with wolves.[19] The European wolf differs slightly from that of North America, and has been ranked by many naturalists as a distinct species. The common wolf of India is also by some esteemed as a third species, and here again we find a marked resemblance between the pariah dogs of certain districts of India and the Indian wolf.[20]

With respect to Jackals, Isidore Geoffroy Saint Hilaire[21] says that not one constant difference can be pointed out between their structure and that of the smaller races of dogs. They agree closely in habits: jackals, when tamed and called by their

[19] Paget's 'Travels in Hungary and Transylvania,' vol. i. p. 501. Jeitteles, 'Fauna Hungariæ Superioris,' 1862, s. 13. *See* Pliny, 'Hist. of the World' (Eng. transl.), 8th book, ch. xl., about the Gauls crossing their dogs. *See* also Aristotle, 'Hist. Animal.' lib. viii. c. 28. For good evidence about wolves and dogs naturally crossing near the Pyrenees, *see* M. Mauduyt, 'Du Loup et de ses Races,' Poitiers, 1851; also Pallas, in 'Acta Acad. St. Petersburgh,' 1780, part ii. p. 94.

[20] I give this on excellent authority, namely, Mr. Blyth (under the signature of Zoophilus), in the 'Indian Sporting Review,' Oct. 1856, p. 134. Mr. Blyth states that he was struck with the resemblance between a brush-tailed race of pariah-dogs, north-west of Cawnpore, and the Indian wolf. He gives corroborative evidence with respect to the dogs of the valley of the Nerbudda.

[21] For numerous and interesting details on the resemblance of dogs and jackals, *see* Isid. Geoffroy St. Hilaire, 'Hist. Nat. Gén.,' 1860, tom. iii. p. 101. *See* also 'Hist. Nat. des Mammifères,' par Prof. Gervais, 1855, tom. ii. p. 60.

master, wag their tails, crouch, and throw themselves on their backs; they smell at the tails of dogs, and void their urine sideways.[22] A number of excellent naturalists, from the time of Güldenstädt to that of Ehrenberg, Hemprich, and Cretzschmar, have expressed themselves in the strongest terms with respect to the resemblance of the half-domestic dogs of Asia and Egypt to jackals. M. Nordmann, for instance, says, "Les chiens d'Awhasie ressemblent étonnamment à des chacals." Ehrenberg[23] asserts that the domestic dogs of Lower Egypt, and certain mummied dogs, have for their wild type a species of wolf (*C. lupaster*) of the country; whereas the domestic dogs of Nubia and certain other mummied dogs have the closest relation to a wild species of the same country, viz. *C. sabbar*, which is only a form of the common jackal. Pallas asserts that jackals and dogs sometimes naturally cross in the East; and a case is on record in Algeria.[24] The greater number of naturalists divide the jackals of Asia and Africa into several species, but some few rank them all as one.

I may add that the domestic dogs on the coast of Guinea are fox-like animals, and are dumb.[25] On the east coast of Africa, between lat. 4° and 6° south, and about ten days' journey in the interior, a semi-domestic dog, as the Rev. S. Erhardt informs me, is kept, which the natives assert is derived from a similar wild animal. Lichtenstein[26] says that the dogs of the Bosjemans present a striking resemblance even in colour (excepting the black stripe down the back) with the *C. mesomelas* of South Africa. Mr. E. Layard informs me that he has seen a Caffre dog which closely resembled an Esquimaux dog. In Australia the Dingo is both domesticated and wild; though this animal may have been introduced aboriginally by man, yet it must be considered as almost an endemic form, for its remains have been found in a similar state of preservation and associated with

[22] Güldenstädt, 'Nov. Comment. Acad. Petrop.,' tom. xx., pro anno 1775, p. 449.

[23] Quoted by De Blainville in his 'Ostéographie, Canidæ,' pp. 79, 98.

[24] *See* Pallas, in 'Act. Acad. St. Petersburgh,' 1780, part ii. p. 91. For Algeria, *see* Isid. Geoffroy St. Hilaire, 'Hist. Nat. Gén.,' tom. iii. p. 177. In both countries it is the male jackal which pairs with female domestic dogs.

[25] John Barbut's 'Description of the Coast of Guinea in 1746.'

[26] 'Travels in South Africa,' vol. ii. p. 272.

extinct mammals, so that its introduction must have been ancient.[27]

From this resemblance in several countries of the half-domesticated dogs to the wild species still living there,—from the facility with which they can often be crossed together,—from even half-tamed animals being so much valued by savages,—and from the other circumstances previously remarked on which favour their domestication, it is highly probable that the domestic dogs of the world have descended from two good species of wolf (viz. *C. lupus* and *C. latrans*), and from two or three other doubtful species of wolves (namely, the European, Indian, and North African forms); from at least one or two South American canine species; from several races or species of the jackal; and perhaps from one or more extinct species. Those authors who attribute great influence to the action of climate by itself may thus account for the resemblance of the domesticated dogs and native animals in the same countries; but I know of no facts supporting the belief in so powerful an action of climate.

It cannot be objected to the view of several canine species having been anciently domesticated, that these animals are tamed with difficulty: facts have been already given on this head, but I may add that the young of the *Canis primævus* of India were tamed by Mr. Hodgson,[28] and became as sensible to caresses, and manifested as much intelligence, as any sporting dog of the same age. There is not much difference, as we have already shown and shall immediately further see, in habits between the domestic dogs of the North American Indians and the wolves of that country, or between the Eastern pariah dogs and jackals, or between the dogs which have run wild in various countries and the several natural species of the family. The habit of barking, however, which is almost universal with domesticated

[27] Selwyn, Geology of Victoria; 'Journal of Geolog. Soc.,' vol. xiv., 1858, p. 536, and vol. xvi., 1860, p. 148; and Prof. M'Coy, in 'Annals and Mag. of Nat. Hist.' (3rd series), vol. ix., 1862, p. 147. The Dingo differs from the dogs of the central Polynesian islands. Dieffenbach remarks ('Travels,' vol. ii. p. 45) that the native New Zealand dog also differs from the Dingo.

[28] 'Proceedings Zoolog. Soc.,' 1833, p. 112. *See*, also, on the taming of the common wolf, L. Lloyd, 'Scandinavian Adventures,' vol. i. p. 460, 1854. With respect to the jackal, *see* Prof. Gervais, 'Hist. Nat. Mamm.,' tom. ii. p. 61. With respect to the aguara of Paraguay, *see* Rengger's work.

dogs, and which does not characterise a single natural species of the family, seems an exception; but this habit is soon lost and soon reacquired. The case of the wild dogs on the island of Juan Fernandez having become dumb has often been quoted, and there is reason to believe[29] that the dumbness ensued in the course of thirty-three years; on the other hand, dogs taken from this island by Ulloa slowly reacquired the habit of barking. The Mackenzie-river dogs, of the *Canis latrans* type, when brought to England, never learned to bark properly; but one born in the Zoological Gardens[30] "made his voice sound as loudly as any other dog of the same age and size." According to Professor Nillson,[31] a wolf-whelp reared by a bitch barks. I. Geoffroy Saint Hilaire exhibited a jackal which barked with the same tone as any common dog.[32] An interesting account has been given by Mr. G. Clarke[33] of some dogs run wild on Juan de Nova, in the Indian Ocean; "they had entirely lost the faculty of barking; they had no inclination for the company of other dogs, nor did they acquire their voice," during a captivity of several months. On the island they "congregate in vast packs, and catch sea-birds with as much address as foxes could display." The feral dogs of La Plata have not become dumb; they are of large size, hunt single or in packs, and burrow holes for their young.[34] In these habits the feral dogs of La Plata resemble wolves and jackals; both of which hunt either singly or in packs, and burrow holes.[35] These feral dogs have not become uniform in colour on Juan Fernandez, Juan de Nova, or La Plata.[36] In Cuba the feral dogs are described by Poeppig as nearly all mouse-coloured, with short ears and light-blue eyes.

[29] Roulin, in 'Mém. présent. par divers Savans,' tom. vi. p. 341.

[30] Martin, 'History of the Dog,' p. 14.

[31] Quoted by L. Lloyd in 'Field Sports of North of Europe,' vol. i. p. 387.

[32] Quatrefages, 'Soc. d'Acclimat.,' May 11th, 1863, p. 7.

[33] 'Annals and Mag. of Nat. Hist.,' vol. xv., 1845, p. 140.

[34] Azara, Voyages dans l'Amér. Mérid.,' tom. i. p. 381; his account is fully confirmed by Rengger. Quatrefages gives an account of a bitch brought from Jerusalem to France which burrowed a hole and littered in it. *See* 'Discours, Exposition des Races Canines,' 1865, p. 3.

[35] With respect to wolves burrowing holes, *see* Richardson, 'Fauna Boreali-Americana,' p. 64; and Bechstein, 'Naturgesch. Deutschlands,' b. i. s. 617.

[36] *See* Poeppig, 'Reise in Chile,' b. i. s. 290; Mr. G. Clarke, as above; and Rengger, s. 155.

In St. Domingo, Col. Ham. Smith says [37] that the feral dogs are very large, like greyhounds, of a uniform pale blue-ash, with small ears, and large light-brown eyes. Even the wild Dingo, though so anciently naturalised in Australia, "varies considerably in colour," as I am informed by Mr. P. P. King: a half-bred Dingo reared in England [38] showed signs of wishing to burrow.

From the several foregoing facts we see that reversion in the feral state gives no indication of the colour or size of the aboriginal parent-species. One fact, however, with respect to the colouring of domestic dogs, I at one time hoped might have thrown some light on their origin; and it is worth giving, as showing how colouring follows laws, even in so anciently and thoroughly domesticated an animal as the dog. Black dogs with tan-coloured feet, whatever breed they may belong to, almost invariably have a tan-coloured spot on the upper and inner corners of each eye, and their lips are generally thus coloured. I have seen only two exceptions to this rule, namely, in a spaniel and terrier. Dogs of a light-brown colour often have a lighter, yellowish-brown spot over the eyes; sometimes the spot is white, and in a mongrel terrier the spot was black. Mr. Waring kindly examined for me a stud of fifteen greyhounds in Suffolk: eleven of them were black, or black and white, or brindled, and these had no eye-spots; but three were red and one slaty-blue, and these four had dark-coloured spots over their eyes. Although the spots thus sometimes differ in colour, they strongly tend to be tan-coloured; this is proved by my having seen four spaniels, a setter, two Yorkshire shepherd dogs, a large mongrel, and some fox-hounds, coloured black and white, with not a trace of tan-colour, excepting the spots over the eyes, and sometimes a little on the feet. These latter cases, and many others, show plainly that the colour of the feet and the eye-spots are in some way correlated. I have noticed, in various breeds, every gradation, from the whole face being tan-coloured, to a complete ring round the eyes, to a minute spot over the inner and upper corners. The spots occur in various sub-breeds of terriers and spaniels; in setters; in hounds of various kinds, including the turnspit-like German badger-hound; in shepherd dogs; in a mongrel, of which neither parent had the spots; in one pure bulldog, though the spots were in this case almost white; and in greyhounds,—but true black-and-tan greyhounds are excessively rare; nevertheless I have been assured by Mr. Warwick, that one ran at the Caledonian Champion meeting of April, 1860, and was "marked precisely like a black-and-tan terrier." Mr. Swinhoe at my request looked at the dogs in China, at Amoy, and he soon noticed a brown dog with yellow spots over the eyes. Colonel H. Smith [39] figures the magnificent black mastiff of Thibet with a

[37] Dogs, 'Nat. Library,' vol. x. p. 121: an endemic South American dog seems also to have become feral in this island. *See* Gosse's 'Jamaica,' p. 340.

[38] Low, 'Domesticated Animals,' p. 650.

[39] 'The Naturalist Library,' Dogs, vol. x. pp. 4, 19.

tan-coloured stripe over the eyes, feet, and chaps; and what is more singular, he figures the Alco, or native domestic dog of Mexico, as black and white, with narrow tan-coloured rings round the eyes; at the Exhibition of dogs in London, May, 1863, a so-called forest-dog from North-West Mexico was shown, which had pale tan-coloured spots over the eyes. The occurrence of these tan-coloured spots in dogs of such extremely different breeds, living in various parts of the world, makes the fact highly remarkable.

We shall hereafter see, especially in the chapter on Pigeons, that coloured marks are strongly inherited, and that they often aid us in discovering the primitive forms of our domestic races. Hence, if any wild canine species had distinctly exhibited the tan-coloured spots over the eyes, it might have been argued that this was the parent-form of nearly all our domestic races. But after looking at many coloured plates, and through the whole collection of skins in the British Museum, I can find no species thus marked. It is no doubt possible that some extinct species was thus coloured. On the other hand, in looking at the various species, there seems to be a tolerably plain correlation between tan-coloured legs and face; and less frequently between black legs and a black face; and this general rule of colouring explains to a certain extent the above-given cases of correlation between the eye-spots and the colour of the feet. Moreover, some jackals and foxes have a trace of a white ring round their eyes, as in *C. mesomelas*, *C. aureus*, and (judging from Colonel Ham. Smith's drawing) in *C. alopex* and *C. thaleb*. Other species have a trace of a black line over the corners of the eyes, as in *C. variegatus*, *cinereo-variegatus*, and *fulvus*, and the wild Dingo. Hence I am inclined to conclude that a tendency for tan-coloured spots to appear over the eyes in the various breeds of dogs, is analogous to the case observed by Desmarest, namely, that when any white appears on a dog the tip of the tail is always white, "de manière à rappeler la tache terminale de même couleur, qui caractérise la plupart des Canidés sauvages."[40]

It has been objected that our domestic dogs cannot be descended from wolves or jackals, because their periods of gestation are different. The supposed difference rests on statements made by Buffon, Gilibert, Bechstein, and others; but these are now known to be erroneous; and the period is found to agree in the wolf, jackal, and dog, as closely as could be expected, for it is often in some degree variable.[41] Tessier, who

[40] Quoted by Prof. Gervais, 'Hist. Nat. Mamm.,' tom. ii. p. 66.

[41] J. Hunter shows that the long period of seventy-three days given by Buffon is easily explained by the bitch having received the dog many times during a period of sixteen days ('Phil. Transact.,' 1787, p. 253). Hunter found that the gestation of a mongrel from wolf and dog ('Phil. Transact.,' 1789, p. 160) apparently was sixty-three days, for she received the dog more than once. The period of a mongrel dog and jackal was fifty-nine days. Fred. Cuvier found the period of gestation of the wolf to be ('Dict. Class. d'Hist. Nat.,' tom. iv. p.

has closely attended to this subject, allows a difference of four days in the gestation of the dog. The Rev. W. D. Fox has given me three carefully recorded cases of retrievers, in which the bitch was put only once to the dog; and not counting this day, but counting that of parturition, the periods were fifty-nine, sixty-two, and sixty-seven days. The average period is sixty-three days; but Bellingeri states that this holds good only with large dogs; and that for small races it is from sixty to sixty-three days; Mr. Eyton of Eyton, who has had much experience with dogs, also informs me that the time is apt to be longer with large than with small dogs.

F. Cuvier has objected that the jackal would not have been domesticated on account of its offensive smell; but savages are not sensitive in this respect. The degree of odour, also, differs in the different kinds of jackal;[42] and Colonel H. Smith makes a sectional division of the group with one character dependent on not being offensive. On the other hand, dogs—for instance, rough and smooth terriers—differ much in this respect; and M. Godron states that the hairless so-called Turkish dog is more odoriferous than other dogs. Isidore Geoffroy[43] gave to a dog the same odour as that from a jackal by feeding it on raw flesh.

The belief that our dogs are descended from wolves, jackals, South American Canidæ, and other species, suggests a far more important difficulty. These animals in their undomesticated state, judging from a widely-spread analogy, would have been in some degree sterile if intercrossed; and such sterility will be admitted as almost certain by all those who believe that the lessened fertility of crossed forms is an infallible criterion of specific distinctness. Anyhow these animals keep distinct in the countries which they inhabit in common. On the other hand, all domestic dogs, which are here supposed to be descended

8) two months and a few days, which agrees with the dog. Isid. G. St. Hilaire, who has discussed the whole subject, and from whom I quote Bellingeri, states ('Hist. Nat. Gén.,' tom. iii. p. 112) that in the Jardin des Plantes the period of the jackal has been found to be from sixty to sixty-three days, exactly as with the dog.

42 *See* Isid. Geoffroy St. Hilaire, 'Hist. Nat. Gén.,' tom. iii. p. 112, on the odour of jackals. Col. Ham. Smith, in 'Nat. Hist. Lib.,' vol. x. p. 289.

43 Quoted by Quatrefages in 'Bull. Soc. d'Acclimat.,' May 11th, 1863.

from several distinct species, are, as far as is known, mutually fertile together. But, as Broca has well remarked,[44] the fertility of successive generations of mongrel dogs has never been scrutinised with that care which is thought indispensable when species are crossed. The few facts leading to the conclusion that the sexual feelings and reproductive powers differ in the several races of the dog when crossed are (passing over mere size as rendering propagation difficult) as follows: the Mexican Alco[45] apparently dislikes dogs of other kinds, but this perhaps is not strictly a sexual feeling; the hairless endemic dog of Paraguay, according to Rengger, mixes less with the European races than these do with each other; the Spitz-dog in Germany is said to receive the fox more readily than do other breeds; and Dr. Hodgkin states that a female Dingo in England attracted the male wild foxes. If these latter statements can be trusted, they prove some degree of sexual difference in the breeds of the dog. But the fact remains that our domestic dogs, differing so widely as they do in external structure, are far more fertile together than we have reason to believe their supposed wild parents would have been. Pallas assumes[46] that a long course of domestication eliminates that sterility which the parent-species would have exhibited if only lately captured; no distinct facts are recorded in support of this hypothesis; but the evidence seems to me so strong (independently of the evidence derived from other domesticated animals) in favour of our domestic dogs having descended from several wild stocks, that I am led to admit the truth of this hypothesis.

There is another and closely allied difficulty consequent on the doctrine of the descent of our domestic dogs from several wild species, namely, that they do not seem to be perfectly fertile with their supposed parents. But the experiment has not been quite fairly tried; the Hungarian dog, for instance,

[44] 'Journal de la Physiologie,' tom. ii. p. 385.

[45] *See* Mr. R. Hill's excellent account of this breed in Gosse's 'Jamaica,' p. 338; Rengger's 'Saeugethiere von Paraguay,' s. 153. With respect to Spitz dogs, *see* Bechstein's 'Naturgesch. Deutschlands,' 1801, b. i. s. 638. With respect to Dr. Hodgkin's statement made before Brit. Assoc., *see* 'The Zoologist,' vol. iv., for 1845-46, p. 1097.

[46] 'Acta Acad. St. Petersburgh,' 1780, part ii. pp. 84, 100.

which in external appearance so closely resembles the European wolf, ought to be crossed with this wolf; and the pariah-dogs of India with Indian wolves and jackals; and so in other cases. That the sterility is very slight between certain dogs and wolves and other Canidæ is shown by savages taking the trouble to cross them. Buffon got four successive generations from the wolf and dog, and the mongrels were perfectly fertile together.[47] But more lately M. Flourens states positively as the result of his numerous experiments that hybrids from the wolf and dog, crossed *inter se*, become sterile at the third generation, and those from the jackal and dog at the fourth generation.[48] But these animals were closely confined; and many wild animals, as we shall see in a future chapter, are rendered by confinement in some degree or even utterly sterile. The Dingo, which breeds freely in Australia with our imported dogs, would not breed though repeatedly crossed in the Jardin des Plantes.[49] Some hounds from Central Africa, brought home by Major Denham, never bred in the Tower of London;[50] and a similar tendency to sterility might be transmitted to the hybrid offspring of a wild animal. Moreover, it appears that in M. Flourens' experiments the hybrids were closely bred in and in for three or four generations; but this circumstance, although it would almost certainly increase the tendency to sterility, would hardly account for the final result, even though aided by close confinement, unless there had been some original tendency to lessened fertility. Several years ago I saw confined in the Zoological Gardens of London a female hybrid from an English dog and jackal, which even in this the first generation was so sterile that, as I was assured by

[47] M. Broca has shown ('Journal de Physiologie,' tom. ii. p. 353) that Buffon's experiments have been often misrepresented. Broca has collected (pp. 390-395) many facts on the fertility of crossed dogs, wolves, and jackals.

[48] 'De la Longévité Humaine,' par M. Flourens, 1855, p. 143. Mr. Blyth says ('Indian Sporting Review,' vol. ii. p. 137) that he has seen in India several hybrids from the pariah-dog and jackal; and between one of these hybrids and a terrier. The experiments of Hunter on the jackal are well known. *See* also Isid. Geoffroy St. Hilaire, 'Hist. Nat. Gén.,' tom. iii. p. 217, who speaks of the hybrid offspring of the jackal as perfectly fertile for three generations.

[49] On authority of F. Cuvier, quoted in Bronn's 'Geschichte der Natur,' B. ii. s. 164.

[50] W. C. L. Martin, 'History of the Dog,' 1845, p. 203. Mr. Philip P. King, after ample opportunities of observation, informs me that the Dingo and European dogs often cross in Australia.

her keeper, she did not fully exhibit her proper periods; but this case, from the numerous instances of fertile hybrids from these two animals, was certainly exceptional. In almost all experiments on the crossing of animals there are so many causes of doubt, that it is extremely difficult to come to any positive conclusion. It would, however, appear, that those who believe that our dogs are descended from several species will have not only to admit that their offspring after a long course of domestication generally lose all tendency to sterility when crossed together; but that between certain breeds of dogs and some of their supposed aboriginal parents a certain degree of sterility has been retained or possibly even acquired.

Notwithstanding the difficulties in regard to fertility given in the last two paragraphs, when we reflect on the inherent improbability of man having domesticated throughout the world one single species alone of so widely distributed, so easily tamed, and so useful a group as the Canidæ; when we reflect on the extreme antiquity of the different breeds; and especially when we reflect on the close similarity, both in external structure and habits, between the domestic dogs of various countries and the wild species still inhabiting these same countries, the balance of evidence is strongly in favour of the multiple origin of our dogs.

Differences between the several Breeds of the Dog.—If the several breeds have descended from several wild stocks, their difference can obviously in part be explained by that of their parent-species. For instance, the form of the greyhound may be partly accounted for by descent from some such animal as the slim Abyssinian *Canis simensis*,[51] with its elongated muzzle; that of the larger dogs from the larger wolves, and the smaller and slighter dogs from jackals: and thus perhaps we may account for certain constitutional and climatal differences. But it would be a great error to suppose that there has not been in addition [52] a large amount of variation. The intercrossing of the several aboriginal wild stocks, and of the subsequently formed

[51] Rüppel, 'Neue Wirbelthiere von Abyssinien,' 1835-40; 'Mammif.,' s. 39, pl. xiv. There is a specimen of this fine animal in the British Museum.

[52] Even Pallas admits this: *see* 'Act. Acad. St. Petersburgh,' 1780, p. 93.

races, has probably increased the total number of breeds, and, as we shall presently see, has greatly modified some of them. But we cannot explain by crossing the origin of such extreme forms as thoroughbred greyhounds, bloodhounds, bulldogs, Blenheim spaniels, terriers, pugs, &c., unless we believe that forms equally or more strongly characterised in these different respects once existed in nature. But hardly any one has been bold enough to suppose that such unnatural forms ever did or could exist in a wild state. When compared with all known members of the family of Canidæ they betray a distinct and abnormal origin. No instance is on record of such dogs as bloodhounds, spaniels, true greyhounds having been kept by savages: they are the product of long-continued civilization.

The number of breeds and sub-breeds of the dog is great: Youatt, for instance, describes twelve kinds of greyhounds. I will not attempt to enumerate or describe the varieties, for we cannot discriminate how much of their difference is due to variation, and how much to descent from different aboriginal stocks. But it may be worth while briefly to mention some points. Commencing with the skull, Cuvier has admitted[53] that in form the differences are "plus fortes que celles d'aucunes espèces sauvages d'un même genre naturel." The proportions of the different bones; the curvature of the lower jaw, the position of the condyles with respect to the plane of the teeth (on which F. Cuvier founded his classification), and in mastiffs the shape of its posterior branch; the shape of the zygomatic arch, and of the temporal fossæ; the position of the occiput—all vary considerably.[54] The dog has properly six pairs of molar teeth in the upper jaw, and seven in the lower; but several naturalists have seen not rarely an additional pair in the upper jaw;[55] and Professor Gervais says that there are dogs "qui ont sept paires de dents supérieures et huit inférieures." De Blainville[56] has given full particulars on the frequency of these deviations in the number of the teeth, and has shown that it is not always the same tooth which is supernumerary. In short-muzzled races, according to H. Müller,[57] the molar teeth stand obliquely, whilst in long-muzzled races they are placed longitudinally, with open spaces between them. The naked, so-called Egyptian or Turkish dog is extremely deficient in its teeth,[58]—

[53] Quoted by I. Geoffroy, 'Hist. Nat. Gén.,' tom. iii. p. 453.

[54] F. Cuvier, in 'Annales du Muséum,' tom. xviii. p. 337; Godron, 'De l'Espèce,' tom. i. p. 342; and Col. Ham. Smith, in 'Naturalist's Library,' vol. ix. p. 101.

[55] Isid. Geoffroy Saint Hilaire, 'Hist. des Anomalies,' 1832, tom. i. p. 660. Gervais, 'Hist. Nat. des Mammifères,' tom. ii., 1855, p. 66. De Blainville ('Ostéographie, Canidæ,' p. 137) has also seen an extra molar on both sides.

[56] 'Ostéographie, Canidæ,' p. 137.

[57] Würzburger, 'Medecin, Zeitschrift,' 1860, B. i. s. 265.

[58] Mr. Yarrell, in 'Proc. Zoolog. Soc.,' Oct. 8th, 1833. Mr. Waterhouse showed me a skull of one of these dogs, which had only a single molar on each side and some imperfect incisors.

sometimes having none except one molar on each side; but this, though characteristic of the breed, must be considered as a monstrosity. M. Girard,[59] who seems to have attended closely to the subject, says that the period of the appearance of the permanent teeth differs in different dogs, being earlier in large dogs; thus the mastiff assumes its adult teeth in four or five months, whilst in the spaniel the period is sometimes more than seven or eight months.

With respect to minor differences little need be said. Isidore Geoffroy has shown[60] that in size some dogs are six times as long (the tail being excluded) as others; and that the height relatively to the length of the body varies from between one to two, and one to nearly four. In the Scotch deer-hound there is a striking and remarkable difference in the size of the male and female.[61] Every one knows how the ears vary in size in different breeds, and with their great development their muscles become atrophied. Certain breeds of dogs are described as having a deep furrow between the nostrils and lips. The caudal vertebræ, according to F. Cuvier, on whose authority the two last statements rest, vary in number; and the tail in shepherd dogs is almost absent. The mammæ vary from seven to ten in number; Daubenton, having examined twenty-one dogs, found eight with five mammæ on each side; eight with four on each side; and the others with an unequal number on the two sides.[62] Dogs have properly five toes in front and four behind, but a fifth toe is often added; and F. Cuvier states that, when a fifth toe is present, a fourth cuneiform bone is developed; and, in this case, sometimes the great cuneiform bone is raised, and gives on its inner side a large articular surface to the astragalus; so that even the relative connection of the bones, the most constant of all characters, varies. These modifications, however, in the feet of dogs are not important, because they ought to be ranked, as De Blainville has shown,[63] as monstrosities. Nevertheless they are interesting from being correlated with the size of the body, for they occur much more frequently with mastiffs and other large breeds than with small dogs. Closely allied varieties, however, sometimes differ in this respect; thus Mr. Hodgson states that the black-and-tan Lassa variety of the Thibet mastiff has the fifth digit, whilst the Mustang sub-variety is not thus characterised. The extent to which the skin is developed between the toes varies much; but we shall return to this point. The degree to which the various breeds differ in the perfection of their senses, dispositions, and inherited habits is notorious to every one. The breeds present some constitutional differences: the pulse, says Youatt,[64] "varies materially according to the breed, as well

---

[59] Quoted in 'The Veterinary,' London, vol. viii. p. 415.

[60] 'Hist. Nat. Général,' tom. iii. p. 448.

[61] W. Scrope, 'Art of Deer-Stalking,' p. 354.

[62] Quoted by Col. Ham. Smith in 'Naturalist's Library,' vol. x. p. 79.

[63] De Blainville, 'Ostéographie, Canidæ,' p. 134. F. Cuvier, 'Annales du Muséum,' tom. xviii. p. 342. In regard to mastiffs, *see* Col. Ham. Smith, 'Nat. Lib.,' vol. x. p. 218. For the Thibet mastiff, *see* Mr. Hodgson in 'Journal of As. Soc. of Bengal,' vol. i., 1832, p. 342.

[64] 'The Dog,' 1845, p. 186. With respect to diseases, Youatt asserts (p.

as to the size of the animal." Different breeds of dogs are subject in different degrees to various diseases. They certainly become adapted to different climates under which they have long existed. It is notorious that most of our best European breeds deteriorate in India.[65] The Rev. R. Everest[66] believes that no one has succeeded in keeping the Newfoundland dog long alive in India; so it is, according to Lichtenstein,[67] even at the Cape of Good Hope. The Thibet mastiff degenerates on the plains of India, and can live only on the mountains.[68] Lloyd [69] asserts that our bloodhounds and bulldogs have been tried, and cannot withstand the cold of the northern European forests.

Seeing in how many characters the races of the dog differ from each other, and remembering Cuvier's admission that their skulls differ more than do those of the species of any natural genus, and bearing in mind how closely the bones of wolves, jackals, foxes, and other Canidæ agree, it is remarkable that we meet with the statement, repeated over and over again, that the races of the dog differ in no important characters. A highly competent judge, Prof. Gervais,[70] admits, "si l'on prenait sans contrôle les altérations dont chacun de ces organes est susceptible, on pourrait croire qu'il y a entre les chiens domestiques des différences plus grandes que celles qui séparent ailleurs les espèces, quelquefois même les genres." Some of the differences above enumerated are in one respect of comparatively little value, for they are not characteristic of distinct breeds: no one pretends that such is the case with the additional molar teeth or with the number of mammæ; the additional digit is generally present with mastiffs, and some of the more important differences in the skull and lower jaw are more or less characteristic of various breeds. But we must not forget that the predominant power of selection has not been applied in any of these cases; we have variability in important parts, but the differences have not been fixed by selection. Man

167) that the Italian greyhound is "strongly subject" to polypi in the matrix or vagina. The spaniel and pug (p. 182) are most liable to bronchocele. The liability to distemper (p. 232) is extremely different in different breeds. On the distemper, *see* also Col. Hutchinson on 'Dog Breaking,' 1850, p. 279.

[65] *See* Youatt on the Dog, p. 15; 'The Veterinary,' London, vol. xi. p. 235.

[66] 'Journal of As. Soc. of Bengal,' vol. iii. p. 19.

[67] 'Travels,' vol. ii. p. 15.

[68] Hodgson, in 'Journal of As. Soc. of Bengal,' vol. i. p. 342.

[69] 'Field Sports of the North of Europe,' vol. ii. p. 165.

[70] 'Hist. Nat. des Mammif.,' 1855, tom. ii. pp. 66, 67.

cares for the form and fleetness of his greyhounds, for the size of his mastiffs, for the strength of the jaw in his bulldogs, &c.; but he cares nothing about the number of their molar teeth or mammæ or digits; nor do we know that differences in these organs are correlated with, or owe their development to, differences in other parts of the body about which man does care. Those who have attended to the subject of selection will admit that, nature having given variability, man, if he so chose, could fix five toes to the hinder feet of certain breeds of dogs, as certainly as to the feet of his Dorking-fowls: he could probably fix, but with much more difficulty, an additional pair of molar teeth in either jaw, in the same way as he has given additional horns to certain breeds of sheep; if he wished to produce a toothless breed of dogs, having the so-called Turkish dog with its imperfect teeth to work on, he could probably do so, for he has succeeded in making hornless breeds of cattle and sheep.

With respect to the precise causes and steps by which the several races of dogs have come to differ so greatly from each other, we are, as in most other cases, profoundly ignorant. We may attribute part of the difference in external form and constitution to inheritance from distinct wild stocks, that is to changes effected under nature before domestication. We must attribute something to the crossing of the several domestic and natural races. I shall, however, soon recur to the crossing of races. We have already seen how often savages cross their dogs with wild native species; and Pennant gives a curious account[71] of the manner in which Fochabers, in Scotland, was stocked "with a multitude of curs of a most wolfish aspect" from a single hybrid-wolf brought into that district.

It would appear that climate to a certain extent directly modifies the forms of dogs. We have lately seen that several of our English breeds cannot live in India, and it is positively asserted that when bred there for a few generations they degenerate not only in their mental faculties, but in form. Captain Williamson,[72] who carefully attended to this subject, states that "hounds are the most rapid in their decline;" "greyhounds and

[71] 'History of Quadrupeds,' 1793, vol. i. p. 238.

[72] 'Oriental Field Sports,' quoted by Youatt, 'The Dog,' p. 15.

pointers, also, rapidly decline." But spaniels, after eight or nine generations, and without a cross from Europe, are as good as their ancestors. Dr. Falconer informs me that bulldogs, which have been known, when first brought into the country, to pin down even an elephant by its trunk, not only fall off after two or three generations in pluck and ferocity, but lose the under-hung character of their lower jaws; their muzzles become finer and their bodies lighter. English dogs imported into India are so valuable that probably due care has been taken to prevent their crossing with native dogs; so that the deterioration cannot be thus accounted for. The Rev. R. Everest informs me that he obtained a pair of setters, born in India, which perfectly resembled their Scotch parents: he raised several litters from them in Delhi, taking the most stringent precautions to prevent a cross, but he never succeeded, though this was only the second generation in India, in obtaining a single young dog like its parents in size or make; their nostrils were more contracted, their noses more pointed, their size inferior, and their limbs more slender. This remarkable tendency to rapid deterioration in European dogs subjected to the climate of India, may perhaps partly be accounted for by the tendency to reversion to a primordial condition which many animals exhibit, as we shall see in a future chapter, when exposed to new conditions of life.

Some of the peculiarities characteristic of the several breeds of the dog have probably arisen suddenly, and, though strictly inherited, may be called monstrosities; for instance, the shape of the legs and body in the turnspit of Europe and India; the shape of the head and the under-hanging jaw in the bull and pug-dog, so alike in this one respect and so unlike in all others. A peculiarity suddenly arising, and therefore in one sense deserving to be called a monstrosity, may, however, be increased and fixed by man's selection. We can hardly doubt that long-continued training, as with the greyhound in coursing hares, as with water-dogs in swimming—and the want of exercise, in the case of lapdogs—must have produced some direct effect on their structure and instincts. But we shall immediately see that the most potent cause of change has probably been the selection, both methodical and unconscious, of slight individual differences,—the latter kind of selection resulting from the occasional preservation, during hundreds of generations, of those individual dogs which were the most useful to man for certain purposes and under certain conditions of life. In a future chapter on Selection I shall show that even barbarians attend closely to the qualities of their dogs. This unconscious selection by man would be aided by a kind of natural selection; for the dogs of savages have partly to gain their own subsistence; for instance, in Australia, as we hear from Mr. Nind,[73] the dogs are sometimes compelled by want to leave their masters and provide for themselves; but in a few days they generally return. And we may infer that dogs of different shapes, sizes, and habits, would have the best chance of surviving under different circumstances,—on open, sterile plains, where they have to run down their own prey,—on rocky coasts, where they have to feed on crabs and fish left in the tidal pools, as in the case of New Guinea and Tierra del Fuego. In this latter country, as I am informed by Mr. Bridges, the Catechist to the Mission, the dogs turn over the stones on the shore to catch the crustaceans which lie beneath, and they "are clever enough to knock off the shell-fish at a first blow;" for if this be not done, shell-fish are well known to have an almost invincible power of adhesion.

It has already been remarked that dogs differ in the degree to which their feet are webbed. In dogs of the Newfoundland breed, which are eminently aquatic in their habits, the skin, according to Isidore Geoffroy,[74] extends to the third phalanges, whilst in ordinary dogs it extends only to the second. In two Newfoundland dogs which I examined, when the toes were stretched apart and viewed on the under side, the skin extended in a nearly straight line between the outer margins of the balls of the toes; whereas, in two terriers of distinct sub-breeds, the skin viewed in the same manner was deeply scooped out. In Canada there is a dog which is peculiar to the country and common there, and this has "half-webbed feet and is fond of the water."[75] English otter-hounds are said to have webbed feet: a friend examined for me the feet of two, in comparison

[73] Quoted by Mr. Galton, 'Domestication of Animals,' p. 13.

[74] 'Hist. Nat. Gén.,' tom. iii. p. 450.

[75] Mr. Greenhow on the Canadian Dog, in Loudon's 'Mag. of Nat. Hist.,' vol. vi., 1833, p. 511.

with the feet of some harriers and bloodhounds; he found the skin variable in extent in all, but more developed in the otter than in the other hounds.[76] As aquatic animals which belong to quite different orders have webbed feet, there can be no doubt that this structure would be serviceable to dogs that frequent the water. We may confidently infer that no man ever selected his water-dogs by the extent to which the skin was developed between their toes; but what he does, is to preserve and breed from those individuals which hunt best in the water, or best retrieve wounded game, and thus he unconsciously selects dogs with feet slightly better webbed. Man thus closely imitates Natural Selection. We have an excellent illustration of this same process in North America, where, according to Sir J. Richardson,[77] all the wolves, foxes, and aboriginal domestic dogs have their feet broader than in the corresponding species of the Old World, and "well calculated for running on the snow." Now, in these Arctic regions, the life or death of every animal will often depend on its success in hunting over the snow when softened; and this will in part depend on the feet being broad; yet they must not be so broad as to interfere with the activity of the animal when the ground is sticky, or with its power of burrowing holes, or with other habits of life.

As changes in domestic breeds which take place so slowly as not to be noticed at any one period, whether due to the selection of individual variations or of differences resulting from crosses, are most important in understanding the origin of our domestic productions, and likewise in throwing indirect light on the changes effected under nature, I will give in detail such cases as I have been able to collect. Lawrence,[78] who paid particular attention to the history of the foxhound, writing in 1829, says that between eighty and ninety years before "an entirely new foxhound was raised through the breeder's art," the ears of the old southern hound being reduced, the bone and bulk lightened, the waist increased in length, and the stature

[76] *See* Mr. C. O. Groom-Napier on the webbing of the hind feet of Otter-hounds, in 'Land and Water,' Oct. 13th, 1866, p. 270.

[77] 'Fauna Boreali-Americana,' 1829, p. 62.

[78] 'The Horse in all his Varieties,' &c., 1829, pp. 230, 234.

somewhat added to. It is believed that this was effected by a cross with the greyhound. With respect to this latter dog, Youatt,[79] who is generally cautious in his statements, says that the greyhound within the last fifty years, that is before the commencement of the present century, "assumed a somewhat different character from that which he once possessed. He is now distinguished by a beautiful symmetry of form, of which he could not once boast, and he has even superior speed to that which he formerly exhibited. He is no longer used to struggle with deer, but contends with his fellows over a shorter and speedier course." An able writer[80] believes that our English greyhounds are the descendants, *progressively improved*, of the large rough greyhounds which existed in Scotland so early as the third century. A cross at some former period with the Italian greyhound has been suspected; but this seems hardly probable, considering the feebleness of this latter breed. Lord Orford, as is well known, crossed his famous greyhounds, which failed in courage, with a bulldog—this breed being chosen from being deficient in the power of scent; "after the sixth or seventh generation," says Youatt, "there was not a vestige left of the form of the bulldog, but his courage and indomitable perseverance remained."

Youatt infers, from a comparison of an old picture of King Charles's spaniels with the living dog, that "the breed of the present day is materially altered for the worse:" the muzzle has become shorter, the forehead more prominent, and the eyes larger: the changes in this case have probably been due to simple selection. The setter, as this author remarks in another place, "is evidently the large spaniel improved to his present péculiar size and beauty, and taught another way of marking his game. If the form of the dog were not sufficiently satisfactory on this point, we might have recourse to history:" he then refers to a document dated 1685 bearing on this subject, and adds that the pure Irish setter shows no signs of a cross with the pointer, which some authors suspect has been the case with the English setter. Another writer[81] remarks

[79] 'The Dog,' 1845, pp. 31, 35; with respect to King Charles's spaniel, p. 45; for the setter, p. 90.

[80] In the 'Encyclop. of Rural Sports,' p. 557.

[81] 'The Farrier,' 1828, vol. i. p. 337.

that, if the mastiff and English bulldog had formerly been as distinct as they are at the present time (*i.e.* 1828), so accurate an observer as the poet Gay (who was the author of 'Rural Sports' in 1711) would have spoken in his Fable of the *Bull and the Bulldog*, and not of the *Bull and the Mastiff*. There can be no doubt that the fancy bulldogs of the present day, now that they are not used for bull-baiting, have become greatly reduced in size, without any express intention on the part of the breeder. Our pointers are certainly descended from a Spanish breed, as even their names, Don, Ponto, Carlos, &c., would show: it is said that they were not known in England before the Revolution in 1688;[82] but the breed since its introduction has been much modified, for Mr. Borrow, who is a sportsman and knows Spain intimately well, informs me that he has not seen in that country any breed "corresponding in figure with the English pointer; but there are genuine pointers near Xeres which have been imported by English gentlemen." A nearly parallel case is offered by the Newfoundland dog, which was certainly brought into England from that country, but which has since been so much modified that, as several writers have observed, it does not now closely resemble any existing native dog in Newfoundland.[83]

These several cases of slow and gradual changes in our English dogs possess some interest; for though the changes have generally, but not invariably, been caused by one or two crosses with a distinct breed, yet we may feel sure, from the well-known extreme variability of crossed breeds, that rigorous and long-continued selection must have been practised, in order to improve them in a definite manner. As soon as any strain or family became slightly improved or better adapted to altered circumstances, it would tend to supplant the older and less improved strains. For instance, as soon as the old foxhound was improved by a cross with the greyhound, or by simple selection, and assumed its present character—and the change was probably required by

[82] *See* Col. Hamilton Smith on the antiquity of the Pointer, in 'Naturalist's Library,' vol. x. p. 196.

[83] The Newfoundland dog is believed to have originated from a cross between the Esquimaux dog and a large French hound. *See* Dr. Hodgkin, 'Brit. Assoc.,' 1844; Bechstein's 'Naturgesch. Deutschlands,' Band i. s. 574; 'Naturalist's Library,' vol. x. p. 132; also Mr. Jukes' 'Excursion in and about Newfoundland.'

the increased fleetness of our hunters—it rapidly spread throughout the country, and is now everywhere nearly uniform. But the process of improvement is still going on, for every one tries to improve his strain by occasionally procuring dogs from the best kennels. Through this process of gradual substitution the old English hound has been lost; and so it has been with the old Irish greyhound and apparently with the old English bulldog. But the extinction of former breeds is apparently aided by another cause; for whenever a breed is kept in scanty numbers, as at present with the bloodhound, it is reared with difficulty, and this apparently is due to the evil effects of long-continued close interbreeding. As several breeds of the dog have been slightly but sensibly modified within so short a period as the last one or two centuries, by the selection of the best individual dogs, modified in many cases by crosses with other breeds; and as we shall hereafter see that the breeding of dogs was attended to in ancient times, as it still is by savages, we may conclude that we have in selection, even if only occasionally practised, a potent means of modification.

DOMESTIC CATS.

CATS have been domesticated in the East from an ancient period; Mr. Blyth informs me that they are mentioned in a Sanskrit writing 2000 years old, and in Egypt their antiquity is known to be even greater, as shown by monumental drawings and their mummied bodies. These mummies, according to De Blainville,[84] who has particularly studied the subject, belong to no less than three species, namely, *F. caligulata*, *bubastes*, and *chaus*. The two former species are said to be still found, both wild and domesticated, in parts of Egypt. *F. caligulata* presents a difference in the first inferior milk molar tooth, as compared with the domestic cats of Europe, which makes De Blainville conclude that it is not one of the parent-forms of our cats. Several naturalists, as Pallas, Temminck, Blyth, believe that domestic cats are the descendants of several species com-

[84] De Blainville, 'Ostéographie, Felis,' p. 65, on the character of *F. caligulata*; pp. 85, 89, 90, 175, on the other mummied species. He quotes Ehrenberg on *F. maniculata* being mummied.

mingled: it is certain that cats cross readily with various wild species, and it would appear that the character of the domestic breeds has, at least in some cases, been thus affected. Sir W. Jardine has no doubt that, "in the north of Scotland, there has been occasional crossing with our native species (*F. sylvestris*), and that the result of these crosses has been kept in our houses. I have seen," he adds, "many cats very closely resembling the wild cat, and one or two that could scarcely be distinguished from it." Mr. Blyth[85] remarks on this passage, "but such cats are never seen in the southern parts of England; still, as compared with any Indian tame cat, the affinity of the ordinary British cat to *F. sylvestris* is manifest; and due I suspect to frequent intermixture at a time when the tame cat was first introduced into Britain and continued rare, while the wild species was far more abundant than at present." In Hungary, Jeitteles[86] was assured on trustworthy authority that a wild male cat crossed with a female domestic cat, and that the hybrids long lived in a domesticated state. In Algiers the domestic cat has crossed with the wild cat (*F. Lybica*) of that country.[87] In South Africa, as Mr. E. Layard informs me, the domestic cat intermingles freely with the wild *F. caffra;* he has seen a pair of hybrids which were quite tame and particularly attached to the lady who brought them up; and Mr. Fry has found that these hybrids are fertile. In India the domestic cat, according to Mr. Blyth, has crossed with four Indian species. With respect to one of these species, *F. chaus*, an excellent observer, Sir W. Elliot, informs me that he once killed, near Madras, a wild brood, which were evidently hybrids from the domestic cat; these young animals had a thick lynx-like tail and the broad brown bar on the inside of the forearm characteristic of *F. chaus.* Sir W. Elliot adds that he has often observed this same mark on the forearms of domestic cats in India. Mr. Blyth states that domestic cats coloured nearly like *F. chaus*, but not resembling that species in shape, abound in

[85] Asiatic Soc. of Calcutta; Curator's Report, Aug. 1856. The passage from Sir W. Jardine is quoted from this Report. Mr. Blyth, who has especially attended to the wild and domestic cats of India, has given in this Report a very interesting discussion on their origin.

[86] 'Fauna Hungariæ Sup.,' 1862, s. 12.

[87] Isid. Geoffroy Saint Hilaire, 'Hist. Nat. Gén.,' tom. iii. p. 177.

Bengal; he adds, "such a colouration is utterly unknown in European cats, and the proper tabby markings (pale streaks on a black ground, peculiarly and symmetrically disposed), so common in English cats, are never seen in those of India." Dr. D. Short has assured Mr. Blyth [88] that at Hansi hybrids between the common cat and *F. ornata* (or *torquata*) occur, "and that many of the domestic cats of that part of India were undistinguishable from the wild *F. ornata*." Azara states, but only on the authority of the inhabitants, that in Paraguay the cat has crossed with two native species. From these several cases we see that in Europe, Asia, Africa, and America, the common cat, which lives a freer life than most other domesticated animals, has crossed with various wild species; and that in some instances the crossing has been sufficiently frequent to affect the character of the breed.

Whether domestic cats have descended from several distinct species, or have only been modified by occasional crosses, their fertility, as far as is known, is unimpaired. The large Angora or Persian cat is the most distinct in structure and habits of all the domestic breeds; and is believed by Pallas, but on no distinct evidence, to be descended from the *F. manul* of middle Asia; but I am assured by Mr. Blyth that this cat breeds freely with Indian cats, which, as we have already seen, have apparently been much crossed with *F. chaus*. In England half-bred Angora cats are perfectly fertile with the common cat; I do not know whether the half-breeds are fertile one with another; but as they are common in some parts of Europe, any marked degree of sterility could hardly fail to have been noticed.

Within the same country we do not meet with distinct races of the cat, as we do of dogs and of most other domestic animals; though the cats of the same country present a considerable amount of fluctuating variability. The explanation obviously is that, from their nocturnal and rambling habits, indiscriminate crossing cannot without much trouble be prevented. Selection cannot be brought into play to produce distinct breeds, or to keep those distinct which have been imported from foreign lands. On the other hand, in islands and

[88] 'Proc. Zoolog. Soc.,' 1863, p. 184.

in countries completely separated from each other, we meet with breeds more or less distinct; and these cases are worth giving as showing that the scarcity of distinct races in the same country is not caused by a deficiency of variability in the animal. The tail-less cats of the Isle of Man are said to differ from common cats not only in the want of a tail, but in the greater length of their hind legs, in the size of their heads, and in habits. The Creole cat of Antigua, as I am informed by Mr. Nicholson, is smaller, and has a more elongated head, than the British cat. In Ceylon, as Mr. Thwaites writes to me, every one at first notices the different appearance of the native cat from the English animal; it is of small size, with closely lying hairs; its head is small, with a receding forehead; but the ears are large and sharp; altogether it has what is there called a "low-caste" appearance. Rengger[89] says that the domestic cat, which has been bred for 300 years in Paraguay, presents a striking difference from the European cat; it is smaller by a fourth, has a more lanky body, its hair is short, shining, scanty, and lies close, especially on the tail: he adds that the change has been less at Ascension, the capital of Paraguay, owing to the continual crossing with newly imported cats; and this fact well illustrates the importance of separation. The conditions of life in Paraguay appear not to be highly favourable to the cat, for, though they have run half-wild, they do not become thoroughly feral, like so many other European animals. In another part of South America, according to Roulin,[90] the introduced cat has lost the habit of uttering its hideous nocturnal howl. The Rev. W. D. Fox purchased a cat in Portsmouth, which he was told came from the coast of Guinea; its skin was black and wrinkled, fur bluish-grey and short, its ears rather bare, legs long, and whole aspect peculiar. This "negro" cat was fertile with common cats. On the opposite coast of Africa, at Mombas, Captain Owen, R.N.,[91] states that all the cats are covered with short stiff hair instead of fur: he gives a curious account of a cat from Algoa Bay, which had been kept for some time on board and could be identified with certainty; this

[89] 'Saeugethiere von Paraguay,' 1830, s. 212.

[90] 'Mem. présentés par divers Savans: Acad. Roy. des Sciences,' tom. vi. p. 346. Gomara first noticed this fact in 1554.

[91] 'Narrative of Voyages,' vol. ii. p. 180.

animal was left for only eight weeks at Mombas, but during that short period it "underwent a complete metamorphosis, having parted with its sandy-coloured fur." A cat from the Cape of Good Hope has been described by Desmarest as remarkable from a red stripe extending along the whole length of its back. Throughout an immense area, namely, the Malayan archipelago, Siam, Pegu, and Burmah, all the cats have truncated tails about half the proper length,[92] often with a sort of knot at the end. In the Caroline archipelago the cats have very long legs, and are of a reddish-yellow colour.[93] In China a breed has drooping ears. At Tobolsk, according to Gmelin, there is a red-coloured breed. In Asia, also, we find the well-known Angora or Persian breed.

The domestic cat has run wild in several countries, and everywhere assumes, as far as can be judged by the short recorded descriptions, a uniform character. Near Maldonado, in La Plata, I shot one which seemed perfectly wild; it was carefully examined by Mr. Waterhouse,[94] who found nothing remarkable in it, excepting its great size. In New Zealand, according to Dieffenbach, the feral cats assume a streaky grey colour like that of wild cats; and this is the case with the half-wild cats of the Scotch Highlands.

We have seen that distant countries possess distinct domestic races of the cat. The differences may be in part due to descent from several aboriginal species, or at least to crosses with them. In some cases, as in Paraguay, Mombas, and Antigua, the differences seem due to the direct action of different conditions of life. In other cases some slight effect may possibly be attributed to natural selection, as cats in many cases have largely to support themselves and to escape diverse dangers. But man, owing to the difficulty of pairing cats, has done nothing by methodical selection; and probably very little by unintentional selection; though in each litter he generally saves the prettiest,

[92] J. Crawfurd, 'Descript. Dict. of the Indian Islands,' p. 255. The Madagascar cat is said to have a twisted tail: *see* Desmarest, in 'Encyclop. Nat. Mamm.,' 1820, p. 233, for some of the other breeds.

[93] Admiral Lutké's Voyage, vol. iii. p. 308.

[94] 'Zoology of the Voyage of the Beagle, Mammalia,' p. 20. Dieffenbach, 'Travels in New Zealand,' vol. ii. p. 185. Ch. St. John, 'Wild Sports of the Highlands,' 1846, p. 40.

and values most a good breed of mouse or rat-catchers. Those cats which have a strong tendency to prowl after game, generally get destroyed by traps. As cats are so much petted, a breed bearing the same relation to other cats, that lapdogs bear to larger dogs, would have been much valued; and if selection could have been applied, we should certainly have had many breeds in each long-civilized country, for there is plenty of variability to work upon.

We see in this country considerable diversity in size, some in the proportions of the body, and extreme variability in colouring. I have only lately attended to this subject, but have already heard of some singular cases of variation; one of a cat born in the West Indies toothless, and remaining so all its life. Mr. Tegetmeier has shown me the skull of a female cat with its canines so much developed that they protruded uncovered beyond the lips; the tooth with the fang being 95, and the part projecting from the gum ·6 of an inch in length. I have heard of a family of six-toed cats. The tail varies greatly in length; I have seen a cat which always carried its tail flat on its back when pleased. The ears vary in shape, and certain strains, in England, inherit a pencil-like tuft of hairs, above a quarter of an inch in length, on the tips of their ears; and this same peculiarity, according to Mr. Blyth, characterises some cats in India. The great variability in the length of the tail and the lynx-like tufts of hairs on the ears are apparently analogous to differences in certain wild species of the genus. A much more important difference, according to Daubenton,[95] is that the intestines of domestic cats are wider, and a third longer, than in wild cats of the same size; and this apparently has been caused by their less strictly carnivorous diet.

[95] Quoted by Isid. Geoffroy, 'Hist. Nat. Gén.,' tom. iii. p. 427.

CHAPTER II.

HORSES AND ASSES.

HORSE. — DIFFERENCES IN THE BREEDS — INDIVIDUAL VARIABILITY OF — DIRECT EFFECTS OF THE CONDITIONS OF LIFE — CAN WITHSTAND MUCH COLD — BREEDS MUCH MODIFIED BY SELECTION — COLOURS OF THE HORSE —DAPPLING — DARK STRIPES ON THE SPINE, LEGS, SHOULDERS, AND FOREHEAD — DUN-COLOURED HORSES MOST FREQUENTLY STRIPED — STRIPES PROBABLY DUE TO REVERSION TO THE PRIMITIVE STATE OF THE HORSE.

ASSES. — BREEDS OF — COLOUR OF — LEG- AND SHOULDER- STRIPES — SHOULDER-STRIPES SOMETIMES ABSENT, SOMETIMES FORKED.

THE history of the Horse is lost in antiquity. Remains of this animal in a domesticated condition have been found in the Swiss lake-dwellings, belonging to the latter part of the Stone period.[1] At the present time the number of breeds is great, as may be seen by consulting any treatise on the Horse.[2] Looking only to the native ponies of Great Britain, those of the Shetland Isles, Wales, the New Forest, and Devonshire are distinguishable; and so it is with each separate island in the great Malay archipelago.[3] Some of the breeds present great differences in size, shape of ears, length of mane, proportions of the body, form of the withers and hind quarters, and especially in the head. Compare the race-horse, dray-horse, and a Shetland pony in size, configuration, and disposition; and see how much greater the difference is than between the six or seven other living species of the genus Equus.

[1] Rütimeyer, 'Fauna der Pfalbauten,' 1861, s. 122.

[2] *See* Youatt on the Horse: J. Lawrence on the Horse, 1829: W. C. L. Martin, 'History of the Horse,' 1845: Col. Ham. Smith, in 'Naturalist's Library, Horses,' 1841, vol. xii.: Prof. Veith, 'Die Naturgesch. Haussäugethiere,' 1856.

[3] Crawfurd, 'Descript. Dict. of Indian Islands,' 1856, p. 153. "There are many different breeds, every island having at least one peculiar to it." Thus in Sumatra there are at least two breeds; in Achin and Batubara one; in Java several breeds; one in Bali, Lomboc, Sumbawa (one of the best breeds), Tambora, Bima, Gunung-api, Celebes, Sumba, and Philippines. Other breeds are specified by Zollinger in the 'Journal of the Indian Archipelago,' vol. v. p. 343, &c.

Of individual variations not known to characterise particular breeds, and not great or injurious enough to be called monstrosities, I have not collected many cases. Mr. G. Brown, of the Cirencester Agricultural College, who has particularly attended to the dentition of our domestic animals, writes to me that he has "several times noticed eight permanent incisors instead of six in the jaw." Male horses alone properly have canines, but they are occasionally found in the mare, though of small size.[4] The number of ribs is properly eighteen, but Youatt[5] asserts that not unfrequently there are nineteen on each side, the additional one being always the posterior rib. I have seen several notices of variations in the bones of the leg; thus Mr. Price[6] speaks of an additional bone in the hock, and of certain abnormal appearances between the tibia and astragalus, as quite common in Irish horses, and not due to disease. Horses have often been observed, according to M. Gaudry,[7] to possess a trapezium and a rudiment of a fifth metacarpal bone, so that "one sees appearing by monstrosity, in the foot of the horse, structures which normally exist in the foot of the Hipparion," —an allied and extinct animal. In various countries horn-like projections have been observed on the frontal bones of the horse: in one case described by Mr. Percival they arose about two inches above the orbital processes, and were "very like those in a calf from five to six months old," being from half to three-quarters of an inch in length.[8] Azara has described two cases in South America in which the projections were between three and four inches in length: other instances have occurred in Spain.

That there has been much inherited variation in the horse cannot be doubted, when we reflect on the number of the breeds existing throughout the world or even within the same country, and when we know that they have largely increased in number

[4] 'The Horse,' &c., by John Lawrence, 1829, p. 14.

[5] 'The Veterinary,' London, vol. v. p. 543.

[6] Proc. Veterinary Assoc., in 'The Veterinary,' vol. xiii. p. 42.

[7] 'Bulletin de la Soc. Géolog.,' tom. xxii., 1866, p. 22.

[8] Mr. Percival, of the Enniskillen Dragoons, in 'The Veterinary,' vol. i. p. 224: *see* Azara, 'Des Quadrupèdes du Paraguay,' tom. ii. p. 313. The French translator of Azara refers to other cases mentioned by Huzard as occurring in Spain.

since the earliest known records.[9] Even in so fleeting a character as colour, Hofacker[10] found that, out of two hundred and sixteen cases in which horses of the same colour were paired, only eleven pairs produced foals of a quite different colour. As Professor Low[11] has remarked, the English race-horse offers the best possible evidence of inheritance. The pedigree of a race-horse is of more value in judging of its probable success than its appearance: "King Herod" gained in prizes 201,505*l.* sterling, and begot 497 winners; "Eclipse" begot 334 winners.

Whether the whole amount of difference between the various breeds be due to variation is doubtful. From the fertility of the most distinct breeds[12] when crossed, naturalists have generally looked at all the breeds as having descended from a single species. Few will agree with Colonel H. Smith, who believes that they have descended from no less than five primitive and differently coloured stocks.[13] But as several species and varieties of the horse existed[14] during the later tertiary periods, and as Rütimeyer found differences in the size and form of the skull in the earliest known domesticated horses,[15] we ought not to feel sure that all our breeds have descended from a single species. As we see that the savages of North and South America easily reclaim the feral horses, there is no improbability in savages in various quarters of the world having domesticated more than one native species or natural race. No aboriginal or truly wild horse is positively known now to exist; for it is thought by some authors that the wild horses of the East are escaped domestic animals.[16] If our domestic breeds have descended from several

[9] Godron, 'De l'Espèce,' tom. i. p. 378.

[10] 'Ueber die Eigenschaften,' &c., 1828, s. 10.

[11] 'Domesticated Animals of the British Islands,' pp. 527, 532. In all the veterinary treatises and papers which I have read, the writers insist in the strongest terms on the inheritance by the horse of all good and bad tendencies and qualities. Perhaps the principle of inheritance is not really stronger in the horse than in any other animal; but, from its value, the tendency has been more carefully observed.

[12] Andrew Knight crossed breeds so different in size as a dray-horse and Norwegian pony: *see* A. Walker on 'Intermarriage,' 1838, p. 205.

[13] 'Naturalist's Library,' Horses, vol. xii. p. 208.

[14] Gervais, 'Hist. Nat. Mamm.,' tom. ii. p. 143. Owen, 'British Fossil Mammals,' p. 383.

[15] 'Kenntniss der fossilen Pferde,' 1863, s. 131.

[16] Mr. W. C. L. Martin ('The Horse,' 1845, p. 34), in arguing against the belief that the wild Eastern horses are merely feral, has remarked on the improbability of man in ancient times having extirpated a species in a region where it can now exist in numbers.

species or natural races, these apparently have all become extinct in the wild state. With our present knowledge, the common view that all have descended from a single species is, perhaps, the most probable.

With respect to the causes of the modifications which horses have undergone, the conditions of life seem to produce a considerable direct effect. Mr. D. Forbes, who has had excellent opportunities of comparing the horses of Spain with those of South America, informs me that the horses of Chile, which have lived under nearly the same conditions as their progenitors in Andalusia, remain unaltered, whilst the Pampas horses and the Puno ponies are considerably modified. There can be no doubt that horses become greatly reduced in size and altered in appearance by living on mountains and islands; and this apparently is due to want of nutritious or varied food. Every one knows how small and rugged the ponies are on the Northern islands and on the mountains of Europe. Corsica and Sardinia have their native ponies; and there were,[17] or still are, on some islands on the coast of Virginia, ponies like those of the Shetland Islands, which are believed to have originated through exposure to unfavourable conditions. The Puno ponies, which inhabit the lofty regions of the Cordillera, are, as I hear from Mr. D. Forbes, strange little creatures, very unlike their Spanish progenitors. Further south, in the Falkland Islands, the offspring of the horses imported in 1764 have already so much deteriorated in size[18] and strength that they are unfitted for catching wild cattle with the lasso; so that fresh horses have to be brought for this purpose from La Plata at a great expense. The reduced size of the horses bred on both southern and northern islands, and on several mountain-chains, can hardly have been caused by the cold, as a similar reduction has occurred on the Virginian and Mediterranean islands. The horse can withstand intense cold, for wild troops live on the plains of Siberia under lat. 56°,[19] and aboriginally the horse must

[17] 'Transact. Maryland Academy,' vol. i. part i. p. 28.

[18] Mr. Mackinnon on 'The Falkland Islands,' p. 25. The average height of the Falkland horses is said to be 14 hands 2 inches. *See* also my 'Journal of Researches.'

[19] Pallas, 'Act. Acad. St. Petersburgh,' 1777, part ii. p. 265. With respect to the tarpans scraping away the snow, *see* Col. Hamilton Smith in 'Nat. Lib.,' vol. xii. p. 165.

have inhabited countries annually covered with snow, for he long retains the instinct of scraping it away to get at the herbage beneath. The wild tarpans in the East have this instinct; and, as I am informed by Admiral Sulivan, this is likewise the case with the horses which have run wild on the Falkland Islands; now this is the more remarkable as the progenitors of these horses could not have followed this instinct during many generations in La Plata: the wild cattle of the Falklands never scrape away the snow, and perish when the ground is long covered. In the northern parts of America the horses, descended from those introduced by the Spanish conquerors of Mexico, have the same habit, as have the native bisons, but not so the cattle introduced from Europe.[20]

The horse can flourish under intense heat as well as under intense cold, for he is known to come to the highest perfection, though not attaining a large size, in Arabia and northern Africa. Much humidity is apparently more injurious to the horse than heat or cold. In the Falkland Islands, horses suffer much from the dampness; and this same circumstance may perhaps partly account for the singular fact that to the eastward of the Bay of Bengal,[21] over an enormous and humid area, in Ava, Pegu, Siam, the Malayan archipelago, the Loo Choo Islands, and a large part of China, no full-sized horse is found. When we advance as far eastward as Japan, the horse reacquires his full size.[22]

With most of our domesticated animals, some breeds are kept on account of their curiosity or beauty; but the horse is valued almost solely for its utility. Hence semi-monstrous breeds are not preserved; and probably all the existing breeds have been slowly formed either by the direct action of the conditions of life, or through the selection of individual differences. No doubt semi-monstrous breeds might have been formed: thus Mr. Waterton records[23] the case of a mare which produced suc-

[20] Franklin's 'Narrative,' vol. i. p. 87; note by Sir J. Richardson.

[21] Mr. J. H. Moor, 'Notices of the Indian Archipelago:' Singapore, 1837, p. 189. A pony from Java was sent ('Athenæum,' 1842, p. 718) to the Queen only 28 inches in height. For the Loo Choo Islands, *see* Beechey's 'Voyage,' 4th edit., vol. i. p. 499.

[22] J. Crawford, 'History of the Horse;' 'Journal of Royal United Service Institution,' vol. iv.

[23] 'Essays on Natural History,' 2nd series, p. 161.

cessively three foals without tails; so that a tailless race might have been formed like the tailless races of dogs and cats. A Russian breed of horses is said to have frizzled hair, and Azara[24] relates that in Paraguay horses are occasionally born, but are generally destroyed, with hair like that on the head of a negro; and this peculiarity is transmitted even to half-breeds: it is a curious case of correlation that such horses have short manes and tails, and their hoofs are of a peculiar shape like those of a mule.

It is scarcely possible to doubt that the long-continued selection of qualities serviceable to man has been the chief agent in the formation of the several breeds of the horse. Look at a dray-horse, and see how well adapted he is to draw heavy weights, and how unlike in appearance to any allied wild animal. The English race-horse is known to have proceeded from the commingled blood of Arabs, Turks, and Barbs; but selection and training have together made him a very different animal from his parent-stocks. As a writer in India, who evidently knows the pure Arab well, asks, who now, "looking at our present breed of race-horses, could have conceived that they were the result of the union of the Arab horse and African mare?" The improvement is so marked that in running for the Goodwood Cup "the first descendants of Arabian, Turkish, and Persian horses, are allowed a discount of 18 lbs. weight; and when both parents are of these countries a discount of 36 lbs.[25] It is notorious that the Arabs have long been as careful about the pedigree of their horses as we are, and this implies great and continued care in breeding. Seeing what has been done in England by careful breeding, can we doubt that the Arabs must likewise have produced during the course of centuries a marked effect on the qualities of their horses? But we may go much farther back in time, for in the most ancient known book, the Bible, we hear of studs carefully kept for breeding,

[24] 'Quadrupèdes du Paraguay,' tom. ii. p. 333.

[25] Prof. Low, 'Domesticated Animals,' p. 546. With respect to the writer in India, *see* 'India Sporting Review,' vol. ii. p. 181. As Lawrence has remarked ('The Horse,' p. 9), "perhaps no instance has ever occurred of a three-part bred horse (*i.e.* a horse, one of whose grand-parents was of impure blood) saving his distance in running two miles with thoroughbred racers." Some few instances are on record of seven-eighths racers having been successful.

and of horses imported at high prices from various countries.[26] We may therefore conclude that, whether or not the various existing breeds of the horse have proceeded from one or more aboriginal stocks, yet that a great amount of change has resulted from the direct action of the conditions of life, and probably a still greater amount from the long-continued selection by man of slight individual differences.

With several domesticated quadrupeds and birds, certain coloured marks are either strongly inherited or tend to reappear after having long been lost. As this subject will hereafter be seen to be of importance, I will give a full account of the colouring of horses. All English breeds, however unlike in size and appearance, and several of those in India and the Malay archipelago, present a similar range and diversity of colour. The English race-horse, however, is said[27] never to be dun-coloured; but as dun and cream-coloured horses are considered by the Arabs as worthless, "and fit only for Jews to ride,"[28] these tints may have been removed by long-continued selection. Horses of every colour, and of such widely different kinds as dray-horses, cobs, and ponies, are all occasionally dappled,[29] in the same manner as is so conspicuous with grey horses. This fact does not throw any clear light on the colouring of the aboriginal horse, but is a case of analogous variation, for even asses are sometimes dappled, and I have seen, in the British Museum, a hybrid from the ass and zebra dappled on its hinder quarters. By the expression analogous variation (and it is one that I shall often have occasion to use) I mean a variation occurring in a species or variety which resembles a normal character in another and distinct species or variety. Analogous variations may arise, as will be explained in a future chapter,

[26] Prof. Gervais (in his 'Hist. Nat. Mamm.,' tom. ii. p. 144) has collected many facts on this head. For instance, Solomon (Kings, b. i. ch. x. v. 28) bought horses in Egypt at a high price.

[27] 'The Field,' July 13th, 1861, p. 42.

[28] E. Vernon Harcourt, 'Sporting in Algeria,' p. 26.

[29] I state this from my own observations made during several years on the colours of horses. I have seen cream-coloured, light-dun and mouse-dun horses dappled, which I mention because it has been stated (Martin, 'History of the Horse,' p. 134) that duns are never dappled. Martin (p. 205) refers to dappled asses. In 'The Farrier' (London, 1828, pp. 453, 455) there are some good remarks on the dappling of horses; and likewise in Col. Hamilton Smith on 'The Horse.'

from two or more forms with a similar constitution having been exposed to similar conditions,—or from one of two forms having reacquired through reversion a character inherited by the other form from their common progenitor,—or from both forms having reverted to the same ancestral character. We shall immediately see that horses occasionally exhibit a tendency to become striped over a large part of their bodies; and as we know that stripes readily pass into spots and cloudy marks in the varieties of the domestic cat and in several feline species—even the cubs of the uniformly-coloured lion being spotted with dark marks on a lighter ground—we may suspect that the dappling of the horse, which has been noticed by some authors with surprise, is a modification or vestige of a tendency to become striped.

This tendency in the horse to become striped is in several respects an interesting fact. Horses of all colours, of the most diverse breeds, in various parts of the world, often have a dark stripe extending along the spine, from the mane to the tail; but this is so common that I need enter into no particulars.[30] Occasionally horses are transversely barred on the legs, chiefly on the under side; and more rarely they have a distinct stripe on the shoulder, like that on the shoulder of the ass, or a broad dark patch representing a stripe. Before entering on any details I must premise that

Fig. 1.—Dun Devonshire Pony, with shoulder, spinal, and leg stripes.

[30] Some details are given in 'The Farrier,' 1828, pp. 452, 455. One of the least ponies I ever saw, of the colour of a mouse, had a conspicuous spinal stripe. A small Indian chesnut pony had the same stripe, as had a remarkably heavy chesnut cart-horse. Race-horses often have the spinal stripe.

the term dun-coloured is vague, and includes three groups of colour, viz. that between cream-colour and reddish-brown, which graduates into light-bay or light-chesnut—this, I believe, is often called fallow-dun; secondly, leaden or slate-colour or mouse-dun, which graduates into an ash-colour; and, lastly, dark-dun, between brown and black. In England I have examined a rather large, lightly-built, fallow-dun Devonshire pony (fig. 1), with a conspicuous stripe along the back, with light transverse stripes on the under sides of its front legs, and with four parallel stripes on each shoulder. Of these four stripes the posterior one was very minute and faint; the anterior one, on the other hand, was long and broad, but interrupted in the middle, and truncated at its lower extremity, with the anterior angle produced into a long tapering point. I mention this latter fact because the shoulder-stripe of the ass occasionally presents exactly the same appearance. I have had an outline and description sent to me of a small, purely-bred, light fallow-dun Welch pony, with a spinal stripe, a single transverse stripe on each leg, and three shoulder-stripes; the posterior stripe corresponding with that on the shoulder of the ass was the longest, whilst the two anterior parallel stripes, arising from the mane, decreased in length, in a reversed manner as compared with the shoulder-stripes on the above-described Devonshire pony. I have seen a bright fallow-dun, strong cob, with its front legs transversely barred on the under sides in the most conspicuous manner; also a dark-leaden mouse-coloured pony with similar leg stripes, but much less conspicuous; also a bright fallow-dun colt, fully three-parts thoroughbred, with very plain transverse stripes on the legs; also a chesnut-dun cart-horse with a conspicuous spinal stripe, with distinct traces of shoulder-stripes, but none on the legs; I could add other cases. My son made a sketch for me of a large, heavy, Belgian cart-horse, of a fallow-dun, with a conspicuous spinal stripe, traces of leg-stripes, and with two parallel (three inches apart) stripes about seven or eight inches in length on both shoulders. I have seen another rather light cart-horse, of a dirty dark cream-colour, with striped legs, and on one shoulder a large ill-defined dark cloudy patch, and on the opposite shoulder two parallel faint stripes. All the cases yet mentioned are duns of various tints; but Mr. W. W. Edwards has seen a nearly thoroughbred chesnut horse which had the spinal stripe, and distinct bars on the legs; and I have seen two bay carriage-horses with black spinal stripes; one of these horses had on each shoulder a light shoulder-stripe, and the other had a broad black ill-defined stripe, running obliquely half-way down each shoulder; neither had leg-stripes.

The most interesting case which I have met with occurred in a colt of my own breeding. A bay mare (descended from a dark-brown Flemish mare by a light grey Turcoman horse) was put to Hercules, a thoroughbred dark bay, whose sire (Kingston) and dam were both bays. The colt ultimately turned out brown; but when only a fortnight old it was a dirty bay, shaded with mouse-grey, and in parts with a yellowish tint: it had only a trace of the spinal stripe, with a few obscure transverse bars on the legs; but almost the whole body was marked with very narrow dark stripes, in most parts so obscure as to be visible only in certain lights, like the stripes which may be seen on black kittens. These stripes were distinct on the hind-quarters, where they diverged from the spine, and pointed a little forwards; many of them as they diverged from the spine became a little branched, exactly in the same manner as in some zebrine species. The stripes were plainest on the forehead between the ears, where they formed a set of pointed arches, one under the other, decreasing in size downwards towards the muzzle; exactly similar marks may be seen on the forehead of the quagga and Burchell's zebra. When this foal was two or three months old all the stripes entirely disappeared. I have seen similar marks on the forehead of a fully grown, fallow-dun, cob-like horse, having a conspicuous spinal stripe, and with its front legs well barred.

In Norway the colour of the native horse or pony is dun, varying from almost cream-colour to dark mouse-dun; and an animal is not considered purely bred unless it has the spinal and leg stripes.[31] In one part of the country my son estimated that about a third of the ponies had striped legs; he counted seven stripes on the fore-legs and two on the hind-legs of one pony; only a few of them exhibited traces of shoulder-stripes; but I have heard of a cob imported from Norway which had the shoulder as well as the other stripes well developed. Colonel Ham. Smith[32] alludes to dun-horses with the spinal stripe in the Sierras of Spain; and the horses originally derived from Spain, in some parts of South America, are now duns. Sir W. Elliot informs me that he inspected a herd of 300 South American horses imported into Madras, and many of these had transverse stripes on the legs and short shoulder-stripes; the most strongly marked individual, of which a coloured drawing was sent me, was a mouse-dun, with the shoulder-stripes slightly forked.

In the North-Western parts of India striped horses of more than one breed are apparently commoner than in any other part of the world; and I have received information respecting them from several officers, especially from Colonel Poole, Colonel Curtis, Major Campbell, Brigadier St. John, and others. The Kattywar horses are often fifteen or sixteen hands in height, and are well but lightly built. They are of all colours, but the several kinds of duns prevail; and these are so generally striped, that a horse without stripes is not considered pure. Colonel Poole believes that all the duns have the spinal stripe, the leg-stripes are generally present, and he thinks that about half the horses have the shoulder-stripe; this stripe is sometimes double or treble on both shoulders. Colonel Poole has often seen stripes on the cheeks and sides of the nose. He has seen stripes on the grey and bay Kattywars when first foaled, but they soon faded away. I have received other accounts of cream-coloured, bay, brown, and grey Kattywar horses being striped. Eastward of India, the Shan (north of Burmah) ponies, as I am informed by Mr. Blyth, have spinal, leg, and shoulder stripes. Sir W. Elliot informs me that he saw two bay Pegu ponies with

---

[31] I have received information, through the kindness of the Consul-General, Mr. J. R. Crowe, from Prof. Boeck, Rasck, and Esmarck, on the colours of the Norwegian ponies. *See*, also, 'The Field,' 1861, p. 431.

[32] Col. Ham. Smith, 'Nat. Lib.,' vol. xii. p. 275.

leg-stripes. Burmese and Javanese ponies are frequently dun-coloured, and have the three kinds of stripes, "in the same degree as in England."[33] Mr. Swinhoe informs me that he examined two light-dun ponies of two Chinese breeds, viz. those of Shangai and Amoy; both had the spinal stripe, and the latter an indistinct shoulder-stripe.

We thus see that in all parts of the world breeds of the horse as different as possible, when of a dun-colour (including under this term a wide range of tint from cream to dusky black), and rarely when of bay, grey, and chesnut shades, have the several above-specified stripes. Horses which are of a yellow colour with white mane and tail, and which are sometimes called duns, I have never seen with stripes.[34]

From reasons which will be apparent in the chapter on Reversion, I have endeavoured, but with poor success, to discover whether duns, which are so much oftener striped than other coloured horses, are ever produced from the crossing of two horses, neither of which are duns. Most persons to whom I have applied believe that one parent must be a dun; and it is generally asserted, that, when this is the case, the dun-colour and the stripes are strongly inherited.[35] One case has fallen under my own observation of a foal from a black mare by a bay horse, which when fully grown was a dark fallow-dun and had a narrow but plain spinal stripe. Hofacker[36] gives two instances of mouse-duns (Mausrapp) being produced from two parents of different colours and neither duns.

I have also endeavoured with little success to find out whether the stripes are generally plainer or less plain in the foal than in the adult horse. Colonel Poole informs me that, as he believes, "the stripes are plainest when the colt is first foaled; they then become less and less distinct till after the first coat is shed, when they come out as strongly as before; but certainly often fade away as the age of the horse increases." Two other accounts confirm this fading of the stripes in old horses in India. One writer, on the other hand, states that colts are often born without stripes, but that they appear as the colt grows older. Three authorities affirm that in Norway the stripes are less plain in the foal than in the adult. Perhaps there is no fixed rule. In the case described by me of the young foal which was narrowly striped over nearly all its body, there was no doubt about the early and complete disappearance of the stripes. Mr. W. W. Edwards examined for me twenty-two foals of race-horses, and twelve had the spinal stripe more or less plain; this fact, and some other accounts which I have received, lead me to believe that the spinal stripe often disappears in the English race-horse when old. On the whole I infer that the stripes are generally plainest in the foal, and tend to disappear in old age.

The stripes are variable in colour, but are always darker than the rest of the body. They do not by any means always

[33] Mr. G. Clark, in 'Annal and Mag. of Nat. History,' 2nd series, vol. ii., 1848, p. 363. Mr. Wallace informs me that he saw in Java a dun and clay-coloured horse with spinal and leg stripes.

[34] *See*, also, on this point, 'The Field,' July 27th, 1861, p. 91.

[35] 'The Field,' 1861, pp. 431, 493, 545.

[36] 'Ueber die Eigenschaften,' &c., 1828, s. 13, 14.

coexist on the different parts of the body: the legs may be striped without any shoulder-stripe, or the converse case, which is rarer, may occur; but I have never heard of either shoulder or leg-stripes without the spinal stripe. The latter is by far the commonest of all the stripes, as might have been expected, as it characterises the other seven or eight species of the genus. It is remarkable that so trifling a character as the shoulder-stripe being double or triple should occur in such different breeds as Welch and Devonshire ponies, the Shan pony, heavy cart-horses, light South American horses, and the lanky Kattywar breed. Colonel Hamilton Smith believes that one of his five supposed primitive stocks was dun-coloured and striped; and that the stripes in all the other breeds result from ancient crosses with this one primitive dun; but it is extremely improbable that different breeds living in such distant quarters of the world should all have been crossed with any one aboriginally distinct stock. Nor have we any reason to believe that the effects of a cross at a very remote period could be propagated for so many generations as is implied on this view.

With respect to the primitive colour of the horse having been dun, Colonel Hamilton Smith[37] has collected a large body of evidence showing that this tint was common in the East as far back as the time of Alexander, and that the wild horses of Western Asia and Eastern Europe now are, or recently were, of various shades of dun. It seems that not very long ago a wild breed of dun-coloured horses with a spinal stripe was preserved in the royal parks in Prussia. I hear from Hungary that the inhabitants of that country look at the duns with a spinal stripe as the aboriginal stock, and so it is in Norway. Dun-coloured ponies are not rare in the mountainous parts of Devonshire, Wales, and Scotland, where the aboriginal breed would have had the best chance of being preserved. In South America in the time of Azara, when the horse had been feral for about 250 years, 90 out of 100 horses were "bai-châtains," and the remaining ten were "zains," and not more than one in 2000

[37] 'Naturalist's Library,' vol. xii. (1841), pp. 109, 156 to 163, 280, 281. Cream-colour, passing into Isabella (*i.e.* the colour of the dirty linen of Queen Isabella), seems to have been common in ancient times. *See* also Pallas's account of the wild horses of the East, who speaks of dun and brown as the prevalent colours.

black. Zain is generally translated as dark without any white; but as Azara speaks of mules being "zain-clair," I suspect that zain must have meant dun-coloured. In some parts of the world feral horses show a strong tendency to become roans.[38]

In the following chapters on the Pigeon we shall see that in pure breeds of various colours, when a blue bird is occasionally produced, certain black marks invariably appear on the wings and tail; so again, when variously coloured breeds are crossed, blue birds with the same black marks are frequently produced. We shall further see that these facts are explained by, and afford strong evidence in favour of, the view that all the breeds are descended from the rock-pigeon, or *Columba livia*, which is thus coloured and marked. But the appearance of the stripes on the various breeds of the horse, when of a dun-colour, does not afford nearly such good evidence of their descent from a single primitive stock as in the case of the pigeon; because no certainly wild horse is known as a standard of comparison; because the stripes when they do appear are variable in character; because there is far from sufficient evidence of the appearance of the stripes from the crossing of distinct breeds; and lastly, because all the species of the genus Equus have the spinal stripe, and several have shoulder and leg stripes. Nevertheless the similarity in the most distinct breeds in their general range of colour, in their dappling, and in the occasional appearance, especially in duns, of leg-stripes and of double or triple shoulder-stripes, taken together, indicate the probability of the descent of all the existing races from a single, dun-coloured, more or less striped, primitive stock, to which our horses still occasionally revert.

[38] Azara, 'Quadrupèdes du Paraguay,' tom. ii. p. 307; for the colour of mules, *see* p. 350. In North America, Catlin (vol. ii. p. 57) describes the wild horses, believed to have descended from the Spanish horses of Mexico, as of all colours, black, grey, roan, and roan pied with sorrel. F. Michaux ('Travels in North America,' Eng. translat., p. 235) describes two wild horses from Mexico as roan. In the Falkland Islands, where the horse has been feral only between 60 and 70 years, I was told that roans and iron-greys were the prevalent colours. These several facts show that horses do not generally revert to any uniform colour.

THE ASS.

FOUR species of Asses, besides three of zebras, have been described by naturalists; but there can now be little doubt that our domesticated animal is descended from one alone, namely, the *Asinus tæniopus* of Abyssinia.[39] The ass is sometimes advanced as an instance of an animal domesticated, as we know by the Old Testament, from an ancient period, which has varied only in a very slight degree. But this is by no means strictly true; for in Syria alone there are four breeds;[40] first, a light and graceful animal, with an agreeable gait, used by ladies; secondly, an Arab breed reserved exclusively for the saddle; thirdly, a stouter animal used for ploughing and various purposes; and lastly, the large Damascus breed, with a peculiarly long body and ears. In this country, and generally in Central Europe, though the ass is by no means uniform in appearance, it has not given rise to distinct breeds like those of the horse. This may probably be accounted for by the animal being kept chiefly by poor persons, who do not rear large numbers, nor carefully match and select the young. For, as we shall see in a future chapter, the ass can with ease be greatly improved in size and strength by careful selection, combined no doubt with good food; and we may infer that all its other characters would be equally amenable to selection. The small size of the ass in England and Northern Europe is apparently due far more to want of care in breeding than to cold; for in Western India, where the ass is used as a beast of burden by some of the lower castes, it is not much larger than a Newfoundland dog, "being generally not more than from twenty to thirty inches high."[41]

The ass varies greatly in colour; and its legs, especially the fore-legs, both in England and other countries—for instance, in China—are occasionally barred transversely more plainly than those of dun-coloured horses. With the horse the occasional appearance of leg-stripes was accounted for, through the principle of reversion, by the supposition that the primitive horse was

[39] Dr. Sclater, in 'Proc. Zoolog. Soc.,' 1862, p. 164.

[40] W. C. Martin, 'History of the Horse,' 1845, p. 207.

[41] Col. Sykes' Cat. of Mammalia, 'Proc. Zoolog. Soc.,' July 12th, 1831. Williamson, 'Oriental Field Sports,' vol. ii., quoted by Martin, p. 206.

thus striped; with the ass we may confidently advance this explanation, for the parent-form, the *A. tæniopus*, is known to be barred, though only in a slight degree, across the legs. The stripes are believed to occur most frequently and to be plainest on the legs of the domestic ass during early youth,[42] as is apparently likewise the case with the horse. The shoulder-stripe, which is so eminently characteristic of the species, is nevertheless variable in breadth, length, and manner of termination. I have measured a shoulder-stripe four times as broad as another; and some more than twice as long as others. In one light-grey ass the shoulder-stripe was only six inches in length, and as thin as a piece of string; and in another animal of the same colour there was only a dusky shade representing a stripe. I have heard of three white asses, not albinoes, with no trace of shoulder or spinal stripes;[43] and I have seen nine other asses with no shoulder-stripe, and some of them had no spinal stripe. Three of the nine were light-greys, one a dark-grey, another grey passing into reddish-roan, and the others were brown, two being tinted on parts of their bodies with a reddish or bay shade. Hence we may conclude that, if grey and reddish-brown asses had been steadily selected and bred from, the shoulder-stripe would have been almost as generally and as completely lost as in the case of the horse.

The shoulder-stripe on the ass is sometimes double, and Mr. Blyth has seen even three or four parallel stripes.[44] I have observed in ten cases shoulder-stripes abruptly truncated at the lower end, with the anterior angle produced into a tapering point, precisely as has been figured in the dun Devonshire pony. I have seen three cases of the terminal portion abruptly and angularly bent; and two cases of a distinct though slight forking. In Syria, Dr. Hooker and his party observed for me no less than five instances of the shoulder-stripe being plainly forked over the fore leg. In the common mule it is likewise sometimes forked. When I first noticed the forking and angular bending of the shoulder-stripe, I had seen enough of the stripes

[42] Blyth, in 'Charlesworth's Mag. of Nat. Hist.,' vol. iv., 1840, p. 83. I have also been assured by a breeder that this is the case.

[43] One case is given by Martin, 'The Horse,' p. 205.

[44] 'Journal As. Soc. of Bengal,' vol. xxviii. 1860, p. 231. Martin on the Horse, p. 205.

in the various equine species to feel convinced that even a character so unimportant as this had a distinct meaning, and was thus led to attend to the subject. I now find that in the *Asinus Burchellii* and *quagga*, the stripe which corresponds with the shoulder-stripe of the ass, as well as some of the stripes on the neck, bifurcate, and that some of those near the shoulder have their extremities angularly bent backwards. The forking and angular bending of the stripes on the shoulders apparently stand in relation with the changed direction of the nearly upright stripes on the sides of the body and neck to the transverse bars on the legs. Finally we see that the presence of shoulder, leg, and spinal stripes in the horse,—their occasional absence in the ass,—the occurrence of double and triple shoulder-stripes in both animals, and the similar manner in which these stripes terminate at their lower extremities,—are all cases of analogous variation in the horse and ass. These cases are probably not due to similar conditions acting on similar constitutions, but to a partial reversion in colour to the common progenitor of these two species, as well as of the other species of the genus. We shall hereafter have to return to this subject, and discuss it more fully.

CHAPTER III.

PIGS — CATTLE — SHEEP — GOATS.

PIGS BELONG TO TWO DISTINCT TYPES, SUS SCROFA AND INDICA — TORF-SCHWEIN — JAPAN PIG — FERTILITY OF CROSSED PIGS — CHANGES IN THE SKULL OF THE HIGHLY CULTIVATED RACES — CONVERGENCE OF CHARACTER — GESTATION — SOLID-HOOFED SWINE — CURIOUS APPENDAGES TO THE JAWS — DECREASE IN SIZE OF THE TUSKS — YOUNG PIGS LONGITUDINALLY STRIPED — FERAL PIGS — CROSSED BREEDS.

CATTLE. — ZEBU A DISTINCT SPECIES — EUROPEAN CATTLE PROBABLY DESCENDED FROM THREE WILD FORMS — ALL THE RACES NOW FERTILE TOGETHER — BRITISH PARK CATTLE — ON THE COLOUR OF THE ABORIGINAL SPECIES — CONSTITUTIONAL DIFFERENCES — SOUTH AFRICAN RACES — SOUTH AMERICAN RACES — NIATA CATTLE — ORIGIN OF THE VARIOUS RACES OF CATTLE.

SHEEP. — REMARKABLE RACES OF — VARIATIONS ATTACHED TO THE MALE SEX — ADAPTATIONS TO VARIOUS CONDITIONS — GESTATION OF — CHANGES IN THE WOOL —SEMI-MONSTROUS BREEDS.

GOATS. — REMARKABLE VARIATIONS OF.

THE breeds of the pig have recently been more closely studied, though much still remains to be done, than those of almost any other domesticated animal. This has been effected by Hermann von Nathusius in two admirable works, especially in the later one on the Skulls of the several races, and by Rütimeyer in his celebrated Fauna of the ancient Swiss lake-dwellings.[1] Nathusius has shown that all the known breeds may be divided in two great groups: one resembling in all important respects and no doubt descended from the common wild boar; so that this may be called the *Sus scrofa* group. The other group differs in several important and constant osteological charcters; its wild parent-form is unknown; the name given to it by Nathusius, according to the law of priority, is *Sus Indica* of Pallas. This name must now be followed, though an unfortunate one, as the wild aboriginal does not inhabit India, and the best-known domesticated breeds have been imported from Siam and China.

[1] Hermann von Nathusius, 'Die Racen des Schweines,' Berlin, 1860; and 'Vorstudien fur Geschichte,' &c., 'Schweineschädel,' Berlin, 1864. Rütimeyer, 'Die Fauna der Pfahlbauten,' Basel, 1861.

Firstly, the *Sus scrofa* breeds, or those resembling the common wild boar. These still exist, according to Nathusius (Schweineschädel, s. 75), in various parts of central and northern Europe; formerly every kingdom,[2] and almost every province in Britain, possessed its own native breed; but these are now everywhere rapidly disappearing, being replaced by improved breeds crossed with the *S. Indica* form. The skull in the breeds of the *S. scrofa* type resembles, in all important respects, that of the European wild boar; but it has become (Schweineschädel, s. 63-68) higher and broader relatively to its length; and the hinder part is more upright. The differences, however, are all variable in degree. The breeds which thus resemble *S. scrofa* in their essential skull-characters differ conspicuously from each other in other respects, as in the length of the ears and legs, curvature of the ribs, colour, hairiness, size and proportions of the body.

The wild *Sus scrofa* has a wide range, namely, Europe, North Africa, as identified by osteological characters by Rütimeyer, and Hindostan, as similarly identified by Nathusius. But the wild boars inhabiting these several countries differ so much from each other in external characters, that they have been ranked by some naturalists as specifically distinct. Even within Hindostan these animals, according to Mr. Blyth, form very distinct races in the different districts; in the N. Western provinces, as I am informed by the Rev. R. Everest, the boar never exceeds 36 inches in height, whilst in Bengal one has been measured 44 inches in height. In Europe, Northern Africa, and Hindostan, domestic pigs have been known to cross with the wild native species;[3] and in Hindostan an accurate observer,[4] Sir Walter Elliot, after describing the differences between wild Indian and wild German boars, remarks that "the same differences are perceptible in the domesticated

[2] Nathusius, 'Die Racen des Schweines,' Berlin, 1860. An excellent appendix is given with references to published and trustworthy drawings of the breeds of each country.

[3] For Europe, *see* Bechstein, 'Naturgesch. Deutschlands,' 1801, b. i., s. 505. Several accounts have been published on the fertility of the offspring from wild and tame swine. *See* Burdach's 'Physiology,' and Godron, 'De l'Espèce,' tom. i. p. 370. For Africa, 'Bull. de la Soc. d'Acclimat.,' tom. iv. p. 389. For India, *see* Nathusius, 'Schweineschädel,' s. 148.

[4] Sir W. Elliot, Catalogue of Mammalia, 'Madras Journal of Lit. and Science,' vol. x. p. 219.

individuals of the two countries." We may therefore conclude that the breeds of the *Sus scrofa* type have either descended from, or been modified by crossing with, forms which may be ranked as geographical races, but which are, according to some naturalists, distinct species.

Pigs of the *Sus Indica* type are best known to Englishmen under the form of the Chinese breed. The skull of *S. Indica*, as described by Nathusius, differs from that of *S. scrofa* in several minor respects, as in its greater breadth and in some details in the teeth; but chiefly in the shortness of the lachrymal bones, in the greater width of the fore part of the palate-bones, and in the divergence of the premolar teeth. It deserves especial notice that these latter characters are not gained, even in the least degree, by the domesticated forms of *S. scrofa*. After reading the remarks and descriptions given by Nathusius, it seems to me to be merely playing with words to doubt whether *S. Indica* ought to be ranked as a species; for the above-specified differences are more strongly marked than any that can be pointed out between, for instance, the fox and the wolf, or the ass and the horse. As already stated, *S. Indica* is not known in a wild state; but its domesticated forms, according to Nathusius, come near to *S. vittatus* of Java and some allied species. A pig found wild in the Aru islands (Schweineschädel, s. 169) is apparently identical with *S. Indica;* but it is doubtful whether this is a truly native animal. The domesticated breeds of China, Cochin-China, and Siam belong to this type. The Roman or Neapolitan breed, the Andalusian, the Hungarian, and the "Krause" swine of Nathusius, inhabiting south-eastern Europe and Turkey, and having fine curly hair, and the small Swiss "Bündtnerschwein" of Rütimeyer, all agree in their more important skull characters with *S. Indica*, and, as is supposed, have all been largely crossed with this form. Pigs of this type have existed during a long period on the shores of the Mediterranean, for a figure (Schweineschädel, s. 142) closely resembling the existing Neapolitan pig has been found in the buried city of Herculaneum.

Rütimeyer has made the remarkable discovery that there lived contemporaneously in Switzerland, during the later Stone or Neolithic period, two domesticated forms, the *S. scrofa*, and the *S. scrofa palustris* or Torfschwein. Rütimeyer perceived that the latter approached the Eastern breeds, and, according to Nathusius, it certainly belongs to the *S. Indica* group; but Rütimeyer has subsequently shown that it differs in some well-marked characters. This author was formerly convinced that his Torfschwein existed as a wild animal during the first part of the Stone period, and was domesticated during a later part of the same period.[5] Nathusius, whilst he fully admits the curious fact first observed by Rütimeyer, that the bones of domesticated and wild animals can be distinguished by their different aspect, yet, from special difficulties in the case of the bones of the pig (Schweineschädel, s. 147), is not convinced of the truth of this conclusion; and Rütimeyer himself seems now to feel some doubt. As the Torfschwein was domesticated at so early a period, and as its remains have been found in several parts of Europe, belonging to various historic and prehistoric ages,[6] and as closely allied forms still exist in Hungary and on the shores of the Mediterranean, one is led to suspect that the wild *S. Indica* formerly ranged from Europe to China, in the same manner as *S. scrofa* now ranges from Europe to Hindostan. Or, as Rütimeyer apparently suspects, a third allied species may formerly have lived in Europe and Eastern Asia.

Several breeds, differing in the proportions of the body, in the length of the ears, in the nature of the hair, in colour, &c., come under the *S. Indica* type. Nor is this surprising, considering how ancient the domestication of this form has been both in Europe and in China. In this latter country the date is believed by an eminent Chinese scholar[7] to go back at least 4900 years from the present time. This same scholar alludes to the existence of many local varieties of the pig in China; and at the present time the Chinese take extraordinary pains in feeding and tending their pigs, not even allowing them to walk from place to place.[8] Hence the Chinese breed, as Nathusius has remarked,[9] displays in an eminent degree the characters of a highly-cultivated race, and hence, no doubt, its

[5] 'Pfahlbauten,' s. 163 et passim.

[6] *See* Rütimeyer's Neue Beitrage, ... Torfschweine, Verh. Naturfor. Gesell. in Basel, iv. i., 1865, s. 139.

[7] Stan. Julien, quoted by De Blainville, 'Ostéographie,' p. 163.

[8] Richardson, 'Pigs, their Origin,' &c., p. 26.

[9] 'Die Racen des Schweines,' s. 47, 64.

high value in the improvement of our European breeds. Nathusius makes a remarkable statement (Schweineschädel, s. 138), that the infusion of the $\frac{1}{32}$nd, or even of the $\frac{1}{64}$th, part of the blood of *S. Indica* into a breed of *S. scrofa,* is sufficient plainly to modify the skull of the latter species. This singular fact may perhaps be accounted for by several of the chief distinctive characters of *S. Indica,* such as the shortness of the lachrymal bones, &c., being common to several of the species of the genus; for in crosses the characters which are common to many species apparently tend to be prepotent over those appertaining to only a few species.

The Japan pig (*S. pliciceps* of Gray), which has been recently exhibited in the Zoological Gardens, has an extraordinary appearance from its short head, broad forehead and nose, great fleshy ears, and deeply furrowed skin. The following woodcut is copied from that given by Mr. Bartlett.[10] Not only

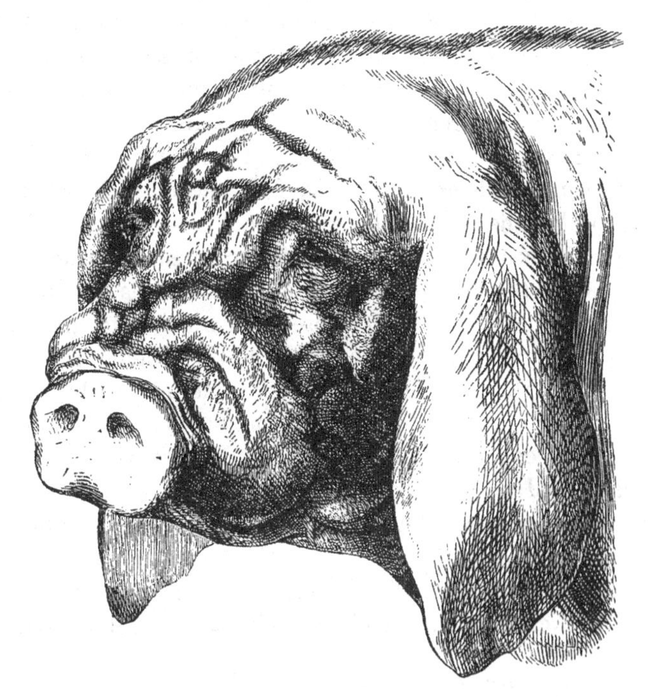

Fig. 2.—Head of Japan or Masked Pig. (Copied from Mr. Bartlett's paper in Proc. Zoolog. Soc. 1861, p. 263.)

[10] 'Proc. Zoolog. Soc.,' 1861, p. 263.

is the face furrowed, but thick folds of skin, which are harder than the other parts, almost like the plates on the Indian rhinoceros, hang about the shoulders and rump. It is coloured black, with white feet, and breeds true. That it has long been domesticated there can be little doubt; and this might have been inferred even from the fact that its young are not longitudinally striped; for this is a character common to all the species included within the genus *Sus* and the allied genera whilst in their natural state.[11] Dr. Gray[12] has described the skull of this animal, which he ranks not only as a distinct species, but places it in a distinct section of the genus. Nathusius, however, after his careful study of the whole group, states positively (Schweineschädel, s. 153-158) that the skull in all essential characters closely resembles that of the short-eared Chinese breed of the *S. Indica* type. Hence Nathusius considers the Japan pig as only a domesticated variety of *S. Indica:* if this really be the case, it is a wonderful instance of the amount of modification which can be effected under domestication.

Formerly there existed in the central islands of the Pacific Ocean a singular breed of pigs. These are described by the Rev. D. Tyerman and G. Bennett[13] as of small size, humpbacked, with a disproportionately long head, with short ears turned backwards, with a bushy tail not more than two inches in length, placed as if it grew from the back. Within half a century after the introduction into these islands of European and Chinese pigs, the native breed, according to the above authors, became almost completely lost by being repeatedly crossed with them. Secluded islands, as might have been expected, seem favourable for the production or retention of peculiar breeds; thus, in the Orkney Islands, the hogs have been described as very small, with erect and sharp ears, and "with an appearance altogether different from the hogs brought from the south."[14]

Seeing how different the Chinese pigs, belonging to the *Sus Indica* type, are in their osteological characters and in external

[11] Sclater, in 'Proc. Zoolog. Soc.,' Feb. 26th, 1861.

[12] 'Proc. Zoolog. Soc.,' 1862, p. 13.

[13] 'Journal of Voyages and Travels from 1821 to 1829,' vol. i. p. 300.

[14] Rev. G. Low, 'Fauna Orcadensis,' p. 10. *See* also Dr. Hibbert's account of the pig of the Shetland Islands.

appearance from the pigs of the *S. scrofa* type, so that they must be considered specifically distinct, it is a fact well deserving attention, that Chinese and common pigs have been repeatedly crossed in various manners, with unimpaired fertility. One great breeder who had used pure Chinese pigs assured me that the fertility of the half-breeds *inter se* and of their recrossed progeny was actually increased; and this is the general belief of agriculturists. Again, the Japan pig or *S. pliciceps* of Gray is so distinct in appearance from all common pigs, that it stretches one's belief to the utmost to admit that it is simply a domestic variety; yet this breed has been found perfectly fertile with the Berkshire breed; and Mr. Eyton informs me that he paired a half-bred brother and sister and found them quite fertile together.

The modifications of the skull in the most highly cultivated races are wonderful. To appreciate the amount of change, Nathusius' work, with its excellent figures, should be studied. The whole of the exterior of the skull in all its parts has been altered; the hinder surface, instead of sloping backwards, is directed forwards, entailing many changes in other parts; the front of the head is deeply concave; the orbits have a different shape; the auditory meatus has a different direction and shape; the incisors of the upper and lower jaws do not touch each other, and they stand in both jaws above the plane of the molars; the canines of the upper jaw stand in front of those of the lower jaw, and this is a remarkable anomaly: the articular surfaces of the occipital condyles are so greatly changed in shape, that, as Nathusius remarks (s. 133), no naturalist, seeing this important part of the skull by itself, would suppose that it belonged to the genus Sus. These and various other modifications, as Nathusius observes, can hardly be considered as monstrosities, for they are not injurious, and are strictly inherited. The whole head is much shortened; thus, whilst in common breeds its length to that of the body is as 1 to 6, in the "cultur-races" the proportion is as 1 to 9, and even recently as 1 to 11.[15] The following woodcut[16]

[15] 'Die Racen des Schweines,' s. 70.

[16] These woodcuts are copied from engravings given in Mr. S. Sidney's excellent edition of 'The Pig,' by Youatt, 1860. *See* pp. 1, 16, 19.

of the head of a wild boar and of a sow from a photograph of the Yorkshire Large Breed, may aid in showing how greatly the head in a highly cultivated race has been modified and shortened.

Nathusius has well discussed the causes of the remarkable changes in the skull and shape of the body which the highly cultivated races have undergone. These modifications occur chiefly in the pure and crossed races of the *S. Indica* type; but their commencement may be clearly detected in the slightly improved breeds of the *S. scrofa* type.[17] Nathusius states positively (s. 99, 103), as the result of common experience and of his experiments, that rich and abundant food, given during youth, tends by some direct action to make the head broader and shorter; and that poor food works a contrary result. He lays much stress on the fact that all wild and semi-domesticated pigs, in ploughing up the ground with their muzzles, have, whilst young, to exert the powerful muscles fixed to the hinder part of the head. In highly cultivated races this habit is no longer followed, and consequently the back of the skull becomes modified in shape, entailing other changes in other parts. There can hardly be a doubt that so great a change in habits would

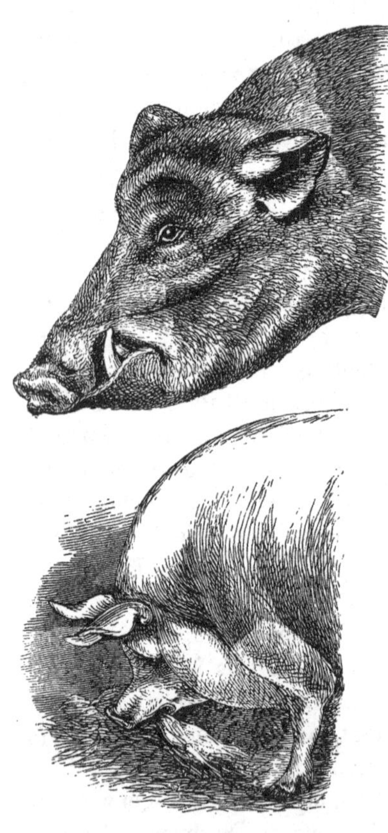

Fig. 3.—Head of Wild Boar, and of "Golden Days," a pig of the Yorkshire Large Breed; the latter from a photograph. (Copied from Sidney's edit. of 'The Pig,' by Youatt.)

[17] 'Schweineschädel,' s. 74, 135.

affect the skull; but it seems rather doubtful how far this will account for the greatly reduced length of the skull and for its concave front. It is well known (Nathusius himself advancing many cases, s. 104) that there is a strong tendency in many domestic animals—in bull- and pug- dogs, in the niata cattle, in sheep, in Polish fowls, short-faced tumbler pigeons, and in one variety of the carp—for the bones of the face to become greatly shortened. In the case of the dog, as H. Müller has shown, this seems caused by an abnormal state of the primordial cartilage. We may, however, readily admit that abundant and rich food supplied during many generations would give an inherited tendency to increased size of body, and that, from disuse, the limbs would become finer and shorter.[18] We shall in a future chapter also see that the skull and limbs are apparently in some manner correlated, so that any change in the one tends to affect the other.

Nathusius has remarked, and the observation is an interesting one, that the peculiar form of the skull and body in the most highly cultivated races is not characteristic of any one race, but is common to all when improved up to the same standard. Thus the large-bodied, long-eared, English breeds with a convex back, and the small-bodied, short-eared, Chinese breeds with a concave back, when bred to the same state of perfection, nearly resemble each other in the form of the head and body. This result, it appears, is partly due to similar causes of change acting on the several races, and partly to man breeding the pig for one sole purpose, namely, for the greatest amount of flesh and fat; so that selection has always tended towards one and the same end. With most domestic animals the result of selection has been divergence of character, here it has been convergence.[19]

The nature of the food supplied during many generations has apparently affected the length of the intestines; for, according to Cuvier,[20] their length to that of the body in the wild boar is as 9 to 1,—in the common domestic boar as 13·5 to 1,—and in the Siam breed as 16 to 1. In this latter breed the greater

[18] Nathusius, 'Die Racen des Schweines,' s. 71.

[19] 'Die Racen des Schweines,' s. 47. 'Schweineschädel,' s. 104. Compare, also, the figures of the old Irish and the improved Irish breeds in Richardson on 'The Pig,' 1847.

[20] Quoted by Isid. Geoffroy, 'Hist. Nat. Gén.,' tom. iii. p. 441.

length may be due either to descent from a distinct species or to more ancient domestication. The number of mammæ vary, as does the period of gestation. The latest authority says[21] that "the period averages from 17 to 20 weeks," but I think there must be some error in this statement: in M. Tessier's observations on 25 sows it varied from 109 to 123 days. The Rev. W. D. Fox has given me ten carefully recorded cases with well-bred pigs, in which the period varied from 101 to 116 days. According to Nathusius the period is shortest in the races which come early to maturity; but in these latter the course of development does not appear to be actually shortened, for the young animal is born, judging from the state of the skull, less fully developed, or in a more embryonic condition,[22] than in the case of common swine, which arrive at maturity at a later age. In the highly cultivated and early matured races, the teeth, also, are developed earlier.

The difference in the number of the vertebræ and ribs in different kinds of pigs, as observed by Mr. Eyton,[23] and as given in the following table, has often been quoted. The African sow probably belongs to the *S. scrofa* type; and Mr. Eyton informs me that, since the publication of his paper, cross-bred animals from the African and English races were found by Lord Hill to be perfectly fertile.

	English Long-legged Male.	African Female.	Chinese Male.	Wild Boar, from Cuvier.	French Domestic Boar, from Cuvier.
Dorsal vertebræ ..	15	13	15	14	14
Lumbar	6	6	4	5	5
Dorsal and lumbar together	21	19	19	19	19
Sacral	5	5	4	4	4
Total number of vertebræ	26	24	23	23	23

[21] S. Sidney, 'The Pig,' p. 61.

[22] 'Schweineschädel,' s. 2, 20.

[23] 'Proc. Zoolog. Soc.,' 1837, p. 23. I have not given the caudal vertebræ, as Mr. Eyton says some might possibly have been lost. I have added together the dorsal and lumbar vertebræ, owing to Prof. Owen's remarks ('Journal Linn. Soc.,' vol. ii. p. 28) on the difference between dorsal and lumbar vertebræ depending only on the development of the ribs. Nevertheless the difference in the number of the ribs in pigs deserves notice.

Some semi-monstrous breeds deserve notice. From the time of Aristotle to the present time solid-hoofed swine have occasionally been observed in various parts of the world. Although this peculiarity is strongly inherited, it is hardly probable that all the animals with solid hoofs have descended from the same parents; it is more probable that the same peculiarity has reappeared at various times and places. Dr. Struthers has lately described and figured[24] the structure of the feet; in both front and hind feet the distal phalanges of the two greater toes are represented by a single, great, hoof-bearing phalanx; and in the front feet, the middle phalanges are represented by a bone which is single towards the lower end, but bears two separate articulations towards the upper end. From other accounts it appears that an intermediate toe is likewise sometimes superadded.

Another curious anomaly is offered by the appendages, described by M. Eudes-Deslongchamps as often characterizing the Normandy pigs. These appendages are always attached to the same spot, to the corners of the jaw; they are cylindrical, about three inches in length, covered with bristles, and with a pencil of bristles rising out of a sinus on one side: they have a cartilaginous centre, with two small longitudinal muscles; they occur either symmetrically on both sides of the face or on one

Fig. 4.—Old Irish Pig, with jaw-appendages. (Copied from H. D. Richardson on Pigs.)

[24] 'Edinburgh New Philosoph. Journal,' April 1863. *See* also De Blainville's 'Ostéographie,' p. 128, for various authorities on this subject.

side alone. Richardson figures them on the gaunt old "Irish Greyhound pig;" and Nathusius states that they occasionally appear in all the long-eared races, but are not strictly inherited, for they occur or fail in animals of the same litter.[25] As no wild pigs are known to have analogous appendages, we have at present no reason to suppose that their appearance is due to reversion; and if this be so, we are forced to admit that somewhat complex, though apparently useless, structures may be suddenly developed without the aid of selection. This case perhaps throws some little light on the manner of appearance of the hideous fleshy protuberances, though of an essentially different nature from the above-described appendages, on the cheeks of the wart-hog or Phascochœrus Africanus.

It is a remarkable fact that the boars of all domesticated breeds have much shorter tusks than wild boars. Many facts show that with all animals the state of the hair is much affected by exposure to, or protection from, climate; and as we see that the state of the hair and teeth are correlated in Turkish dogs (other analogous facts will be hereafter given), may we not venture to surmise that the reduction of the tusks in the domestic boar is related to his coat of bristles being diminished from living under shelter? On the other hand, as we shall immediately see, the tusks and bristles reappear with feral boars, which are no longer protected from the weather. It is not surprising that the tusks should be more affected than the other teeth; as parts developed to serve as secondary sexual characters are always liable to much variation.

It is a well-known fact that the young of wild European and Indian pigs,[26] for the first six months, are longitudinally banded with light-coloured stripes. This character generally disappears under domestication. The Turkish domestic pigs, however, have striped young, as have those of Westphalia, "whatever may be their hue;"[27] whether these latter pigs belong to the

[25] Eudes-Deslongchamps, 'Mémoires de la Soc. Linn. de Normandie,' vol. vii., 1842, p. 41. Richardson, 'Pigs, their Origin, &c.,' 1847, p. 30. Nathusius, 'Die Racen des Schweines,' 1860, s. 54.

[26] D. Johnson's 'Sketches of Indian Field Sports,' p. 272. Mr. Crawfurd informs me that the same fact holds good with the wild pigs of the Malay peninsula.

[27] For Turkish pigs, *see* Desmarest, 'Mammalogie,' 1820, p. 391. For those of Westphalia, *see* Richardson's 'Pigs, their Origin,' &c., 1847, p. 41.

same curly-haired race with the Turkish swine, I do not know. The pigs which have run wild in Jamaica and the semi-feral pigs of New Granada, both those which are black and those which are black with a white band across the stomach, often extending over the back, have resumed this aboriginal character and produce longitudinally-striped young. This is likewise the case, at least occasionally, with the neglected pigs in the Zambesi settlement on the coast of Africa.[28]

The common belief that all domesticated animals, when they run wild, revert completely to the character of their parent-stock, is chiefly founded, as far as I can discover, on feral pigs. But even in this case the belief is not grounded on sufficient evidence; for the two main types of *S. scrofa* and *Indica* have never been distinguished in a feral state. The young, as we have just seen, reacquire their longitudinal stripes, and the boars invariably reassume their tusks. They revert also in the general shape of their bodies, and in the length of their legs and muzzles, to the state of the wild animal, as might have been expected from the amount of exercise which they are compelled to take in search of food. In Jamaica the feral pigs do not acquire the full size of the European wild boar, "never attaining a greater height than 20 inches at the shoulder." In various countries they reassume their original bristly covering, but in different

[28] With respect to the several foregoing and following statements on feral pigs, *see* Roulin, in 'Mém. présentés par divers Savans à l'Acad.,' &c., Paris, tom. vi., 1835, p. 326. It should be observed that his account does not apply to truly feral pigs; but to pigs long introduced into the country and living in a half-wild state. For the truly feral pigs of Jamaica, *see* Gosse's 'Sojourn in Jamaica,' 1851, p. 386; and Col. Hamilton Smith, in 'Nat. Library,' vol. ix. p. 93. With respect to Africa, *see* Livingstone's 'Expedition to the Zambesi,' 1865, p. 153. The most precise statement with respect to the tusks of the West Indian feral boars is by P. Labat (quoted by Roulin); but this author attributes the state of these pigs to descent from a domestic stock which he saw in Spain. Admiral Sulivan, R.N., had ample opportunities of observing the wild pigs on Eagle Islet in the Falklands; and he informs me that they resembled wild boars with bristly ridged backs and large tusks. The pigs which have run wild in the province of Buenos Ayres (Rengger, 'Säugethiere,' s. 331) have not reverted to the wild type. De Blainville ('Ostéographie,' p. 132) refers to two skulls of domestic pigs sent from Patagonia by Al. d'Orbigny, and he states that they have the occipital elevation of the wild European boar, but that the head altogether is "plus courte et plus ramassée." He refers, also, to the skin of a feral pig from North America, and says, "il ressemble tout à fait à un petit sanglier, mais il est presque tout noir, et peut-être un peu plus ramassé dans ses formes."

degrees, dependent on the climate; thus, according to Roulin, the semi-feral pigs in the hot valleys of New Granada are very scantily clothed; whereas, on the Paramos, at the height of 7000 to 8000 feet, they acquire a thick covering of wool lying under the bristles, like that on the truly wild pigs of France. These pigs on the Paramos are small and stunted. The wild boar of India is said to have the bristles at the end of its tail arranged like the plumes of an arrow, whilst the European boar has a simple tuft; and it is a curious fact that many, but not all, of the feral pigs in Jamaica, derived from a Spanish stock, have a plumed tail.[29] With respect to colour, feral pigs generally revert to that of the wild boar; but in certain parts of S. America, as we have seen, some of the semi-feral pigs have a curious white band across their stomachs; and in certain other hot places the pigs are red, and this colour has likewise occasionally been observed in the feral pigs of Jamaica. From these several facts we see that with pigs when feral there is a strong tendency to revert to the wild type; but that this tendency is largely governed by the nature of the climate, amount of exercise, and other causes of change to which they have been subjected.

The last point worth notice is that we have unusually good evidence of breeds of pigs now keeping perfectly true, which have been formed by the crossing of several distinct breeds. The Improved Essex pigs, for instance, breed very true; but there is no doubt that they largely owe their present excellent qualities to crosses originally made by Lord Western with the Neapolitan race, and to subsequent crosses with the Berkshire breed (this also having been improved by Neapolitan crosses), and likewise, probably, with the Sussex breed.[30] In breeds thus formed by complex crosses, the most careful and unremitting selection during many generations has been found to be indispensable. Chiefly in consequence of so much crossing, some well-known breeds have undergone rapid changes; thus, according to Nathusius,[31] the Berkshire breed of 1780 is quite

[29] Gosse's 'Jamaica,' p. 386, with a quotation from Williamson's 'Oriental Field Sports.' Also Col. Hamilton Smith, in 'Naturalist's Library,' vol. ix. p. 94.

[30] S. Sidney's edition of 'Youatt on the Pig,' 1860, pp. 7, 26, 27, 29, 30.

[31] 'Schweineschädel,' s. 140.

different from that of 1810; and, since this latter period, at least two distinct forms have borne the same name.

CATTLE.

DOMESTIC cattle are almost certainly the descendants of more than one wild form, in the same manner as has been shown to be the case with our dogs and pigs. Naturalists have generally made two main divisions of cattle: the humped kinds inhabiting tropical countries, called in India Zebus, to which the specific name of *Bos Indicus* has been given; and the common non-humped cattle, generally included under the name of *Bos taurus*. The humped cattle were domesticated, as may be seen on the Egyptian monuments, at least as early as the twelfth dynasty, that is 2100 B.C. They differ from common cattle in various osteological characters, even in a greater degree, according to Rütimeyer,[32] than do the fossil species of Europe, namely *Bos primigenius*, *longifrons*, and *frontosus*, from each other. They differ, also, as Mr. Blyth,[33] who has particularly attended to this subject, remarks, in general configuration, in the shape of their ears, in the point where the dewlap commences, in the typical curvature of their horns, in their manner of carrying their heads when at rest, in their ordinary variations of colour, especially in the frequent presence of "nilgau-like markings on their feet," and "in the one being born with teeth protruding through the jaws, and the other not so." They have different habits, and their voice is entirely different. The humped cattle in India "seldom seek shade, and never go into the water and there stand knee-deep, like the cattle of Europe." They have run wild in parts of Oude and Rohilcund, and can maintain themselves in a region infested by tigers. They have given rise to many races differing greatly in size, in the presence

[32] 'Die Fauna der Pfahlbauten,' 1861, s. 109, 149, 222. *See* also Geoffroy Saint Hilaire, in 'Mém. du Mus. d'Hist. Nat.,' tom. x. p. 172; and his son Isidore, in 'Hist. Nat. Gén.,' tom. iii. p. 69. Vasey, in his 'Delineations of the Ox Tribe,' 1851, p. 127, says the zebu has four, and the common ox five, sacral vertebræ. Mr. Hodgson found the ribs either thirteen or fourteen in number; *see* a note in 'Indian Field,' 1858, p. 62.

[33] 'The Indian Field,' 1858, p. 74, where Mr. Blyth gives his authorities with respect to the feral humped cattle. Pickering, also, in his 'Races of Man,' 1850, p. 274, notices the peculiar character of the grunt-like voice of the humped cattle.

of one or two humps, in length of horns, and other respects. Mr. Blyth sums up emphatically that the humped and humpless cattle must be considered as distinct species. When we consider the number of points in external structure and habits, independently of their important osteological differences, in which they differ from each other; and that many of these points are not likely to have been affected by domestication, there can hardly be a doubt, notwithstanding the adverse opinion of some naturalists, that the humped and non-humped cattle must be ranked as specifically distinct.

The European breeds of humpless cattle are numerous. Professor Low enumerates 19 British breeds, only a few of which are identical with those on the Continent. Even the small Channel islands of Guernsey, Jersey, and Alderney, possess their own sub-breeds;[34] and these again differ from the cattle of the other British islands, such as Anglesea, and the western isles of Scotland. Desmarest, who paid attention to the subject, describes 15 French races, excluding sub-varieties and those imported from other countries. In other parts of Europe there are several distinct races, such as the pale-coloured Hungarian cattle, with their light and free step, and their enormous horns sometimes measuring above five feet from tip to tip:[35] the Podolian cattle are remarkable from the height of their forequarters. In the most recent work on Cattle,[36] engravings are given of fifty-five European breeds; it is, however, probable that several of these differ very little from each other, or are merely synonyms. It must not be supposed that numerous breeds of cattle exist only in long-civilized countries, for we shall presently see that several kinds are kept by the savages of Southern Africa.

With respect to the parentage of the several European breeds, we already know much from Nilsson's Memoir,[37] and more especially from Rütimeyer's 'Pfahlbauten' and succeeding works. Two or three species or forms of

[34] Mr. H. E. Marquand, in 'The Times,' June 23rd, 1856.

[35] Vasey, 'Delineations of the Ox-Tribe,' p. 124. Brace's 'Hungary,' 1851, p. 94. The Hungarian cattle descend, according to Rütimeyer ('Zahmen. Europ. Rindes., 1866, s. 13), from *Bos primigenius*.

[36] Moll and Gayot, 'La Connaissance Gén. du Bœuf,' Paris, 1860. Fig 82 is that of the Podolian breed.

[37] A translation appeared in three parts in the 'Annals and Mag. of Nat. Hist.,' 2nd series, vol. iv., 1849.

Bos, closely allied to still living domestic races, have been found fossil in the more recent tertiary deposits of Europe. Following Rütimeyer, we have:—

Bos primigenius.—This magnificent, well-known species was domesticated in Switzerland during the Neolithic period; even at this early period it varied a little, having apparently been crossed with other races. Some of the larger races on the Continent, as the Friesland, &c., and the Pembroke race in England, closely resemble in essential structure *B. primigenius*, and no doubt are its descendants. This is likewise the opinion of Nilsson. *Bos primigenius* existed as a wild animal in Cæsar's time, and is now semi-wild, though much degenerated in size, in the park of Chillingham; for I am informed by Professor Rütimeyer, to whom Lord Tankerville sent a skull, that the Chillingham cattle are less altered from the true primigenius type than any other known breed.[38]

Bos trochoceros.—This form is not included in the three species above mentioned, for it is now considered by Rütimeyer to be the female of an early domesticated form of *B. primigenius*, and as the progenitor of his *frontosus* race. I may add that specific names have been given to four other fossil oxen, now believed to be identical with *B. primigenius*.[39]

Bos longifrons (or *brachyceros*) of Owen.—This very distinct species was of small size, and had a short body with fine legs. It has been found in England associated with the remains of the elephant and rhinoceros.[40] It was the commonest form in a domesticated condition in Switzerland during the earliest part of the Neolithic period. It was domesticated in England during the Roman period, and supplied food to the Roman legionaries.[41] Some remains have been found in Ireland in certain crannoges, of which the dates are believed to be from 843-933 A.D.[42] Professor Owen [43] thinks it probable that the Welsh and Highland cattle are descended from this form; as likewise is the case, according to Rütimeyer, with some of the existing Swiss breeds. These latter are of different shades of colour from light-grey to blackish-brown, with a lighter stripe along the spine, but they have no pure white marks. The cattle of North Wales and the Highlands, on the other hand, are generally black or dark-coloured.

Bos frontosus of Nilsson.—This species is allied to *B. longifrons*, but in the opinion of some good judges is distinct from it. Both co-existed in Scania during the same late geological period,[44] and both have been found in the Irish crannoges.[45] Nilsson believes that his *B. frontosus* may be the

[38] *See*, also, Rütimeyer's 'Beitrage pal. Gesch. der Wiederkauer,' Basel, 1865, s. 54.

[39] Pictet's 'Paléontologie,' tom. i. p. 365 (2nd edit.). With respect to B. trochoceros, *see* Rütimeyer's 'Zahmen Europ. Rindes,' 1866, s. 26.

[40] Owen, 'British Fossil Mammals,' 1846, p. 510.

[41] 'British Pleistocene Mammalia,' by W. B. Dawkins and W. A. Sandford, 1866, p. xv.

[42] W. R. Wilde, 'An Essay on the Animal Remains, &c., Royal Irish Academy,' 1860, p. 29. Also 'Proc. of R. Irish Academy,' 1858, p. 48.

[43] 'Lecture: Royal Institution of G. Britain,' May 2nd, 1856, p. 4. 'British Fossil Mammals,' p. 513.

[44] Nilsson, in 'Annals and Mag. of Nat. Hist.,' 1849, vol. iv. p. 354.

[45] *See* W. R. Wilde, ut supra; and Mr. Blyth, in 'Proc. Irish Academy,' March 5th, 1864.

parent of the mountain cattle of Norway, which have a high protuberance on the skull between the base of the horns. As Professor Owen believes that the Scotch Highland cattle are descended from his *B. longifrons*, it is worth notice that a capable judge[46] has remarked that he saw no cattle in Norway like the Highland breed, but that they more nearly resembled the Devonshire breed.

Hence we see that three forms or species of Bos, originally inhabitants of Europe, have been domesticated; but there is no improbability in this fact, for the genus Bos readily yields to domestication. Besides these three species and the zebu, the yak, the gayal, and the arni[47] (not to mention the buffalo or genus Bubalus) have been domesticated; making altogether seven species of Bos. The zebu and the three European species are now extinct in a wild state, for the cattle of the *B. primigenius* type in the British parks can hardly be considered as truly wild. Although certain races of cattle, domesticated at a very ancient period in Europe, are the descendants of the three above-named fossil species, yet it does not follow that they were here first domesticated. Those who place much reliance on philology argue that our cattle were imported from the East.[48] But as races of men invading any country would probably give their own names to the breeds of cattle which they might there find domesticated, the argument seems inconclusive. There is indirect evidence that our cattle are the descendants of species which originally inhabited a temperate or cold climate, but not a land long covered with snow; for our cattle, as we have seen in the chapter on Horses, apparently have not the instinct of scraping away the snow to get at the herbage beneath. No one could behold the magnificent wild bulls on the bleak Falkland Islands in the southern hemisphere, and doubt about the climate being admirably suited to them. Azara has remarked that in the temperate regions of La Plata the cows conceive when two years old, whilst in the much hotter country of Paraguay they do not conceive till three years old; "from which fact," as he adds, "one may conclude that cattle do not succeed so well in warm countries."[49]

The above-named three fossil forms of Bos have been ranked

46 Laing's 'Tour in Norway,' p. 110.

47 Isid. Geoffroy St. Hilaire, 'Hist. Nat. Gén.,' tom. iii. p. 96.

48 Idem, tom. iii. pp. 82, 91.

49 'Quadrupèdes du Paraguay,' tom. ii. p. 360.

by nearly all palæontologists as distinct species; and it would not be reasonable to change their denomination simply because they are now found to be the parents of several domesticated races. But what is of most importance for us, as showing that they deserve to be ranked as species, is that they co-existed in different parts of Europe during the same period, and yet kept distinct. Their domesticated descendants, on the other hand, if not separated, cross with the utmost freedom and become commingled. The several European breeds have so often been crossed, both intentionally and unintentionally, that, if any sterility ensued from such unions, it would certainly have been detected. As zebus inhabit a distant and much hotter region, and as they differ in so many characters from our European cattle, I have taken pains to ascertain whether the two forms are fertile when crossed. The late Lord Powis imported some zebus and crossed them with common cattle in Shropshire; and I was assured by his steward that the cross-bred animals were perfectly fertile with both parent-stocks. Mr. Blyth informs me that in India hybrids, with various proportions of either blood, are quite fertile; and this can hardly fail to be known, for in some districts[50] the two species are allowed to breed freely together. Most of the cattle which were first introduced into Tasmania were humped, so that at one time thousands of crossed animals existed there; and Mr. B. O'Neile Wilson, M.A., writes to me from Tasmania that he has never heard of any sterility having been observed. He himself formerly possessed a herd of such crossed cattle, and all were perfectly fertile; so much so, that he cannot remember even a single cow failing to calve. These several facts afford an important confirmation of the Pallasian doctrine that the descendants of species which when first domesticated would if crossed probably have been in some degree sterile, become perfectly fertile after a long course of domestication. In a future chapter we shall see that this doctrine throws much light on the difficult subject of Hybridism.

I have alluded to the cattle in Chillingham Park, which, according to Rütimeyer, have been very little changed from the *Bos primigenius* type. This park is so ancient that it is

[50] Walther, 'Das Rindvieh,' 1817, s. 30.

referred to in a record of the year 1220. The cattle in their instincts and habits are truly wild. They are white, with the inside of the ears reddish-brown, eyes rimmed with black, muzzles brown, hoofs black, and horns white tipped with black. Within a period of thirty-three years about a dozen calves were born with "brown and blue spots upon the cheeks or necks; but these, together with any defective animals, were always destroyed." According to Bewick, about the year 1770 some calves appeared with black ears; but these were also destroyed by the keeper, and black ears have not since reappeared. The wild white cattle in the Duke of Hamilton's park, where I have heard of the birth of a black calf, are said by Lord Tankerville to be inferior to those at Chillingham. The cattle kept until the year 1780 by the Duke of Queensberry, but now extinct, had their ears, muzzle, and orbits of the eyes black. Those which have existed from time immemorial at Chartley closely resemble the cattle at Chillingham, but are larger, "with some small difference in the colour of the ears." "They frequently tend to become entirely black; and a singular superstition prevails in the vicinity that, when a black calf is born, some calamity impends over the noble house of Ferrers. All the black calves are destroyed." The cattle at Burton Constable in Yorkshire, now extinct, had ears, muzzle, and the tip of the tail black. Those at Gisburne, also in Yorkshire, are said by Bewick to have been sometimes without dark muzzles, with the inside alone of the ears brown; and they are elsewhere said to have been low in stature and hornless.[51]

The several above-specified differences in the park-cattle, slight though they be, are worth recording, as they show that animals living nearly in a state of nature, and exposed to nearly uniform conditions, if not allowed to roam freely and to cross with other herds, do not keep as uniform as truly

[51] I am much indebted to the present Earl of Tankerville for information about his wild cattle; and for the skull which was sent to Prof. Rütimeyer. The fullest account of the Chillingham cattle is given by Mr. Hindmarsh, together with a letter by the late Lord Tankerville, in 'Annals and Mag. of Nat. Hist.,' vol. ii., 1839, p. 274. *See* Bewick, 'Quadrupeds,' 2nd edit., 1791, p. 35, note. With respect to those of the Duke of Queensberry, *see* Pennant's 'Tour in Scotland,' p. 109. For those of Chartley, *see* Low's 'Domesticated Animals of Britain,' 1845, p. 238. For those of Gisburne, *see* Bewick's 'Quadrupeds, and Encyclop. of Rural Sports,' p. 101.

wild animals. For the preservation of a uniform character, even within the same park, a certain degree of selection—that is, the destruction of the dark-coloured calves—is apparently necessary.

The cattle in all the parks are white; but, from the occasional appearance of dark-coloured calves, it is extremely doubtful whether the aboriginal *Bos primigenius* was white. The following facts, however, show that there is a strong, though not invariable, tendency in wild or escaped cattle, under widely different conditions of life, to become white with coloured ears. If the old writers Boethius and Leslie[52] can be trusted, the wild cattle of Scotland were white and furnished with a great mane; but the colour of their ears is not mentioned. The primæval forest formerly extended across the whole country from Chillingham to Hamilton, and Sir Walter Scott used to maintain that the cattle still preserved in these two parks, at the two extremities of the forest, were remnants of its original inhabitants; and this view certainly seems probable. In Wales,[53] during the tenth century, some of the cattle are described as being white with red ears. Four hundred cattle thus coloured were sent to King John; and an early record speaks of a hundred cattle with red ears having been demanded as a compensation for some offence, but, if the cattle were of a dark or black colour, one hundred and fifty were to be presented. The black cattle of North Wales apparently belong, as we have seen, to the small *longifrons* type: and as the alternative was offered of either 150 dark cattle, or 100 white cattle with red ears, we may presume that the latter were the larger beasts, and probably belonged to the *primigenius* type. Youatt has remarked that at the present day, whenever cattle of the short-horn breed are white, the extremities of their ears are more or less tinged with red.

The cattle which have run wild on the Pampas, in Texas, and in two parts of Africa, have become of a nearly uniform dark

[52] Boethius was born in 1470; 'Annals and Mag. of Nat. Hist.,' vol. ii., 1839, p. 281; and vol. iv. 1849, p. 424.

[53] Youatt on Cattle, 1834, p. 48: See also p. 242, on short-horn cattle. Bell, in his 'British Quadrupeds,' p. 423, states that, after long attending to the subject, he has found that white cattle invariably have coloured ears.

brownish-red.[54] On the Ladrone Islands, in the Pacific Ocean, immense herds of cattle, which were wild in the year 1741, are described as "milk-white, except their ears, which are generally black."[55] The Falkland Islands, situated far south, with all the conditions of life as different as it is possible to conceive from those of the Ladrones, offer a more interesting case. Cattle have run wild there during eighty or ninety years; and in the southern districts the animals are mostly white, with their feet, or whole heads, or only their ears black; but my informant, Admiral Sulivan,[56] who long resided on these islands, does not believe that they are ever purely white. So that in these two archipelagos we see that the cattle tend to become white with coloured ears. In other parts of the Falkland Islands other colours prevail: near Port Pleasant brown is the common tint; round Mount Usborne, about half the animals in some of the herds were lead or mouse-coloured, which elsewhere is an unusual tint. These latter cattle, though generally inhabiting high land, breed about a month earlier than the other cattle; and this circumstance would aid in keeping them distinct and in perpetuating this peculiar colour. It is worth recalling to mind that blue or lead-coloured marks have occasionally appeared on the white cattle of Chillingham. So plainly different were the colours of the wild herds in different parts of the Falkland Islands, that in hunting them, as Admiral Sulivan informs me, white spots in one district, and dark spots in another district, were always looked out for on the distant hills. In the intermediate districts intermediate colours prevailed. Whatever the cause may be, this tendency in the wild cattle of the Falkland Islands, which are all descended from a few brought from La Plata, to break up into herds of three different colours, is an interesting fact.

Returning to the several British breeds, the conspicuous difference in general appearance between Short-horns, Long-horns (now rarely seen), Herefords, Highland cattle, Alderneys, &c., must be familiar to every one. A large part of the differ-

[54] Azara, 'Des Quadrupèdes du Paraguay,' tom. ii. p. 361. Azara quotes Buffon for the feral cattle of Africa. For Texas, *see* 'Times,' Feb. 18th, 1846.

[55] Anson's Voyage. *See* Kerr and Porter's 'Collection,' vol. xii. p. 103.

[56] *See* also Mr. Mackinnon's pamphlet on the Falkland Islands, p. 24.

ence, no doubt, may be due to descent from primordially distinct species; but we may feel sure that there has been in addition a considerable amount of variation. Even during the Neolithic period, the domestic cattle were not actually identical with the aboriginal species. Within recent times most of the breeds have been modified by careful and methodical selection. How strongly the characters thus acquired are inherited, may be inferred from the prices realised by the improved breeds; even at the first sale of Colling's Short-horns, eleven bulls reached an average of 214*l*., and lately Short-horn bulls have been sold for a thousand guineas, and have been exported to all quarters of the world.

Some constitutional differences may be here noticed. The Short-horns arrive at maturity far earlier than the wilder breeds, such as those of Wales or the Highlands. This fact has been shown in an interesting manner by Mr. Simonds,[57] who has given a table of the average period of their dentition, which proves that there is a difference of no less than six months in the appearance of the permanent incisors. The period of gestation, from observations made by Tessier on 1131 cows, varies to the extent of eighty-one days; and what is more interesting, M. Lefour affirms "that the period of gestation is longer in the large German cattle than in the smaller breeds."[58] With respect to the period of conception, it seems certain that Alderney and Zetland cows often become pregnant earlier than other breeds.[59] Lastly, as four fully-developed mammæ is a generic character in the genus Bos,[60] it is worth notice that with our domestic cows the two rudimentary mammæ often become fairly well developed and yield milk.

As numerous breeds are generally found only in long-civilized countries, it may be well to show that in some countries inhabited by barbarous races, who are frequently at war with each other and therefore have little free commu-

[57] 'The Age of the Ox, Sheep, Pig,' &c., by Prof. James Simonds, published by order of the Royal Agricult. Soc.

[58] 'Ann. Agricult. France,' April 1837, as quoted in 'The Veterinary,' vol. xii. p. 725. I quote Tessier's observations from Youatt on Cattle, p. 527.

[59] 'The Veterinary,' vol. viii. p. 681, and vol. x. p. 268. Low's 'Domest. Animals of Great Britain,' p. 297.

[60] Mr. Ogleby, in 'Proc. Zoolog. Soc.,' 1836, p. 138, and 1840, p. 4.

nication, several distinct breeds of cattle now exist or formerly existed. At the Cape of Good Hope Leguat observed, in the year 1720, three kinds.[61] At the present day various travellers have noticed the differences in the breeds in Southern Africa. Sir Andrew Smith several years ago remarked to me that the cattle possessed by the different tribes of Caffres, though living near each other under the same latitude and in the same kind of country, yet differed, and he expressed much surprise at the fact. Mr. Andersson has described[62] the Damara, Bechuana, and Namaqua cattle; and he informs me in a letter that the cattle north of Lake Ngami are likewise different, as Mr. Galton has heard is the case with the cattle of Benguela. The Namaqua cattle in size and shape nearly resemble European cattle, and have short stout horns and large hoofs. The Damara cattle are very peculiar, being big-boned, with slender legs and small hard feet; their tails are adorned with a tuft of long bushy hair nearly touching the ground, and their horns are extraordinarily large. The Bechuana cattle have even larger horns, and there is now a skull in London with the two horns 8 ft. 8¼ in. long, as measured in a straight line from tip to tip, and no less than 13 ft. 5 in. as measured along their curvature! Mr. Andersson in his letter to me says that, though he will not venture to describe the differences between the breeds belonging to the many different sub-tribes, yet such certainly exist, as shown by the wonderful facility with which the natives discriminate them.

That many breeds of cattle have originated through variation, independently of descent from distinct species, we may infer from what we see in South America, where the genus Bos was not endemic, and where the cattle which now exist in such vast numbers are the descendants of a few imported from Spain and Portugal. In Columbia, Roulin[63] describes two peculiar breeds, namely, *pelones*, with extremely thin and fine hair, and *calongos*, absolutely naked. According to Castelnau there are two races in Brazil, one like European cattle, the other different, with remark-

[61] Leguat's Voyage, quoted by Vasey in his 'Delineations of the Ox-tribe,' p. 132.

[62] 'Travels in South Africa,' pp. 317, 336.

[63] 'Mém. de l'Institut présent. par divers Savans,' tom. vi., 1835, p. 333. For Brazil, *see* 'Comptes Rendus,' June 15th, 1846. *See* Azara, 'Quadrupèdes du Paraguay,' tom. ii. pp. 359, 361.

able horns. In Paraguay, Azara describes a breed which certainly originated in S. America, called *chivos*, "because they have straight vertical horns, conical, and very large at the base." He likewise describes a dwarf race in Corrientes, with short legs and a body larger than usual. Cattle without horns, and others with reversed hair, have also originated in Paraguay.

Another monstrous breed, called niatas or natas, of which I saw two small herds on the northern bank of the Plata, is so remarkable as to deserve a fuller description. This breed bears the same relation to other breeds, as bull or pug dogs do to other dogs, or as improved pigs, according to H. von Nathusius, do to common pigs.[64] Rütimeyer believes that these cattle belong to the primigenius type.[65] The forehead is very short and broad, with the nasal end of the skull, together with the whole plane of the upper molar-teeth, curved upwards. The lower jaw projects beyond the upper, and has a corresponding upward curvature. It is an interesting fact that an almost similar conformation characterizes, as I have been informed by Dr. Falconer, the extinct and gigantic Sivatherium of India, and is not known in any other ruminant. The upper lip is much drawn back, the nostrils are seated high up and are widely open, the eyes project outwards, and the horns are large. In walking the head is carried low, and the neck is short. The hind legs appear to be longer, compared with the front legs, than is usual. The exposed incisor teeth, the short head and upturned nostrils, give these cattle the most ludicrous, self-confident air of defiance. The skull which I presented to the College of Surgeons has been thus described by Professor Owen:[66] "It is remarkable from the stunted development of the nasals; premaxillaries, and fore-part of the lower jaw, which is unusually

[64] 'Schweineschädel,' 1864, s. 104. Nathusius states that the form of skull characteristic of the niata cattle occasionally appears in European cattle; but he is mistaken, as we shall hereafter see, in supposing that these cattle do not form a distinct race. Prof. Wyman, of Cambridge, United States, informs me that the common cod-fish presents a similar monstrosity, called by the fishermen the "bulldog cod." Prof. Wyman also concluded, after making numerous inquiries in La Plata, that the niata cattle transmit their peculiarities or form a race.

[65] Ueber Art des Zahmen Europ. Rindes, 1866, s. 28.

[66] 'Descriptive Cat. of Ost. Collect. of College of Surgeons,' 1853, p. 624. Vasey, in his 'Delineations of the Ox-tribe,' has given a figure of this skull; and I sent a photograph of it to Prof. Rütimeyer.

curved upwards to come into contact with the premaxillaries. The nasal bones are about one-third the ordinary length, but retain almost their normal breadth. The triangular vacuity is left between them, the frontal and lachrymal, which latter bone articulates with the premaxillary, and thus excludes the maxillary from any junction with the nasal." So that even the connexion of some of the bones is changed. Other differences might be added: thus the plane of the condyles is somewhat modified, and the terminal edge of the premaxillaries forms an arch. In fact, on comparison with the skull of a common ox, scarcely a single bone presents the same exact shape, and the whole skull has a wonderfully different appearance.

The first brief published notice of this race was by Azara, between the years 1783-96; but Don F. Muniz, of Luxan, who has kindly collected information for me, states that about 1760 these cattle were kept as curiosities near Buenos Ayres. Their origin is not positively known, but they must have originated subsequently to the year 1552, when cattle were first introduced. Signor Muniz informs me that the breed is believed to have originated with the Indians southward of the Plata. Even to this day those reared near the Plata show their less civilized nature in being fiercer than common cattle, and in the cow, if visited too often, easily deserting her first calf. The breed is very true, and a niata bull and cow invariably produce niata calves. The breed has already lasted at least a century. A niata bull crossed with a common cow, and the reverse cross, yield offspring having an intermediate character, but with the niata character strongly displayed. According to Signor Muniz, there is the clearest evidence, contrary to the common belief of agriculturists in analogous cases, that the niata cow when crossed with a common bull transmits her peculiarities more strongly than does the niata bull when crossed with a common cow. When the pasture is tolerably long, these cattle feed as well as common cattle with their tongue and palate; but during the great droughts, when so many animals perish on the Pampas, the niata breed lies under a great disadvantage, and would, if not attended to, become extinct; for the common cattle, like horses, are able just to keep alive by browsing on the twigs of trees and on reeds with their lips: this the niatas cannot so well do, as their lips do not join, and hence they are found to perish before the common cattle. This strikes me as a good illustration of how little we are able to judge from the ordinary habits of an animal, on what circumstances, occurring only at long intervals of time, its rarity or extinction may depend. It shows us, also, how natural selection would have determined the rejection of the niata modification had it arisen in a state of nature.

Having described the semi-monstrous niata breed, I may allude to a white bull, said to have been brought from Africa, which was exhibited in London in 1829, and which has been well figured by Mr. Harvey.[67] It had a hump, and was furnished with a mane. The dewlap was peculiar, being divided between its fore-legs into parallel divisions. Its lateral hoofs were annually shed, and grew to the length of five or six inches. The eye was very peculiar, being remarkably prominent, and "resembled a cup and ball, thus enabling the animal to see on all sides with equal ease; the pupil was small and oval, or rather a parallelogram with the ends cut off, and lying transversely across the ball." A new and strange breed might probably have been formed by careful breeding and selection from this animal.

I have often speculated on the probable causes through which each separate district in Great Britain came to possess in former times its own peculiar breed of cattle; and the question is, perhaps, even more perplexing in the case of Southern Africa. We now know that the differences may be in part attributed to descent from distinct species; but this will not suffice. Have the slight differences in climate and in the nature of the pasture, in the different districts of Britain, directly induced corresponding differences in the cattle? We have seen that the semi-wild cattle in the several British parks are not identical in colouring or size, and that some degree of selection has been requisite to keep them true. It is almost certain that abundant food given during many generations directly affects the size of a breed.[68] That climate directly affects the thickness of the

[67] Loudon's 'Magazine of Nat. Hist.,' vol. i., 1829, p. 113. Separate figures are given of the animal, its hoofs, eye, and dewlap.

[68] Low, 'Domesticated Animals of the British Isles,' p. 264.

skin and the hair is likewise certain: thus Roulin asserts[69] that the hides of the feral cattle on the hot Llanos "are always much less heavy than those of the cattle raised on the high platform of Bogota; and that these hides yield in weight and in thickness of hair to those of the cattle which have run wild on the lofty Paramos." The same difference has been observed in the hides of the cattle reared on the bleak Falkland Islands and on the temperate Pampas. Low has remarked[70] that the cattle which inhabit the more humid parts of Britain have longer hair and thicker skins than other British cattle; and the hair and horns are so closely related to each other, that, as we shall see in a future chapter, they are apt to vary together; thus climate might indirectly affect, through the skin, the form and size of the horns. When we compare highly improved stall-fed cattle with the wilder breeds, or compare mountain and lowland breeds, we cannot doubt that an active life, leading to the free use of the limbs and lungs, affects the shape and proportions of the whole body. It is probable that some breeds, such as the semi-monstrous niata cattle, and some peculiarities, such as being hornless, &c., have appeared suddenly from what we may call a spontaneous variation; but even in this case a rude kind of selection is necessary, and the animals thus characterized must be at least partially separated from others. This degree of care, however, has sometimes been taken even in little-civilized districts, where we should least have expected it, as in the case of the niata, chivo, and hornless cattle in S. America.

That methodical selection has done wonders within a recent period in modifying our cattle, no one doubts. During the process of methodical selection it has occasionally happened that deviations of structure, more strongly pronounced than mere individual differences, yet by no means deserving to be called monstrosities, have been taken advantage of: thus the famous Long-horn Bull, Shakespeare, though of the pure Canley stock, "scarcely inherited a single point of the long-horned breed, his horns excepted;[71] yet in the hands of Mr. Fowler,

[69] 'Mém. de l'Institut présent. par divers Savans,' tom. vi., 1835, p. 332.

[70] Idem, pp. 304, 368, &c.

[71] Youatt on Cattle, p. 193. A full account of this bull is taken from Marshall.

this bull greatly improved his race. We have also reason to believe that selection, carried on so far unconsciously that there was at no one time any distinct intention to improve or change the breed, has in the course of time modified most of our cattle; for by this process, aided by more abundant food, all the lowland British breeds have increased greatly in size and in early maturity since the reign of Henry VII.[72] It should never be forgotten that many animals have to be annually slaughtered; so that each owner must determine which shall be killed and which preserved for breeding. In every district, as Youatt has remarked, there is a prejudice in favour of the native breed; so that animals possessing qualities, whatever they may be, which are most valued in each district, will be oftenest preserved; and this unmethodical selection assuredly will in the long run affect the character of the whole breed. But it may be asked, can this rude kind of selection have been practised by barbarians such as those of southern Africa? In a future chapter on Selection we shall see that this has certainly occurred to some extent. Therefore, looking to the origin of the many breeds of cattle which formerly inhabited the several districts of Britain, I conclude that, although slight differences in the nature of the climate, food, &c., as well as changed habits of life, aided by correlation of growth, and the occasional appearance from unknown causes of considerable deviations of structure, have all probably played their parts; yet that the occasional preservation in each district of those individual animals which were most valued by each owner has perhaps been even more effective in the production of the several British breeds. As soon as two or more breeds had once been formed in any district, or when new breeds descended from distinct species were introduced, their crossing, especially if aided by some selection, will have multiplied the number and modified the characters of the older breeds.

SHEEP.

I SHALL treat this subject briefly.[1] Most authors look at our domestic sheep as descended from several distinct species; but how many still exist is doubtful. Mr. Blyth believes that there

[72] Youatt on Cattle, p. 116. Lord Spencer has written on this same subject.

are in the whole world fourteen species, one of which, the Corsican moufflon, he concludes (as I am informed by him) to be the parent of the smaller, short-tailed breeds, with crescent-shaped horns, such as the old Highland sheep. The larger, long-tailed breeds, having horns with a double flexure, such as the Dorsets, merinos, &c., he believes to be descended from an unknown and extinct species. M. Gervais makes six species of Ovis;[73] but concludes that our domestic sheep form a distinct genus, now completely extinct. A German naturalist[74] believes that our sheep descend from ten aboriginally distinct species, of which only one is still living in a wild state! Another ingenious observer,[75] though not a naturalist, with a bold defiance of everything known on geographical distribution, infers that the sheep of Great Britain alone are the descendants of eleven endemic British forms! Under such a hopeless state of doubt it would be useless for my purpose to give a detailed account of the several breeds; but a few remarks may be added.

Sheep have been domesticated from a very ancient period. Rütimeyer[76] found in the Swiss lake-dwellings the remains of a small breed, with thin and tall legs, and with horns like those of a goat: this race differs somewhat from any one now known. Almost every country has its own peculiar breed; and many countries have many breeds differing greatly from each other. One of the most strongly marked races is an Eastern one with a long tail, including, according to Pallas, twenty vertebræ, and so loaded with fat, that, from being esteemed a delicacy, it is sometimes placed on a truck which is dragged about by the living animal. These sheep, though ranked by Fitzinger as a distinct aboriginal form, seem to bear in their drooping ears the stamp of long domestication. This is likewise the case with those sheep which have two great masses of fat on the rump, with the tail in a rudimentary condition. The Angola variety of

[73] Blyth on the genus Ovis, in 'Annals and Mag. of Nat. History,' vol. vii., 1841, p. 261: with respect to the parentage of the breeds, see Mr. Blyth's excellent articles in 'Land and Water,' 1867, pp. 134, 156. Gervais, 'Hist. Nat. des Mammifères,' 1855, tom. ii. p. 191.

[74] Dr. L. Fitzinger, 'Ueber die Racen des Zahmen Schafes,' 1860, s. 86.

[75] J. Anderson, 'Recreations in Agriculture and Natural History,' vol. ii. p. 164.

[76] 'Pfahlbauten,' s. 127, 193.

the long-tailed race has curious masses of fat on the back of the head and beneath the jaws.[77] Mr. Hodgson in an admirable paper [78] on the sheep of the Himalaya infers from the distribution of the several races, "that this caudal augmentation in most of its phases is an instance of degeneracy in these pre-eminently Alpine animals." The horns present an endless diversity in character; being, especially in the female sex, not rarely absent, or, on the other hand, amounting to four or even eight in number. The horns, when numerous, arise from a crest on the frontal bone, which is elevated in a peculiar manner. It is remarkable that multiplicity of horns "is generally accompanied by great length and coarseness of the fleece." [79] This correlation, however, is not invariable; for I am informed by Mr. D. Forbes, that the Spanish sheep in Chile resemble, in fleece and in all other characters, their parent merino-race, except that instead of a pair they generally bear four horns. The existence of a pair of mammæ is a generic character in the genus Ovis as well as in several allied forms; nevertheless, as Mr. Hodgson has remarked, "this character is not absolutely constant even among the true and proper sheep: for I have more than once met with Cágias (a sub-Himalayan domestic race) possessed of four teats." [80] This case is the more remarkable as, when any part or organ is present in reduced number in comparison with the same part in allied groups, it usually is subject to little variation. The presence of interdigital pits has likewise been considered as a generic distinction in sheep; but Isidore Geoffroy [81] has shown that these pits or pouches are absent in some breeds.

In sheep there is a strong tendency for characters, which have apparently been acquired under domestication, to become attached either exclusively to the male sex, or to be more highly developed in this than in the other sex. Thus in many breeds the horns are deficient in the ewe, though this likewise occurs occasionally with the female of the wild musmon. In the rams of the Wallachian breed "the horns spring almost perpendicularly

77 Youatt on Sheep, p. 120.

78 'Journal of the Asiatic Soc. of Bengal,' vol. xvi. pp. 1007, 1016.

79 Youatt on Sheep, pp. 142-169.

80 'Journal Asiat. Soc. of Bengal,' vol. xvi., 1847, p. 1015.

81 'Hist. Nat. Gén.,' tom. iii. p. 435.

from the frontal bone, and then take a beautiful spiral form; in the ewes they protrude nearly at right angles from the head, and then become twisted in a singular manner." [82] Mr. Hodgson states that the extraordinarily arched nose or chaffron, which is so highly developed in several foreign breeds, is characteristic of the ram alone, and apparently is the result of domestication.[83] I hear from Mr. Blyth that the accumulation of fat in the fat-tailed sheep of the plains of India is greater in the male than in the female; and Fitzinger [84] remarks that the mane in the African maned race is far more developed in the ram than in the ewe.

Different races of sheep, like cattle, present constitutional differences. Thus the improved breeds arrive at maturity at an early age, as has been well shown by Mr. Simonds through their early average period of dentition. The several races have become adapted to different kinds of pasture and climate: for instance, no one can rear Leicester sheep on mountainous regions, where Cheviots flourish. As Youatt has remarked, "in all the different districts of Great Britain we find various breeds of sheep beautifully adapted to the locality which they occupy. No one knows their origin; they are indigenous to the soil, climate, pasturage, and the locality on which they graze; they seem to have been formed for it and by it." [85] Marshall relates [86] that a flock of heavy Lincolnshire and light Norfolk sheep which had been bred together in a large sheep-walk, part of which was low, rich, and moist, and another part high and dry, with benty grass, when turned out, regularly separated from each other; the heavy sheep drawing off to the rich soil, and the lighter sheep to their own soil; so that "whilst there was plenty of grass the two breeds kept themselves as distinct as rooks and pigeons." Numerous sheep from various parts of the world have been brought during a long course of years to the Zoological Gardens of London; but as Youatt, who attended the animals as a vete-

[82] Youatt on Sheep, p. 138.

[83] 'Journal Asiat. Soc. of Bengal,' vol. xvi., 1847, pp. 1015, 1016.

[84] 'Racen des Zahmen Schafes,' s. 77.

[85] 'Rural Economy of Norfolk,' vol. ii. p. 136.

[86] Youatt on Sheep, p. 312. On same subject, *see* excellent remarks in 'Gardener's Chronicle,' 1858, p. 868. For experiments in crossing Cheviot sheep with Leicesters, *see* Youatt, p. 325.

rinary surgeon, remarks, "few or none die of the rot, but they are phthisical; not one of them from a torrid climate lasts out the second year, and when they die their lungs are tuberculated."[87] Even in certain parts of England it has been found impossible to keep certain breeds of sheep; thus on a farm on the banks of the Ouse, the Leicester sheep were so rapidly destroyed by pleuritis[88] that the owner could not keep them; the coarser-skinned sheep never being affected.

The period of gestation was formerly thought to be so unalterable a character, that a supposed difference between the wolf and the dog in this respect was esteemed a sure sign of specific distinction; but we have seen that the period is shorter in the improved breeds of the pig, and in the larger breeds of the ox, than in other breeds of these two animals. And now we know, on the excellent authority of Hermann von Nathusius,[89] that Merino and Southdown sheep, when both have long been kept under exactly the same conditions, differ in their average period of gestation, as is seen in the following Table:—

Merinos	150·3 days.
Southdowns	144·2 ,,
Half-bred Merinos and Southdowns ..	146·3 ,,
¾ blood of Southdown	145·5 ,,
⅞ ,, ,, ...	144·2 ,,

In this graduated difference, in these cross-bred animals having different proportions of Southdown blood, we see how strictly the two periods of gestation have been transmitted. Nathusius remarks that, as Southdowns grow with remarkable rapidity after birth, it is not surprising that their fœtal development should have been shortened. It is of course possible that the difference in these two breeds may be due to their descent from distinct parent-species; but as the early maturity of the Southdowns has long been carefully attended to by breeders, the difference is more probably the result of such attention. Lastly, the fecundity of the several breeds differs much; some generally producing twins or even triplets at a birth, of which fact the curious Shangai sheep (with their truncated and rudimentary

87 Youatt on Sheep, note, p. 491.

88 'The Veterinary,' vol. x. p. 217.

89 A translation of his paper is given in 'Bull. Soc. Imp. d'Acclimat.,' tom. ix., 1862, p. 723.

ears, and great Roman noses), lately exhibited in the Zoological Gardens, offer a remarkable instance.

Sheep are perhaps more readily affected by the direct action of the conditions of life to which they have been exposed than almost any other domestic animal. According to Pallas, and more recently according to Erman, the fat-tailed Kirghisian sheep, when bred for a few generations in Russia, degenerate, and the mass of fat dwindles away, "the scanty and bitter herbage of the steppes seems so essential to their development." Pallas makes an analogous statement with respect to one of the Crimean breeds. Burnes states that the Karakool breed, which produces a fine, curled, black, and valuable fleece, when removed from its own canton near Bokhara to Persia or to other quarters, loses its peculiar fleece.[90] In all such cases, however, it may be that a change of any kind in the conditions of life causes variability and consequent loss of character, and not that certain conditions are necessary for the development of certain characters.

Great heat, however, seems to act directly on the fleece: several accounts have been published of the change which sheep imported from Europe undergo in the West Indies. Dr. Nicholson of Antigua informs me that, after the third generation, the wool disappears from the whole body, except over the loins; and the animal then appears like a goat with a dirty door-mat on its back. A similar change is said to take place on the west coast of Africa.[91] On the other hand, many wool-bearing sheep live on the hot plains of India. Roulin asserts that in the lower and heated valleys of the Cordillera, if the lambs are sheared as soon as the wool has grown to a certain thickness, all goes on afterwards as usual; but if not sheared, the wool detaches itself in flakes, and short shining hair like that

[90] Erman's 'Travels in Siberia' (Eng. trans.), vol. i. p. 228. For Pallas on the fat-tailed sheep, I quote from Anderson's account of the 'Sheep of Russia,' 1794, p. 34. With respect to the Crimean sheep, *see* Pallas' 'Travels' (Eng. trans.), vol. ii. p. 454. For the Karakool sheep, *see* Burnes' 'Travels in Bokhara,' vol. iii. p. 151.

[91] *See* Report of the Directors of the Sierra Leone Company, as quoted in White's 'Gradation of Man,' p. 95. With respect to the change which sheep undergo in the West Indies, *see* also Dr. Davy, in 'Edin. New. Phil. Journal,' Jan. 1852. For the statement made by Roulin, *see* 'Mém. de l'Institut présent. par divers Savans,' tom. vi., 1835, p. 347.

on a goat is produced ever afterwards. This curious result seems merely to be an exaggerated tendency natural to the Merino breed, for as a great authority, namely, Lord Somerville, remarks, "the wool of our Merino sheep after shear-time is hard and coarse to such a degree as to render it almost impossible to suppose that the same animal could bear wool so opposite in quality, compared to that which has been clipped from it: as the cold weather advances, the fleeces recover their soft quality." As in sheep of all breeds the fleece naturally consists of longer and coarser hair covering shorter and softer wool, the change which it often undergoes in hot climates is probably merely a case of unequal development; for even with those sheep which like goats are covered with hair, a small quantity of underlying wool may always be found.[92] In the wild mountain-sheep (*Ovis montana*) of North America there is an annual analogous change of coat; "the wool begins to drop out in early spring, leaving in its place a coat of hair resembling that of the elk, a change of pelage quite different in character from the ordinary thickening of the coat or hair, common to all furred animals in winter,—for instance, in the horse, the cow, &c., which shed their winter coat in the spring."[93]

A slight difference in climate or pasture sometimes slightly affects the fleece, as has been observed even in different districts in England, and as is well shown by the great softness of the wool brought from Southern Australia. But it should be observed, as Youatt repeatedly insists, that the tendency to change may generally be counteracted by careful selection. M. Lasterye, after discussing this subject, sums up as follows: "The preservation of the Merino race in its utmost purity at the Cape of Good Hope, in the marshes of Holland, and under the rigorous climate of Sweden, furnishes an additional support of this my unalterable principle, that fine-woolled sheep may be kept wherever industrious men and intelligent breeders exist."

That methodical selection has effected great changes in several

[92] Youatt on Sheep, p. 69, where Lord Somerville is quoted. *See* p. 117, on the presence of wool under the hair. With respect to the fleeces of Australian sheep, p. 185. On selection counteracting any tendency to change, *see* pp. 70, 117, 120, 168.

[93] Audubon and Bachman, 'The Quadrupeds of North America,' 1846, vol. v. p. 365.

breeds of sheep no one, who knows anything on the subject, entertains a doubt. The case of the Southdowns, as improved by Ellman, offers perhaps the most striking instance. Unconscious or occasional selection has likewise slowly produced a great effect, as we shall see in the chapters on Selection. That crossing has largely modified some breeds, no one who will study what has been written on this subject—for instance, Mr. Spooner's paper—will dispute; but to produce uniformity in a crossed breed, careful selection and "rigorous weeding," as this author expresses it, are indispensable.[94]

In some few instances new breeds have suddenly originated; thus, in 1791, a ram-lamb was born in Massachusetts, having short crooked legs and a long back, like a turnspit-dog. From this one lamb the *otter* or *ancon* semi-monstrous breed was raised; as these sheep could not leap over the fences, it was thought that they would be valuable; but they have been supplanted by merinos, and thus exterminated. These sheep are remarkable from transmitting their character so truly that Colonel Humphreys[95] never heard of "but one questionable case" of an ancon ram and ewe not producing ancon offspring. When they are crossed with other breeds the offspring, with rare exceptions, instead of being intermediate in character, perfectly resemble either parent; and this has occurred even in the case of twins. Lastly, "the ancons have been observed to keep together, separating themselves from the rest of the flock when put into enclosures with other sheep."

A more interesting case has been recorded in the Report of the Juries for the Great Exhibition (1851), namely, the production of a merino ram-lamb on the Mauchamp farm, in 1828, which was remarkable for its long, smooth, straight, and silky wool. By the year 1833 M. Graux had raised rams enough to serve his whole flock, and after a few more years he was able to sell stock of his new breed. So peculiar and valuable is the wool, that it sells at 25 per cent. above the best merino wool: even the fleeces of half-bred animals are valuable, and are known in France as the "Mauchamp-merino." It is interesting, as

[94] 'Journal of R. Agricult. Soc. of England,' vol. xx., part ii. W. C. Spooner on Cross-Breeding.

[95] 'Philosoph. Transactions,' London, 1813, p. 88.

showing how generally any marked deviation of structure is accompanied by other deviations, that the first ram and his immediate offspring were of small size, with large heads, long necks, narrow chests, and long flanks; but these blemishes were removed by judicious crosses and selection. The long smooth wool was also correlated with smooth horns; and as horns and hair are homologous structures, we can understand the meaning of this correlation. If the Mauchamp and ancon breeds had originated a century or two ago, we should have had no record of their birth; and many a naturalist would no doubt have insisted, especially in the case of the Mauchamp race, that they had each descended from, or been crossed with, some unknown aboriginal form.

GOATS.

FROM the recent researches of M. Brandt, most naturalists now believe that all our goats are descended from the *Capra ægagrus* of the mountains of Asia, possibly mingled with the allied Indian species *C. Falconeri* of India.[96] In Switzerland, during the early Stone period, the domestic goat was commoner than the sheep; and this very ancient race differed in no respect from that now common in Switzerland.[97] At the present time, the many races found in several parts of the world differ greatly from each other; nevertheless, as far as they have been tried,[98] they are all quite fertile when crossed. So numerous are the breeds, that Mr. G. Clark[99] has described eight distinct kinds imported into the one island of Mauritius. The ears of one kind were enormously developed, being, as measured by Mr. Clark, no less than 19 inches in length and 4¾ inches in breadth. As with cattle, the mammæ of those breeds which are regularly milked become greatly developed; and, as Mr. Clark remarks, "it is not rare to see their teats touching the ground." The following cases are worth notice as presenting unusual

[96] Isidore Geoffroy St. Hilaire, 'Hist. Nat. Générale,' tom. iii. p. 87. Mr. Blyth ('Land and Water,' 1867, p. 37) has arrived at a similar conclusion, but he thinks that certain Eastern races may perhaps be in part descended from the Asiatic markhor.

[97] Rütimeyer, 'Pfahlbauten,' s. 127.

[98] Godron, 'De l'Espèce,' tom. i. p. 402.

[99] 'Annals and Mag. of Nat. History,' vol. ii. (2nd series), 1848, p. 363.

points of variation. According to Godron,[100] the mammæ differ greatly in shape in different breeds, being elongated in the common goat, hemispherical in the Angora race, and bilobed and divergent in the goats of Syria and Nubia. According to this same author, the males of certain breeds have lost their usual offensive odour. In one of the Indian breeds the males and females have horns of widely-different shapes;[101] and in some breeds the females are destitute of horns.[102] The presence of interdigital pits or glands on all four feet has been thought to characterise the genus Ovis, and their absence to be characteristic of the genus Capra; but Mr. Hodgson has found that they exist in the front feet of the majority of Himalayan goats.[103] Mr. Hodgson measured the intestines in two goats of the Dúgú race, and he found that the proportional length of the great and small intestines differed considerably. In one of these goats the cæcum was thirteen inches, and in the other no less than thirty-six inches in length!

[100] 'De l'Espèce,' tom. i. p. 406. Mr. Clark also refers to differences in the shape of the mammæ. Godron states that in the Nubian race the scrotum is divided into two lobes; and Mr. Clark gives a ludicrous proof of this fact, for he saw in the Mauritius a male goat of the Muscat breed purchased at a high price for a female in full milk. These differences in the scrotum are probably not due to descent from distinct species; for Mr. Clark states that this part varies much in form.

[101] Mr. Clark, 'Annals and Mag. of Nat. Hist.,' vol. ii. (2nd series), 1848, p. 361.

[102] Desmarest, 'Encyclop. Méthod. Mammalogie,' p. 480.

[103] 'Journal of Asiatic Soc. of Bengal,' vol. xvi., 1847, pp. 1020, 1025.

CHAPTER IV.

DOMESTIC RABBITS.

DOMESTIC RABBITS DESCENDED FROM THE COMMON WILD RABBIT — ANCIENT DOMESTICATION — ANCIENT SELECTION — LARGE LOP-EARED RABBITS — VARIOUS BREEDS — FLUCTUATING CHARACTERS — ORIGIN OF THE HIMALAYAN BREED — CURIOUS CASE OF INHERITANCE — FERAL RABBITS IN JAMAICA AND THE FALKLAND ISLANDS — PORTO SANTO FERAL RABBITS — OSTEOLOGICAL CHARACTERS — SKULL — SKULL OF HALF-LOP RABBITS — VARIATIONS IN THE SKULL ANALOGOUS TO DIFFERENCES IN DIFFERENT SPECIES OF HARES — VERTEBRÆ — STERNUM — SCAPULA — EFFECTS OF USE AND DISUSE ON THE PROPORTIONS OF THE LIMBS AND BODY — CAPACITY OF THE SKULL AND REDUCED SIZE OF THE BRAIN — SUMMARY ON THE MODIFICATIONS OF DOMESTICATED RABBITS.

ALL naturalists, with, as far as I know, a single exception, believe that the several domestic breeds of the rabbit are descended from the common wild species; I shall therefore describe them more carefully than in the previous cases. Professor Gervais[1] states "that the true wild rabbit is smaller than the domestic; its proportions are not absolutely the same; its tail is smaller; its ears are shorter and more thickly clothed with hair; and these characters, without speaking of colour, are so many indications opposed to the opinion which unites these animals under the same specific denomination." Few naturalists will agree with this author that such slight differences are sufficient to separate as distinct species the wild and domestic rabbit. How extraordinary it would be, if close confinement, perfect tameness, unnatural food, and careful breeding, all prolonged during many generations, had not produced at least some effect! The tame rabbit has been domesticated from an ancient period. Confucius ranges rabbits among animals worthy to be sacrificed to the gods, and, as he prescribes their multiplication, they were probably at this early period domesticated in China. They are mentioned by several of the classical writers.

[1] M. P. Gervais, 'Hist. Nat des Mammifères,' tom. i., 1854, p. 288.

In 1631 Gervaise Markham writes, "You shall not, as in other cattell, looke to their shape, but to their richnesse, onely elect your buckes, the largest and goodliest conies you can get; and for the richnesse of the skin, that is accounted the richest which hath the equallest mixture of blacke and white haire together, yet the blacke rather shadowing the white; the furre should be thicke, deepe, smooth, and shining; . . . . . they are of body much fatter and larger, and, when another skin is worth two or three pence, they are worth two shillings." From this full description we see that silver-grey rabbits existed in England at this period; and, what is far more important, we see that the breeding or selection of rabbits was then carefully attended to. Aldrovandi, in 1637, describes, on the authority of several old writers (as Scaliger, in 1557), rabbits of various colours, some "like a hare," and he adds that P. Valerianus (who died a very old man in 1558) saw at Verona rabbits four times bigger than ours.[2]

From the fact of the rabbit having been domesticated at an ancient period, we must look to the northern hemisphere of the Old World, and to the warmer temperate regions alone, for the aboriginal parent-form; for the rabbit cannot live without protection in countries as cold as Sweden, and, though it has run wild in the tropical island of Jamaica, it has never greatly multiplied there. It now exists, and has long existed, in the warmer temperate parts of Europe, for fossil remains have been found in several countries.[3] The domestic rabbit readily becomes feral in these same countries, and when variously coloured kinds are turned out they generally revert to the ordinary grey colour.[4] The wild rabbits, if taken young, can be domesticated, though the process is generally very troublesome.[5] The various

[2] U. Aldrovandi, 'De Quadrupedibus digitatis,' 1637, p. 383. For Confucius and G. Markham, *see* a writer who has studied the subject, in 'Cottage Gardener,' Jan. 22nd, 1861, p. 250.

[3] Owen, 'British Fossil Mammals,' p. 212.

[4] 'Pigeons and Rabbits,' by E. S. Delamer, 1854, p. 133. Sir J. Sebright ('Observations on Instinct,' 1836, p. 10) speaks most strongly on the difficulty. But this difficulty is not invariable, as I have received two accounts of perfect success in taming and breeding from the wild rabbit. *See* also Dr. P. Broca, in 'Journal de la Physiologie,' tom. ii. p. 368.

[5] Bechstein, 'Naturgesch. Deutschlands,' 1801, b. i. p. 1133. I have received similar accounts with respect to England and Scotland.

domestic races are often crossed, and are believed to be perfectly fertile together, and a perfect gradation can be shown to exist from the largest domestic kinds, having enormously developed ears, to the common wild kind. The parent-form must have been a burrowing animal, a habit not common, as far as I can discover, to any other species in the large genus Lepus. Only one wild species is known with certainty to exist in Europe; but the rabbit (if it be a true rabbit) from Mount Sinai, and likewise that from Algeria, present slight differences; and these forms have been considered by some authors as specifically distinct.[6] But such slight differences would aid us little in explaining the more considerable differences characteristic of the several domestic races. If the latter are the descendants of two or more closely allied species, all, excepting the common rabbit, have been exterminated in a wild state; and this is very improbable, seeing with what pertinacity this animal holds its ground. From these several reasons we may infer with safety that all the domestic breeds are the descendants of the common wild species. But from what we hear of the late marvellous success in rearing hybrids between the hare and rabbit,[7] it is possible, though not probable, from the great difficulty in making the first cross, that some of the larger races, which are coloured like the hare, may have been modified by crosses with this animal. Nevertheless, the chief differences in the skeletons of the several domestic breeds cannot, as we shall presently see, have been derived from a cross with the hare.

There are many breeds which transmit their characters more or less truly. Every one has seen the enormous lop-eared rabbits exhibited at our shows; various allied sub-breeds are reared on the Continent, such as the so-called Andalusian, which is said to have a large head with a round forehead, and to attain a greater size than any other kind; another large Paris breed is named the Rouennais, and has a square head; the so-called Patagonian rabbit has remarkably short ears and a large round head. Although I have not seen all these breeds, I feel some doubt about there being any marked difference in the

[6] Gervais, 'Hist. Nat. des Mammifères,' tom. i. p. 292.

[7] *See* Dr. P. Broca's interesting memoir on this subject in Brown-Sequard's 'Journ. de Phys.,' vol. ii. p. 367.

shape of their skulls.[8] English lop-eared rabbits often weigh 8 lbs. or 10 lbs., and one has been exhibited weighing 18 lbs.; whereas a full-sized wild rabbit weighs only about 3¼ lbs. The head or skull in all the large lop-eared rabbits examined by me is much longer relatively to its breadth than in the wild rabbit. Many of them have loose transverse folds of skin or dewlaps beneath the throat, which can be pulled out so as to reach nearly to the ends of the jaws. Their ears are prodigiously developed, and hang down on each side of their faces. A rabbit has been exhibited with its two ears, measured from the tip of one to the tip of the other, 22 inches in length, and each ear was 5⅜ inches in breadth. In a common wild rabbit I found that the length of the two ears, from tip to tip, was 7⅝ inches, and the breadth only 1⅞ inch. The great weight of the body in the larger rabbits, and the immense development of their ears, are the qualities which win prizes, and have been carefully selected.

The hare-coloured, or, as it is sometimes called, the Belgian rabbit, differs in nothing except colour from the other large breeds; but Mr. J. Young, of Southampton, a great breeder of this kind, informs me that the females, in all the specimens examined by him, had only six mammæ; and this certainly was the case with two females which came into my possession. Mr. B. P. Brent, however, assures me that the number is variable with other domestic rabbits. The common wild rabbit always has ten mammæ. The Angora rabbit is remarkable from the length and fineness of its fur, which even on the soles of the feet is of considerable length. This breed is the only one which differs in its mental qualities, for it is said to be much more sociable than other rabbits, and the male shows no wish to destroy its young.[9] Two live rabbits were brought to me from Moscow, of about the size of the wild species, but with long soft fur, different from that of the Angora. These Moscow rabbits had pink eyes and were snow-white, excepting the ears, two spots near the nose, the upper and under surface of the tail, and the hinder tarsi, which were blackish-brown. In short, they were

[8] They are briefly described in the 'Journal of Horticulture,' May 7th, 1861, p. 108.

[9] 'Journal of Horticulture,' 1861, p. 380.

coloured nearly like the so-called Himalayan rabbits, presently to be described, and differed from them only in the character of their fur. There are two other breeds which come true to colour, but differ in no other respect, namely silver-greys and chinchillas. Lastly, the Nicard or Dutch rabbit may be mentioned, which varies in colour, and is remarkable from its small size, some specimens weighing only 1¼ lb.; rabbits of this breed make excellent nurses for other and more delicate kinds.[10]

Certain characters are remarkably fluctuating, or are very feebly transmitted by domestic rabbits: thus, one breeder tells me that with the smaller kinds he has hardly ever raised a whole litter of the same colour: with the large lop-eared breeds "it is impossible," says a great judge,[11] "to breed true to colour, but by judicious crossing a great deal may be done towards it. The fancier should know how his does are bred, that is, the colour of their parents." Nevertheless, certain colours, as we shall presently see, are transmitted truly. The dewlap is not strictly inherited. Lop-eared rabbits, with their ears hanging flat down on each side of the face, do not transmit this character at all truly. Mr. Delamer remarks that, "with fancy rabbits, when both the parents are perfectly formed, have model ears, and are handsomely marked, their progeny do not invariably turn out the same." When one parent, or even both, are oar-laps, that is, have their ears sticking out at right angles, or when one parent or both are half-lops, that is, have only one ear dependent, there is nearly as good a chance of the progeny having both ears full-lop, as if both parents had been thus characterized. But I am informed, if both parents have upright ears, there is hardly a chance of a full-lop. In some half-lops the ear that hangs down is broader and longer than the upright ear;[12] so that we have the unusual case of a want of symmetry on the two sides. This difference in the position and size of the two ears probably indicates that the lopping of the ear results

[10] 'Journal of Horticulture,' May 28th, 1861, p. 169.

[11] 'Journal of Horticulture,' 1861, p. 327. With respect to the ears, *see* Delamer on 'Pigeons and Rabbits,' 1854, p. 141; also 'Poultry Chronicle,' vol. ii. p. 499, and ditto for 1854, p. 586.

[12] Delamer, 'Pigeons and Rabbits,' p. 136. *See* also 'Journal of Horticulture,' 1861, p. 375.

Fig. 5.—Half-lop Rabbit. (Copied from E. S. Delamer's work.)

from its great length and weight, favoured no doubt by the weakness of the muscles consequent on disuse. Anderson[13] mentions a breed having only a single ear; and Professor Gervais another breed which is destitute of ears.

The origin of the Himalayan breed (sometimes called Chinese, or Polish, or Russian) is so curious, both in itself, and as throwing some light on the complex laws of inheritance, that it is worth giving in detail. These pretty rabbits are white, except their ears, nose, all four feet, and the upper side of tail, which are all brownish-black; but as they have red eyes, they may be considered as albinoes. I have received several accounts of their breeding perfectly true. From their symmetrical marks, they were at first ranked as specifically distinct, and were provisionally named *L. nigripes*.[14] Some good observers thought that they could detect a difference in their habits, and stoutly maintained that they formed a new species. Their origin is now well known. A writer, in 1857,[15] stated that he had produced Himalayan rabbits in the following manner. But it is first necessary briefly to describe two other breeds: silver-greys or silver-sprigs generally have black heads and legs, and their fine grey fur is interspersed with numerous black and white long hairs.

13 'An Account of the different Kinds of Sheep in the Russian Dominions,' 1794, p. 39.

14 'Proc. Zoolog. Soc.,' June 23rd, 1857, p. 159.

15 'Cottage Gardener,' 1857, p. 141.

They breed perfectly true, and have long been kept in warrens. When they escape and cross with common rabbits, the product, as I hear from Mr. Wyrley Birch, of Wretham Hall, is not a mixture of the two colours, but about half take after the one parent, and the other half after the other parent. Secondly, chinchillas or tame silver-greys (I will use the former name) have short, paler, mouse or slate-coloured fur, interspersed with long, blackish, slate-coloured, and white hairs.[16] These rabbits breed perfectly true. Now, the writer above referred to had a breed of chinchillas which had been crossed with the common black rabbit, and their offspring were either blacks or chinchillas. These latter were again crossed with other chinchillas (which had also been crossed with silver-greys), and from this complicated cross Himalayan rabbits were raised. From these and other similar statements, Mr. Bartlett[17] was led to make a careful trial in the Zoological Gardens, and he found that by simply crossing silver-greys with chinchillas he could always produce some few Himalayans; and the latter, notwithstanding their sudden origin, if kept separate, bred perfectly true.

The Himalayans, when first born, are quite white, and are then true albinoes; but in the course of a few months they gradually assume their dark ears, nose, feet, and tail. Occasionally, however, as I am informed by Mr. W. A. Wooler and the Rev. W. D. Fox, the young are born of a very pale grey colour, and specimens of such fur were sent me by the former gentleman. The grey tint, however, disappears as the animal comes to maturity. So that with these Himalayans there is a tendency, strictly confined to early youth, to revert to the colour of the adult silver-grey parent-stock. Silver-greys and chinchillas, on the other hand, present a remarkable contrast in their colour whilst quite young, for they are born perfectly black, but soon assume their characteristic grey or silver tints. The same thing occurs with grey horses, which, as long as they are foals, are generally of a nearly black colour, but soon become grey, and get whiter and whiter as they grow older. Hence the usual rule is that Himalayans are born white and afterwards become in certain parts of their bodies dark-coloured; whilst

16 'Journal of Horticulture,' April 9th, 1861, p. 35.

17 Mr. Bartlett, in 'Proc. Zoolog. Soc.,' 1861, p. 40.

silver-greys are born black and afterwards become sprinkled with white. Exceptions, however, and of a directly opposite nature, occasionally occur in both cases. For young silver-greys are sometimes born in warrens, as I hear from Mr. W. Birch, of a cream-colour, but these young animals ultimately become black. The Himalayans, on the other hand, sometimes produce, as is stated by an experienced amateur,[18] a single black young one in a litter; but such, before two months elapse, become perfectly white.

To sum up the whole curious case: wild silver-greys may be considered as black rabbits which become grey at an early period of life. When they are crossed with common rabbits, the offspring are said not to have blended colours, but to take after either parent; and in this respect they resemble black and albino varieties of most quadrupeds, which often transmit their colours in this same manner. When they are crossed with chinchillas, that is, with a paler sub-variety, the young are at first pure albinoes, but soon become dark-coloured in certain parts of their bodies, and are then called Himalayans. The young Himalayans, however, are sometimes at first either pale grey or completely black, in either case changing after a time to white. In a future chapter I shall advance a large body of facts showing that, when two varieties are crossed both of which differ in colour from their parent-stock, there is a strong tendency in the young to revert to the aboriginal colour; and what is very remarkable, this reversion occasionally supervenes, not before birth, but during the growth of the animal. Hence, if it could be shown that silver-greys and chinchillas were the offspring of a cross between a black and albino variety with the colours intimately blended—a supposition in itself not improbable, and supported by the circumstance of silver-greys in warrens sometimes producing creamy-white young, which ultimately become black —then all the above-given paradoxical facts on the changes of colour in silver-greys and in their descendants the Himalayans would come under the law of reversion, supervening at different periods of growth and in different degrees, either to the original black or to the original albino parent-variety.

[18] 'Phenomenon in Himalayan Rabbits,' in 'Journal of Horticulture,' 1865, Jan. 27th, p. 102.

It is, also, remarkable that Himalayans, though produced so suddenly, breed true. But as, whilst young, they are albinoes, the case falls under a very general rule; for albinism is well known to be strongly inherited, as with white mice and many other quadrupeds, and even with white flowers. But why, it may be asked, do the ears, tail, nose, and feet, and no other part of the body, revert to a black colour? This apparently depends on a law, which generally holds good, namely, that characters common to many species of a genus—and this, in fact, implies long inheritance in common from the ancient progenitor of the genus—are found to resist variation, or to reappear if lost, more persistently than the characters which are confined to the separate species. Now, in the genus Lepus, a large majority of the species have their ears and the upper surface of the tail tinted black; but the persistence of these marks is best seen in those species which in winter become white: thus, in Scotland the *L. variabilis*[19] in its winter dress has a shade of colour on its nose, and the tips of its ears are black: in the *L. tibetanus* the ears are black, the upper surface of the tail greyish-black, and the soles of the feet brown: in *L. glacialis* the winter fur is pure white, except the soles of the feet and the points of the ears. Even in the variously-coloured fancy rabbits we may often observe a tendency in these same parts to be more darkly tinted than the rest of the body. Thus, as it seems to me, the appearance of the several coloured marks on the Himalayan rabbit, as it grows old, is rendered intelligible. I may add a nearly analogous case: fancy rabbits very often have a white star on their foreheads; and the common English hare, whilst young, generally has, as I have myself observed, a similar white star on its forehead.

When variously coloured rabbits are set free in Europe, and are thus placed under their natural conditions, they generally revert to the aboriginal grey colour; this may be in part due to the tendency in all crossed animals, as lately observed, to revert to their primordial state. But this tendency does not always prevail; thus silver-grey rabbits are kept in warrens, and remain true though living almost in a state of nature; but a

[19] G. R. Waterhouse, 'Natural History of Mammalia: Rodents,' 1846, pp. 52, 60, 105.

warren must not be stocked with both silver-greys and common rabbits; otherwise "in a few years there will be none but common greys surviving."[20] When rabbits run wild in foreign countries, under different conditions of life, they by no means always revert to their aboriginal colour. In Jamaica the feral rabbits are described as "slate-coloured, deeply tinted with sprinklings of white on the neck, on the shoulders, and on the back; softening off to blue-white under the breast and belly."[21] But in this tropical island the conditions were not favourable to their increase, and they never spread widely; and, as I hear from Mr. R. Hill, owing to a great fire which occurred in the woods, they have now become extinct. Rabbits during many years have run wild in the Falkland Islands; they are abundant in certain parts, but do not spread extensively. Most of them are of the common grey colour; a few, as I am informed by Admiral Sulivan, are hare-coloured, and many are black, often with nearly symmetrical white marks on their faces. Hence, M. Lesson described the black variety as a distinct species, under the name of *Lepus magellanicus*, but this, as I have elsewhere shown, is an error.[22] Within recent times the sealers have stocked some of the small outlying islets in the Falkland group with rabbits; and on Pebble Islet, as I hear from Admiral Sulivan, a large proportion are hare-coloured, whereas on Rabbit Islet a large proportion are of a bluish colour which is not elsewhere seen. How the rabbits were coloured which were turned out on these islets is not known.

The rabbits which have become feral on the island of Porto Santo, near Madeira, deserve a fuller account. In 1418 or 1419, J. Gonzales Zarco[23] happened to have a female rabbit on board which had produced young during the voyage, and he turned them all out on the island. These animals soon increased so

[20] Delamer on 'Pigeons and Rabbits,' p. 114.

[21] Gosse's 'Sojourn in Jamaica,' 1851, p. 441, as described by an excellent observer, Mr. R. Hill. This is the only known case in which rabbits have become feral in a hot country. They can be kept, however, at Loanda (*see* Livingstone's 'Travels,' p. 407). In parts of India, as I am informed by Mr. Blyth, they breed well.

[22] Darwin's 'Journal of Researches,' p. 193; and 'Zoology of the Voyage of the Beagle: Mammalia,' p. 92.

[23] Kerr's 'Collection of Voyages,' vol. ii. p. 177; p. 205 for Cada Mosto. According to a work published in Lisbon in 1717, entitled 'Historia Insulana,' written by a Jesuit, the rabbits were turned out in 1420. Some authors believe that the island was discovered in 1413.

rapidly, that they became a nuisance, and actually caused the abandonment of the settlement. Thirty-seven years subsequently, Cada Mosto describes them as innumerable; nor is this surprising, as the island was not inhabited by any beast of prey or by any terrestrial mammal. We do not know the character of the mother-rabbit; but we have every reason to believe that it was the common domesticated kind. The Spanish peninsula, whence Zarco sailed, is known to have abounded with the common wild species at the most remote historical period. As these rabbits were taken on board for food, it is improbable that they should have been of any peculiar breed. That the breed was well domesticated is shown by the doe having littered during the voyage. Mr. Wollaston, at my request, brought home two of these feral rabbits in spirits of wine; and, subsequently, Mr. W. Haywood sent to me three more specimens in brine, and two alive. These seven specimens, though caught at different periods, closely resembled each other. They were full grown, as shown by the state of their bones. Although the conditions of life in Porto Santo are evidently highly favourable to rabbits, as proved by their extraordinarily rapid increase, yet they differ conspicuously in their small size from the wild English rabbit. Four English rabbits, measured from the incisors to the anus, varied between 17 and 17¾ inches in length; whilst two of the Porto Santo rabbits were only 14½ and 15 inches in length. But the decrease in size is best shown by weight; four wild English rabbits averaged 3 lb. 5 oz., whilst one of the Porto Santo rabbits, which had lived for four years in the Zoological Gardens, but had become thin, weighed only 1 lb. 9 oz. A fairer test is afforded by the comparison of the well-cleaned limb-bones of a P. Santo rabbit killed on the island with the same bones of a wild English rabbit of average size, and they differed in the proportion of rather less than five to nine. So that the Porto Santo rabbits have decreased nearly three inches in length, and almost half in weight of body.[24] The head has not decreased in length pro-

[24] Something of the same kind has occurred on the island of Lipari, where, according to Spallanzani ('Voyage dans les deux Siciles,' quoted by Godron sur l'Espèce, p. 364), a countryman turned out some rabbits which multiplied prodigiously, but, says Spallanzani, "les lapins de l'ile de Lipari sont plus petits que ceux qu'on élève en domesticité."

portionally with the body; and the capacity of the brain-case is, as we shall hereafter see, singularly variable. I prepared four skulls, and these resembled each other more closely than do generally the skulls of wild English rabbits; but the only difference in structure which they presented was that the supra-orbital processes of the frontal bones were narrower.

In colour the Porto Santo rabbit differs considerably from the common rabbit; the upper surface is redder, and is rarely interspersed with any black or black-tipped hairs. The throat and certain parts of the under surface, instead of being pure white, are generally pale grey or leaden colour. But the most remarkable difference is in the ears and tail; I have examined many fresh English rabbits, and the large collection of skins in the British Museum from various countries, and all have the upper surface of the tail and the tips of the ears clothed with blackish-grey fur; and this is given in most works as one of the specific characters of the rabbit. Now in the seven Porto Santo rabbits the upper surface of the tail was reddish-brown, and the tips of the ears had no trace of the black edging. But here we meet with a singular circumstance: in June, 1861, I examined two of these rabbits recently sent to the Zoological Gardens, and their tails and ears were coloured as just described; but when one of their dead bodies was sent to me in February, 1865, the ears were plainly edged, and the upper surface of the tail was covered, with blackish-grey fur, and the whole body was much less red; so that under the English climate this individual rabbit had recovered the proper colour of its fur in rather less than four years!

The two little Porto Santo rabbits, whilst alive in the Zoological Gardens, had a remarkably different appearance from the common kind. They were extraordinarily wild and active, so that many persons exclaimed on seeing them that they were more like large rats than rabbits. They were nocturnal to an unusual degree in their habits, and their wildness was never in the least subdued; so that the superintendent, Mr. Bartlett, assured me that he had never had a wilder animal under his charge. This is a singular fact, considering that they are descended from a domesticated breed; I was so much surprised at it, that I requested Mr. Haywood to make inquiries on the spot, whether they were much hunted by the inhabitants, or persecuted by hawks, or cats, or other animals; but this is not the case, and no cause can be assigned for their wildness. They live on the central, higher rocky land and near the sea-cliffs, and, being exceedingly shy and timid, seldom appear in the lower and cultivated districts. They are said to produce from four to six young at a birth, and their breeding season is in July and August. Lastly, and this is a highly remarkable fact, Mr. Bartlett could never succeed in getting these two rabbits, which were both males, to associate or breed with the females of several breeds which were repeatedly placed with them.

If the history of these Porto Santo rabbits had not been known, most naturalists, on observing their much reduced size, their reddish colour above and grey beneath, with neither tail nor ears tipped with black, would have ranked them as a distinct species. They would have been strongly confirmed in this view by seeing them alive in the Zoological Gardens, and hearing that they refused to couple with other rabbits. Yet this rabbit, which there can be little doubt would thus have been ranked as a distinct species, has certainly originated since the year 1420. Finally, from the three cases of the rabbits which have run wild in Porto Santo, Jamaica, and the Falkland Islands, we see that these animals do not, under new conditions of life, revert to or retain their aboriginal character, as is so generally asserted to be the case by most authors.

Osteological Characters.

When we remember, on the one hand, how frequently it is stated that important parts of the structure never vary; and, on the other hand, on what small differences in the skeleton, fossil species have often been founded, the variability of the skull and of some other bones in the domesticated rabbit well deserves attention. It must not be supposed that the more important differences immediately to be described strictly characterise any one breed; all that can be said is, that they are generally present in certain breeds. We should bear in mind that selection has not been applied to fix any character in the skeleton, and that the animals have not had to support themselves under uniform habits of life. We cannot account for most of the differences in the skeleton; but we shall see that the increased size of the body, due to careful nurture and continued selection, has affected the head in a particular manner. Even the elongation and lopping of the ears have influenced in a small degree the form of the whole skull. The want of exercise has apparently modified the proportional length of the limbs in comparison with the body.

As a standard of comparison, I prepared skeletons of two wild rabbits from Kent, one from the Shetland Islands, and one from Antrim in Ireland. As all the bones in these four specimens from such distant localities closely resembled each other, presenting scarcely any appreciable difference, it may be concluded that the bones of the wild rabbit are generally uniform in character.

Skull.—I have carefully examined skulls of ten large lop-eared fancy rabbits, and of five common domestic rabbits, which latter differ from the lop-eared only in not having such large bodies or ears, yet both larger than in the wild rabbit. First for the ten lop-eared rabbits: in all these the skull is remarkably elongated in comparison with its breadth. In a wild rabbit the length was 3·15 inches, in a large fancy rabbit 4·30; whilst the breadth of the cranium enclosing the brain was in both almost exactly the same. Even by taking as the standard of comparison the widest part of the zygomatic arch, the skulls of the lop-eared are proportionally to their breadth three-quarters of an inch too long. The depth of the head has increased almost in the same proportion with the length; it is the breadth alone which has not increased. The parietal and occipital bones enclosing the brain are less arched, both in a longitudinal and transverse line, than in the wild rabbit, so that the shape of the cranium is somewhat different. The surface is rougher, less cleanly sculptured, and the lines of sutures are more prominent.

Although the skulls of the large lop-eared rabbits in comparison with those of the wild rabbit are much elongated relatively to their breadth, yet, relatively to the size of body, they are far from elongated. The lop-eared rabbits which I examined were, though not fat, more than twice as heavy as the wild specimens; but the skull was very far from being twice as long. Even if we take the fairer standard of the length of body, from the nose to the anus, the skull is not on an average as long as it ought to be by a third of an inch. In the small feral P. Santo rabbit, on the other hand, the head relatively to the length of body is about a quarter of an inch too long.

This elongation of the skull relatively to its breadth, I find a universal character, not only with the large lop-eared rabbits, but in all the artificial breeds; as is well seen in the skull of the Angora. I was at first much surprised at the fact, and could not imagine why domestication should produce this uniform result; but the explanation seems to lie in the circumstance that during a number of generations the artificial races have been closely confined, and have had little occasion to exert either their senses, or intellect, or voluntary muscles; consequently the brain, as we shall presently more fully see, has not increased relatively with the size of body. As the brain has not increased, the bony case enclosing it has not increased, and this has evidently affected through correlation the breadth of the entire skull from end to end.

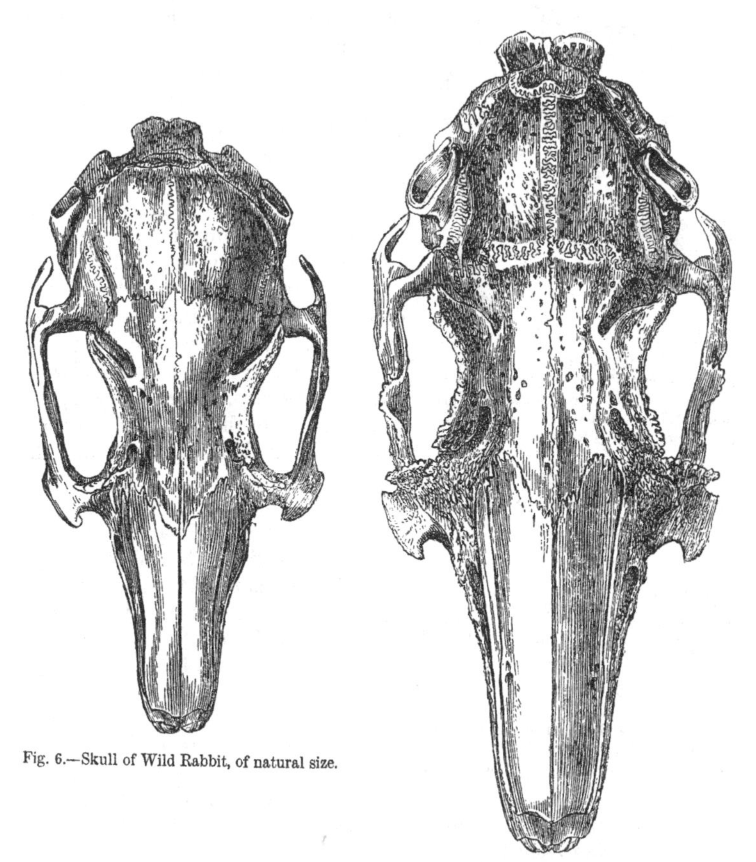

Fig. 6.—Skull of Wild Rabbit, of natural size.

Fig. 7.—Skull of large Lop-eared Rabbit, of natural size.

In all the skulls of the large lop-eared rabbits, the supra-orbital plates or processes of the frontal bones are much broader than in the wild rabbit, and they generally project more upwards. In the zygomatic arch the posterior or projecting point of the malar-bone is broader and blunter; and in the specimen, fig. 8, it is so in a remarkable degree. This point approaches nearer to the auditory meatus than in the wild rabbit, as may be best seen in fig. 8; but this circumstance mainly depends on the changed direction of the meatus. The inter-parietal bone (see fig. 9) differs much in shape in the several skulls; generally it is more oval, or has a greater width in the line of the longitudinal axis of the skull, than in the wild rabbit. The posterior margin of "the square raised platform"[25] of the occiput, instead of being truncated, or projecting slightly as in the wild rabbit, is in most lop-eared rabbits pointed, as in fig. 9, C. The paramastoids relatively to the size of the skull are generally much thicker than in the wild rabbit.

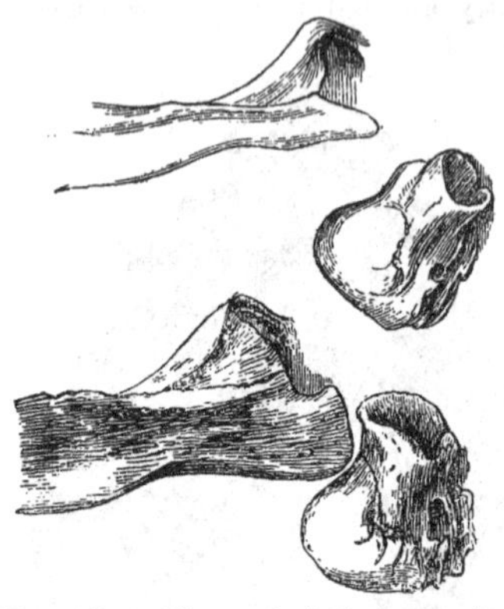

Fig. 8.—Part of Zygomatic Arch, showing the projecting end of the malar bone and the auditory meatus: of natural size. Upper figure, Wild Rabbit. Lower figure, Lop-eared, hare-coloured Rabbit.

The occipital foramen (fig. 10) presents some remarkable differences: in the wild rabbit, the lower edge between the condyles is considerably and almost angularly hollowed out, and the upper edge is deeply and squarely notched; hence the longitudinal axis exceeds the transverse axis. In the skulls of the lop-eared rabbits the transverse axis exceeds the longitudinal; for in none of these skulls was the lower edge between the condyles so deeply hollowed out; in five of them there was no upper square notch, in three there was a trace of the notch, and in two alone it was well developed. These differences in the shape of the foramen are remarkable, considering that it gives passage to so important a structure as the spinal marrow, though apparently the outline of the latter is not affected by the shape of the passage.

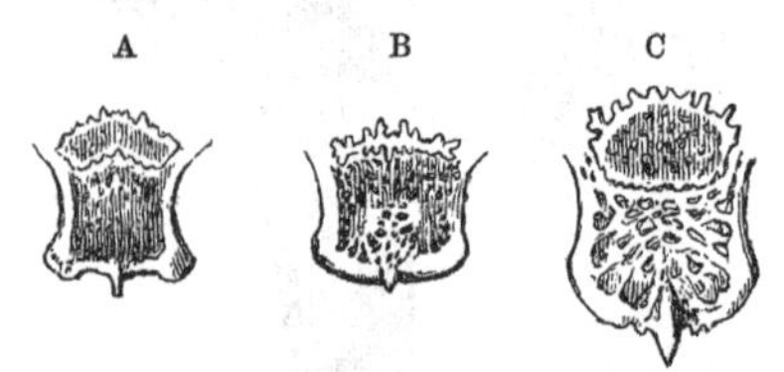

Fig. 9.—Posterior end of Skull, of natural size, showing the inter-parietal bone. A. Wild Rabbit. B. Feral Rabbit from island of P. Santo, near Madeira. C. Large Lop-eared Rabbit.

In all the skulls of the large lop-eared rabbits, the bony auditory meatus is conspicuously larger than in the wild rabbit. In a skull 4·3 inches in length, and which barely exceeded in breadth the skull of a wild rabbit (which was 3·15 inches in length), the longer diameter of the meatus was exactly twice as great. The orifice is more compressed, and its margin on the side nearest the skull stands up higher than the outer side. The whole meatus is directed more forwards. As in breeding lop-eared rabbits the length of the ears, and their consequent lopping and lying flat on the face, are the chief points of excellence, there can hardly be a doubt that the great change in the size, form, and direction of the bony meatus, relatively to this same part in the wild rabbit, is due to the continued selection of individuals having

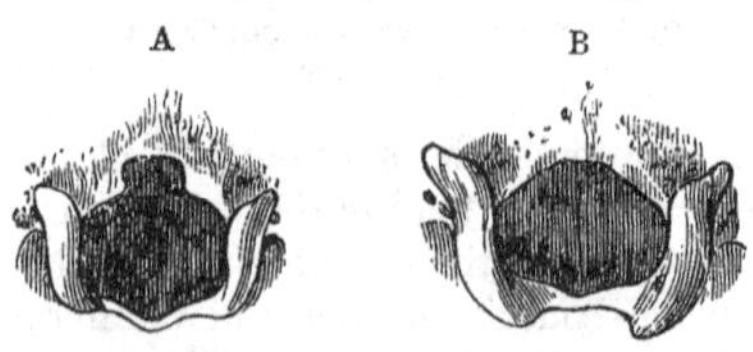

Fig. 10.—Occipital Foramen, of natural size, in—A. Wild Rabbit; B. Large Lop-eared Rabbit.

[25] Waterhouse, 'Nat. Hist. Mammalia,' vol. ii. p. 36.

larger and larger ears. The influence of the external ear on the bony meatus is well shown in the skulls (I have examined three) of half-lops (see fig. 5), in which one ear stands upright, and the other and longer ear hangs down; for in these skulls there was a plain difference in the form and direction of the bony meatus on the two sides. But it is a much more interesting fact, that the changed direction and increased size of the bony meatus have slightly affected on the same side the structure of the whole skull. I here give a drawing of the skull of a half-lop; and it may be observed that the suture between the parietal and frontal bones does not run strictly at right angles to the longitudinal axis of the skull; the left frontal bone projects beyond the right one; both the posterior and anterior margins of the left zygomatic arch on the side of the lopping ear stand a little in advance of the corresponding bones on the opposite side. Even the lower jaw is affected, and the condyles are not quite symmetrical, that on the left standing a little in advance of that on the right. This seems to me a remarkable case of correlation of growth. Who would have surmised that by keeping an animal during many generations under confinement, and so leading to the disuse of the muscles of the ears, and by continually selecting individuals with the longest and largest ears, he would thus indirectly have affected almost every suture in the skull and the form of the lower jaw!

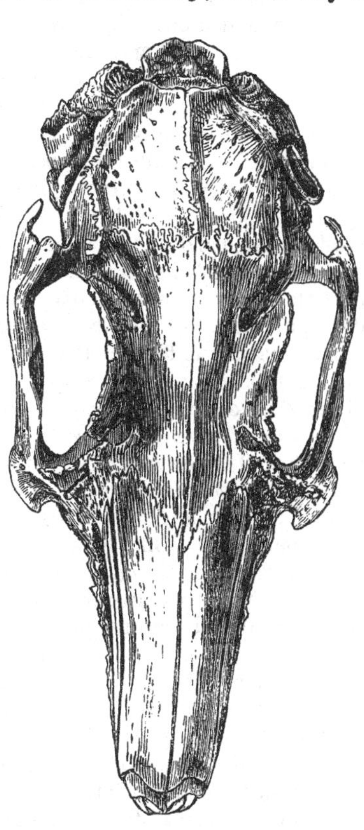

Fig. 11.—Skull, of natural size, of Half-lop Rabbit, showing the different direction of the auditory meatus on the two sides, and the consequent general distortion of the skull. The left ear of the animal (or right side of figure) lopped forwards.

In the large lop-eared rabbits the only difference in the lower jaw, in comparison with that of the wild rabbit, is that the posterior margin of the ascending ramus is broader and more inflected. The teeth in neither jaw present any difference, except that the small incisors, beneath the large ones, are proportionally a little longer. The molar teeth have increased in size proportionally with the increased width of the skull, measured across the zygomatic arch, and not proportionally with its increased length. The inner line of the sockets of the molar teeth in the upper jaw of the wild rabbit forms a perfectly straight line; but in some of the largest skulls of the lop-eared this line was plainly bowed inwards. In one specimen there was an additional molar tooth on each side of the upper jaw, between the molars and premolars; but these two teeth did not correspond in size; and as no rodent has seven molars, this is merely a monstrosity, though a curious one.

The five other skulls of common domestic rabbits, some of which approach in size the above-described largest skulls, whilst the others exceed but little those of the wild rabbit, are only worth notice as presenting a perfect gradation in all the above-specified differences between the skulls of the largest lop-eared and wild rabbits. In all, however, the supra-orbital plates are rather larger, and in all the auditory meatus is larger, in conformity with the increased size of the external ears, than in the wild rabbit. The lower notch in the occipital foramen in some was not so deep as in the wild, but in all five skulls the upper notch was well developed.

The skull of the *Angora* rabbit, like the latter five skulls, is intermediate in general proportions, and in most other characters, between those of the largest lop-eared and wild rabbits. It presents only one singular character: though considerably longer than the skull of the wild, the breadth measured within the posterior supra-orbital fissures is nearly a third less than in the wild. The skulls of the *silver-grey*, and *chinchilla* and *Himalayan* rabbits are more elongated than in the wild, with broader supra-orbital plates, but differ little in any other respect, excepting that the upper and lower notches of the occipital foramen are not so deep or so well developed. The skull of the *Moscow* rabbit scarcely differs in any respect from that of the wild rabbit. In the Porto Santo feral rabbits the supra-orbital plates are generally narrower and more pointed than in our wild rabbits.

As some of the largest lop-eared rabbits of which I prepared skeletons were coloured almost like hares, and as these latter animals and rabbits have, as it is affirmed, been recently crossed in France, it might be thought that some of the above-described characters had been derived from a cross at a remote period with the hare. Consequently I examined skulls of the hare, but no light could thus be thrown on the peculiarities of the skulls of the larger rabbits. It is, however, an interesting fact, as illustrating the law that varieties of one species often assume the characters of other species of the same genus, that I found, on comparing the skulls of ten species of hares in the British Museum, that they differed from each other chiefly in the very same points in which domestic rabbits vary,—namely, in general proportions, in the form and size of the supra-orbital plates, in the form of the free end of the malar bone, and in the line of suture separating the occipital and frontal bones. Moreover two eminently variable characters in the domestic rabbit, namely, the outline of the occipital foramen and the shape of the "raised platform" of the occiput, were likewise variable in two instances in the same species of hare.

Vertebræ.—The number is uniform in all the skeletons which I have examined, with two exceptions, namely, in one of the small feral Porto Santo rabbits and in one of the largest lop-eared kinds; both of these had as usual seven cervical, twelve dorsal with ribs, but, instead of seven lumbar, both had eight lumbar vertebræ. This is remarkable, as Gervais gives seven as the number for the whole genus Lepus. The caudal vertebræ apparently differ by two or three, but I did not attend to them, and they are difficult to count with certainty.

In the first cervical vertebra, or atlas, the anterior margin of the neural arch varies a little in wild specimens, being either nearly smooth, or furnished with a small supra-median atlantoid process; I have figured a specimen with the largest process (*a*) which I have seen; but it will be observed how inferior this is in size and different in shape to that in a large lop-eared rabbit. In the latter, the infra-median process (*b*) is also proportionally much thicker and longer. The alæ are a little squarer in outline.

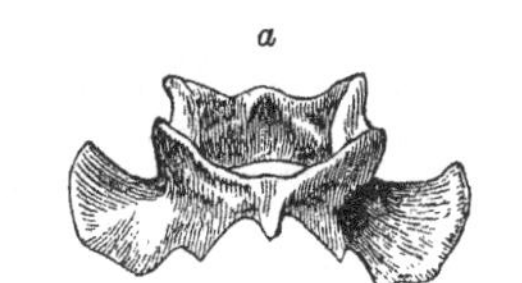

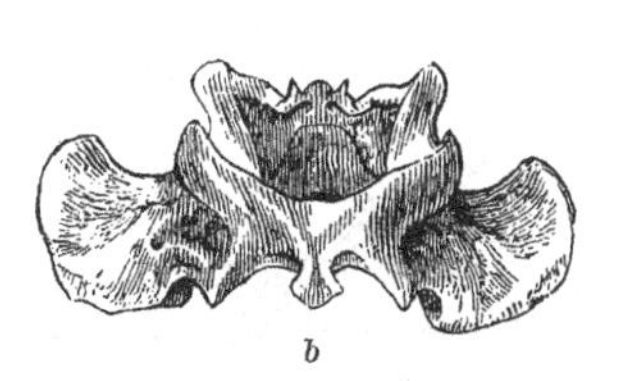

Fig. 12.—Atlas Vertebræ, of natural size; inferior surface viewed obliquely. Upper figure, Wild Rabbit. Lower figure, Hare-coloured, large, Lop-eared Rabbit. *a*, supra-median, atlantoid process; *b*, infra-median process.

Third cervical vertebra.—In the wild rabbit (fig. 13, A *a*) this vertebra, viewed on the inferior surface, has a transverse process, which is directed obliquely backwards, and consists of a single pointed bar; in the fourth vertebra this process is slightly forked in the middle. In the large lop-eared rabbits this process (B *a*) is forked in the third vertebra, as in the fourth of the wild rabbit. But the third cervical vertebræ of the wild and lop-eared (A *b*, B *b*) rabbits differ more conspicuously when their anterior articular surfaces are compared; for the extremities of the antero-dorsal processes in the wild rabbit are simply rounded, whilst in the lop-eared they are trifid, with a deep central pit. The canal for the spinal marrow in the lop-eared (B *b*) is more elongated in a transverse direction than in the wild rabbit; and the passages for the arteries are of a slightly different shape. These several differences in this vertebra seem to me well deserving attention.

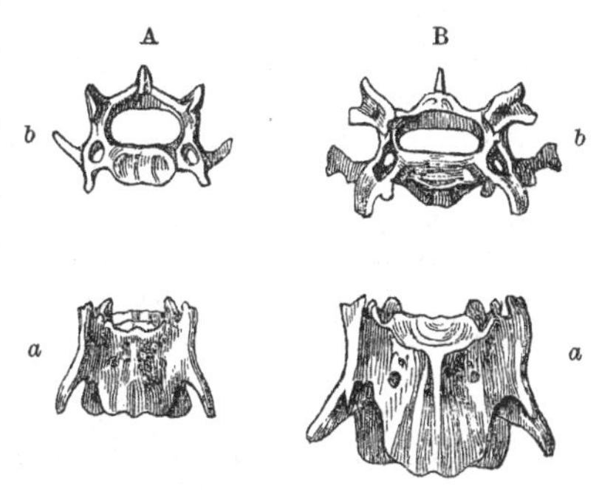

Fig. 13.—Third Cervical Vertebra, of natural size, of—A. Wild Rabbit; B. Hare-coloured, large, Lop-eared Rabbit. *a*, *a*, inferior surface; *b*, *b*, anterior articular surfaces.

First dorsal vertebra.—Its neural spine varies in length in the wild rabbit; being sometimes very short, but generally more than half as long as that of the second dorsal; but I have seen it in two large lop-eared rabbits three-fourths of the length of that of the second dorsal vertebra.

Ninth and tenth dorsal vertebræ.—In the wild rabbit the neural spine of the ninth vertebra is just perceptibly thicker than that of the eighth; and the neural spine of the tenth is plainly thicker and shorter than those of all the anterior vertebræ. In the large lop-eared rabbits the neural spines of the tenth, ninth, eighth, and even in a slight degree that of the seventh vertebra, are very much thicker, and of somewhat different shape, in comparison with those of the wild rabbit. So that this part of the vertebral column differs considerably in appearance from the same part in the wild rabbit, and closely resembles in an interesting manner these same vertebræ in some species of hares. In the Angora, Chinchilla, and Himalayan rabbits, the neural spines of the eighth and ninth vertebræ are in a slight degree thicker than in the wild. On the other hand, in one of the feral Porto Santo rabbits, which in most of its characters deviates in an exactly opposite manner to what the large lop-eared rabbits do from the common wild rabbit, the neural spines of the ninth and tenth vertebræ were not at all larger than those of the several anterior vertebræ. In this same Porto Santo specimen there was no trace in the ninth vertebra of the anterior lateral processes (see woodcut 14), which are plainly developed in all British wild rabbits, and still more plainly developed in the large lop-eared rabbits. In a half-wild rabbit from Sandon Park,[26] a hæmal spine was moderately well developed on the under side of the twelfth dorsal vertebra, and I have seen this in no other specimen.

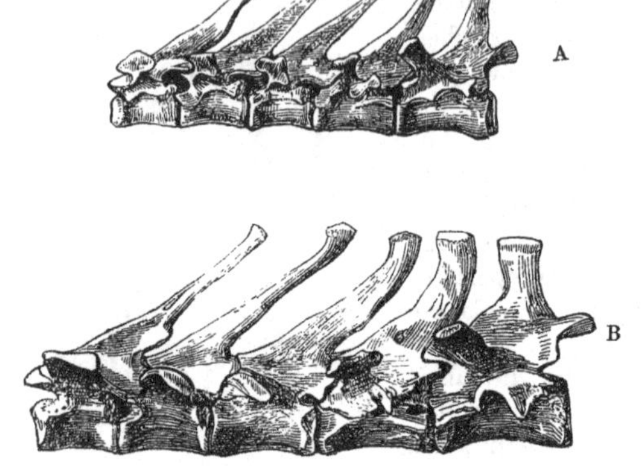

Fig. 14.—Dorsal Vertebræ, from sixth to tenth inclusive, of natural size, viewed laterally. A. Wild Rabbit. B. Large, Hare-coloured, so called Spanish Rabbit.

Lumbar vertebræ.—I have stated that in two cases there were eight instead of seven lumbar vertebræ. The third lumbar vertebra in one skeleton of a wild British rabbit, and in one of the Porto Santo feral rabbits, had a hæmal spine; whilst in four skeletons of large lop-eared rabbits, and in the Himalayan rabbit, this same vertebra had a well-developed hæmal spine.

Pelvis.—In four wild specimens this bone was almost absolutely identical in shape; but in several domesticated breeds shades of differences

[26] These rabbits have run wild for a considerable time in Sandon Park, and in other places in Staffordshire and Shropshire. They originated, as I have been informed by the gamekeeper, from variously-coloured domestic rabbits which had been turned out. They vary in colour; but many are symmetrically coloured, being white with a streak along the spine, and with the ears and certain marks about the head of a blackish-grey tint. They have rather longer bodies than common rabbits.

could be distinguished. In the large lop-eared rabbits the whole upper part of the ilium is straighter, or less splayed outwards, than in the wild rabbit; and the tuberosity on the inner lip of the anterior and upper part of the ilium is proportionally more prominent.

Sternum.—The posterior end of the posterior sternal bone in the wild rabbit (fig. 15, A) is thin and slightly enlarged; in some of the large lop-eared rabbits (B) it is much more enlarged towards the extremity; whilst in other specimens (C) it keeps nearly of the same breadth from end to end, but is much thicker at the extremity.

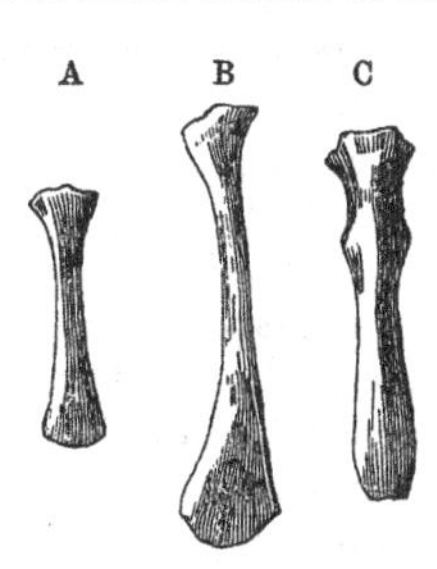

Fig. 15.—Terminal bone of Sternum, of natural size. A. Wild Rabbit. B. Hare-coloured, Lop-eared Rabbit. C. Hare-coloured, Spanish Rabbit. (N.B. The left-hand angle of the upper articular extremity of B was broken, and has been accidentally thus represented.)

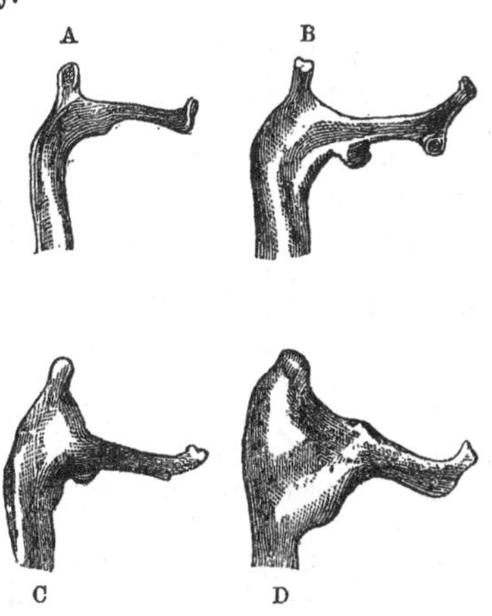

Fig. 16.—Acromion of Scapula, of natural size. A. Wild Rabbit. B, C, D. Large, Lop-eared Rabbits.

Scapula.—The acromion sends out a rectangular bar, ending in an oblique knob, which latter in the wild rabbit (fig. 16, A) varies a little in shape and size, as does the apex of the acromion in sharpness, and the part just below the rectangular bar in breadth. But the variations in these respects in the wild rabbit are very slight; whilst in the large lop-eared rabbits they are considerable. Thus in some specimens (B) the oblique terminal knob is developed into a short bar, forming an obtuse angle with the rectangular bar. In another specimen (C) these two unequal bars form nearly a straight line. The apex of the acromion varies much in breadth and sharpness, as may be seen by comparing figs. B, C, and D.

Limbs.—In these I could detect no variation; but the bones of the feet were too troublesome to compare with much care.

I have now described all the differences in the skeletons which I have observed. It is impossible not to be struck with the high degree of variability or plasticity of many of the bones. We see how erroneous the often-repeated statement is, that only the crests of the bones which give attachment to muscles vary in shape, and that only parts of slight importance become modified under domestication. No one will say, for instance, that the occipital foramen, or the atlas, or the third cervical vertebra is a part of slight importance. If the several vertebræ of the wild and lop-eared rabbits, of which figures have been given, had been found fossil, palæontologists would have declared without hesitation that they had belonged to distinct species.

The effects of the use and disuse of parts.—In the large lop-eared rabbits the relative proportional lengths of the bones of the same leg, and of the front and hind legs compared with each other, have remained nearly the same as in the wild rabbit; but in weight, the bones of the hind legs apparently have not increased in due proportion with the front legs. The weight of the whole body in the large rabbits examined by me was from twice to twice and a half as great as that of the wild rabbit; and the weight of the bones of the front and hind limbs taken together (excluding the feet, on account of the difficulty of perfectly cleaning so many small bones) has increased in the large lop-eared rabbits in nearly the same proportion; consequently in due proportion to the weight of body which they have to support. If we take the length of the body as the standard of comparison, the limbs of the large rabbits have not increased in length in due proportion by one inch, or by one inch and a half. Again, if we take as the standard of comparison the length of the skull, which, as we have before seen, has not increased in length in due proportion to the length of body, the limbs will be found to be, proportionally with those of the wild rabbit, from half to three-quarters of an inch too short. Hence, whatever standard of comparison be taken, the limb-bones of the large lop-eared rabbits have not increased in length, though they have in weight, in full proportion to the other parts of the frame; and this, I presume, may be accounted for by the inactive life which during many generations they have spent. Nor has the scapula increased in length in due proportion to the increased length of the body.

The capacity of the osseous case of the brain is a more interesting point, to which I was led to attend by finding, as previously stated, that with all domesticated rabbits the length of the skull relatively to its breadth has greatly increased in comparison with that of the wild rabbit. If we had possessed a large number of domesticated rabbits of nearly the same size with the wild rabbit, it would have been a simple task to have measured and compared the capacities of their skulls. But this is not the case; almost all the domestic breeds have larger bodies than wild rabbits, and the lop-eared kinds are more than double their weight. As a small animal has to exert its senses, intellect, and instincts equally with a large animal, we ought not by any means to expect an animal twice or thrice as large as another to have a brain of double or treble the size.[27] Now, after weighing

[27] *See* Prof. Owen's remarks on this subject in his paper on the 'Zoological Significance of the Brain, &c., of Man, &c.,' read before Brit. Association, 1862; with respect to Birds, *see* 'Proc. Zoolog. Soc.,' Jan. 11th, 1848, p. 8.

the bodies of four wild rabbits, and of four large but not fattened lop-eared rabbits, I find that on an average the wild are to the lop-eared in weight as 1 to 2·17; in average length of body as 1 to 1 41; whilst in capacity of skull (measured as hereafter to be described) they are only as 1 to 1·15. Hence we see that the capacity of the skull, and consequently the size of the brain, has increased but little, relatively to the increased size of the body; and this fact explains the narrowness of the skull relatively to its length in all domestic rabbits.

In the upper half of the following table I have given the measurements of the skulls of ten wild rabbits; and in the lower half of eleven thoroughly domesticated kinds. As these rabbits differ so greatly in size, it is necessary to have some standard by which to compare the capacities of their skulls. I have selected the length of skull as the best standard, for in the larger rabbits it has not, as already stated, increased in length so much as the body; but as the skull, like every other part, varies in length, neither it nor any other part affords a perfect standard.

In the first column of figures the extreme length of the skull is given in inches and decimals. I am aware that these measurements pretend to greater accuracy than is possible; but I have found it the least trouble to record the exact length which the compass gave. The second and third columns give the length and weight of body, whenever these measurements have been made. The fourth column gives the capacity of the skull by the weight of small shot with which the skulls had been filled; but it is not pretended that these weights are accurate within a few grains. In the fifth column the capacity is given which the skull ought to have had by calculation, according to the length of skull, in comparison with that of the wild rabbit No. 1; in the sixth column the difference between the actual and calculated capacities, and in the seventh the percentage of increase or decrease, are given. For instance, as the wild rabbit No. 5 has a shorter and lighter body than the wild rabbit No. 1, we might have expected that its skull would have had less capacity; the actual capacity, as expressed by the weight of shot, is 875 grains, which is 97 grains less than that of the first rabbit. But comparing these two rabbits by the length of their skulls, we see that in No. 1 the skull is 3·15 inches in length, and in No. 5 2·96 inches in length; according to this ratio, the brain of No. 5 ought to have had a capacity of 913 grains of shot, which is above the actual capacity, but only by 38 grains. Or, to put the case in another way (as in column VII), the brain of this small rabbit, No. 5, for every 100 grains of weight is only 4 per cent. too light,—that is, it ought, according to the standard rabbit No. 1, to have been 4 per cent. heavier. I have taken the rabbit No. 1 as the standard of comparison because, of the skulls having a full average length, this has the least capacity; so that it is the least favourable to the result which I wish to show, namely, that the brain in all long-domesticated rabbits has decreased in size, either actually, or relatively to the length of the head and body, in comparison with the brain of the wild rabbit. Had I taken the Irish rabbit, No. 3, as the standard, the following results would have been somewhat more striking.

Turning to the Table: the first four wild rabbits have skulls of the same length, and these differ but little in capacity. The Sandon rabbit

(No. 4) is interesting, as, though now wild, it is known to be descended from a domesticated breed, as is still shown by its peculiar colouring and longer body; nevertheless the skull has recovered its normal length and full capacity. The next three rabbits are wild, but of small size, and they all have skulls with slightly lessened capacities. The three Porto Santo feral rabbits (Nos. 8 to 10) offer a perplexing case; their bodies are greatly reduced in size, as in a lesser degree are their skulls in length and in actual capacity, in comparison with the skulls of wild English rabbits. But when we compare the capacities of the skull in the three Porto Santo rabbits, we observe a surprising difference, which does not stand in any relation to the slight difference in the length of their skulls, nor, as I believe, to any difference in the size of their bodies; but I neglected to weigh separately their bodies. I can hardly suppose that the medullary matter of the brain in these three rabbits, living under similar conditions, can differ as much as is indicated by the proportional difference of capacity in their skulls; nor do I know whether it is possible that one brain may contain considerably more fluid than another. Hence I can throw no light on this case.

Looking to the lower half of the Table, which gives the measurements of domesticated rabbits, we see that in all the capacity of the skull is less, but in very various degrees, than might have been anticipated according to the length of their skulls, relatively to that of the wild rabbit No. 1. In line 22 the average measurements of seven large lop-eared rabbits are given. Now the question arises, has the average capacity of the skull in these seven large rabbits increased as much as might have been expected from their greatly increased size of body. We may endeavour to answer this question in two ways: in the upper half of the Table we have measurements of the skulls of six small wild rabbits (Nos. 5 to 10), and we find that on an average the skulls are in length ·18 of an inch shorter, and in capacity 91 grains less, than the average length and capacity of the three first wild rabbits on the list. The seven large lop-eared rabbits, on an average, have skulls 4·11 inches in length, and 1136 grains in capacity; so that these skulls have increased in length more than five times as much as the skulls of the six small wild rabbits have decreased in length; hence we might have expected that the skulls of the large lop-eared rabbits would have increased in capacity five times as much as the skulls of the six small rabbits have decreased in capacity; and this would have given an average increased capacity of 455 grains, whilst the real average increase is only 155 grains. Again, the large lop-eared rabbits have bodies of nearly the same weight and size as the common hare, but their heads are longer; consequently, if the lop-eared rabbits had been wild, it might have been expected that their skulls would have had nearly the same capacity as that of the skull of the hare. But this is far from being the case; for the average capacity of the two hare-skulls (Nos. 23, 24) is so much larger than the average capacity of the seven lop-eared skulls, that the latter would have to be increased 21 per cent. to come up to the standard of the hare.[28]

[28] This standard is apparently considerably too low, for Dr. Crisp ('Proc. Zoolog. Soc.,' 1861, p. 86) gives 210 grains as the actual weight of the

Name of Breed.	I. Length of Skull.	II. Length of Body from Incisors to Anus.	III. Weight of whole Body.	IV. Capacity of Skull measured by Small Shot.	V. Capacity calculated according to Length of Skull relatively to that of No. 1.	VI. Difference between actual and calculated capacities of Skulls.	VII. Showing how much per cent. the Brain, by calculation, according to the length of the Skull, is too light or too heavy, relatively to the Brain of the Wild Rabbit No. 1.
WILD AND SEMI-WILD RABBITS.	inches.	inches.	lbs. ozs.	grains.	grains.	grains.	
1. Wild rabbit, Kent	3·15	17·4	3 5	972	..	..	
2. ,, Shetland Islands	3·15	..	..	979	..	..	
3. ,, Ireland	3·15	..	..	992	..	..	[2 per cent. too heavy in comparison with No. 1.]
4. Domestic rabbit, run wild, Sandon	3·15	18·5	..	977			
5. Wild, common variety, small specimen, Kent ..	2·96	17·0	2 14	875	913	38	4 per cent. too light.
6. Wild, fawn-coloured variety, Scotland	3·1	..	..	918	950	32	3 ,, ,,
7. Silver-grey, small specimen, Thetford warren ..	2·95	15·5	2 11	938	910	28	3 ,, too heavy.
8. Feral rabbit, Porto Santo	2 83	..	..	893	873	20	2 ,, ,,
9. ,, ,, ...	2·85	..	..	756	879	123	16 ,, too light.
10. ,, ,, ...	2·95	..	..	835	910	75	9 ,, ,,
Average of the three Porto Santo rabbits ..	2·88	..	..	828	888	60	7 ,, ,,
DOMESTIC RABBITS.							
11. Himalayan	3·5	20·5	..	963	1080	117	12 ,, ,,
12. Moscow	3·25	17·0	3 8	803	1002	199	24 ,, ,,
13. Angora	3·5	19·5	3 1	697	1080	383	54 ,, ,,
14. Chinchilla	3·65	22·0	..	995	1126	131	13 ,, ,,
15. Large lop-eared	4·1	24·5	7 0	1065	1265	200	18 ,, ,,
16. ,, ,, ...	4·1	25·0	7 13	1153	1265	112	9 ,, ,,
17. ,, ,, ...	4·07	..	..	1037	1255	218	21 ,, ,,
18. ,, ,, ...	4·1	25·0	7 4	1208	1265	57	4 ,, ,,
19. ,, ,, ...	4·3	..	..	1232	1326	94	7 ,, ,,
20. ,, ,, ...	4·25	..	..	1124	1311	187	16 ,, ,,
21. Large hare-coloured	3·86	24·0	6 14	1131	1191	60	5 ,, ,,
22. Average of above seven large lop-eared rabbits	4·11	24·62	7 4	1136	1268	132	11 ,, ,,
23. Hare (*L. timidus*) English specimen	3·61		7 0	1315			
24. ,, ,, German specimen	3·82		7 0	1455			

I have previously remarked that, if we had possessed many domestic rabbits of the same average size with the wild rabbit, it would have been easy to compare the capacity of their skulls. Now the Himalayan, Moscow, and Angora rabbits (Nos. 11, 12, 13 of Table) are only a little larger in body, and have skulls only a little longer, than the wild animal, and we see that the actual capacity of their skulls is less than in the wild animal, and considerably less by calculation (column 7), according to the difference in the length of their skulls. The narrowness of the brain-case in these three rabbits could be plainly seen and proved by external measurement. The Chinchilla rabbit (No. 14) is a considerably larger animal than the wild rabbit, yet the capacity of its skull only slightly exceeds that of the wild rabbit. The Angora rabbit, No. 13, offers the most remarkable case; this animal in its pure white colour and length of silky fur bears the stamp of long domesticity. It has a considerably longer head and body than the wild rabbit, but the actual capacity of its skull is less than that of even the little wild Porto Santo rabbits. By the standard of the length of skull the capacity (see column 7) is only half of what it ought to have been! I kept this individual animal alive, and it was not unhealthy nor idiotic. This case of the Angora rabbit so much surprised me, that I repeated all the measurements and found them correct. I have also compared the capacity of the skull of the Angora with that of the wild rabbit by other standards, namely, by the length and weight of the body, and by the weight of the limb-bones; but by all these standards the brain appears to be much too small, though in a less degree when the standard of the limb-bones was used; and this latter circumstance may probably be accounted for by the limbs of this anciently domesticated breed having become much reduced in weight, from its long-continued inactive life. Hence I infer that in the Angora breed, which is said to differ from other breeds in being quieter and more social, the capacity of the skull has really undergone a remarkable amount of reduction.

From the several facts above given,—namely, firstly, that the actual capacity of the skull in the Himalayan, Moscow, and Angora breeds, is less than in the wild rabbit, though they are in all their dimensions rather larger animals; secondly, that the capacity of the skull of the large lop-eared rabbits has not been increased in nearly the same ratio as the capacity of the skull of the smaller wild rabbits has been decreased; and thirdly, that the capacity of the skull in these same large lop-eared rabbits is very inferior to that of the hare, an animal of nearly the same

brain of a hare which weighed 7 lbs., and 125 grains as the weight of the brain of a rabbit which weighed 3 lbs. 5 oz., that is, the same weight as the rabbit No. 1 in my list. Now the contents of the skull of rabbit No. 1 in shot is in my table 972 grains; and according to Dr. Crisp's ratio of 125 to 210, the skull of the hare ought to have contained 1632 grains of shot, instead of only (in the largest hare in my table) 1455 grains.

size,—I conclude, notwithstanding the remarkable differences in capacity in the skulls of the small P. Santo rabbits, and likewise in the large lop-eared kinds, that in all long-domesticated rabbits the brain has either by no means increased in due proportion with the increased length of the head and increased size of the body, or that it has actually decreased in size, relatively to what would have occurred had these animals lived in a state of nature. When we remember that rabbits, from having been domesticated and closely confined during many generations, cannot have exerted their intellect, instincts, senses, and voluntary movements, either in escaping from various dangers or in searching for food, we may conclude that their brains will have been feebly exercised, and consequently have suffered in development. We thus see that the most important and complicated organ in the whole organization is subject to the law of decrease in size from disuse.

Finally, let us sum up the more important modifications which domestic rabbits have undergone, together with their causes as far as we can obscurely see them. By the supply of abundant and nutritious food, together with little exercise, and by the continued selection of the heaviest individuals, the weight of the larger breeds has been more than doubled. The bones of the limbs have increased in weight (but the hind legs less than the front legs), in due proportion with the increased weight of body; but in length they have not increased in due proportion, and this may have been caused by the want of proper exercise. With the increased size of the body the third cervical vertebra has assumed characters proper to the fourth cervical; and the eighth and ninth dorsal vertebræ have similarly assumed characters proper to the tenth and posterior vertebræ. The skull in the larger breeds has increased in length, but not in due proportion with the increased length of body; the brain has not duly increased in dimensions, or has even actually decreased, and consequently the bony case for the brain has remained narrow, and by correlation has affected the bones of the face and the entire length of the skull. The skull has thus acquired its characteristic narrowness. From unknown causes the supra-orbital processes of the frontal bones and the free end of the malar bones have increased in breadth; and in the larger breeds the occipital foramen is generally much less deeply notched than in wild rabbits. Certain parts of the scapula and the terminal sternal bones have become highly variable in shape. The ears have been increased enormously in length and breadth through continued selection; their weight, conjoined probably with the disuse of their muscles, has caused them to lop downwards; and this has affected the position and form of the bony auditory meatus; and this again, by correlation, the position in a slight degree of almost every bone in the upper part of the skull, and even the position of the condyles of the lower jaw.

CHAPTER V.

DOMESTIC PIGEONS.

ENUMERATION AND DESCRIPTION OF THE SEVERAL BREEDS — INDIVIDUAL VARIABILITY — VARIATIONS OF A REMARKABLE NATURE — OSTEOLOGICAL CHARACTERS: SKULL, LOWER JAW, NUMBER OF VERTEBRÆ — CORRELATION OF GROWTH: TONGUE WITH BEAK; EYELIDS AND NOSTRILS WITH WATTLED SKIN — NUMBER OF WING-FEATHERS, AND LENGTH OF WING — COLOUR AND DOWN — WEBBED AND FEATHERED FEET — ON THE EFFECTS OF DISUSE — LENGTH OF FEET IN CORRELATION WITH LENGTH OF BEAK — LENGTH OF STERNUM, SCAPULA, AND FURCULA — LENGTH OF WINGS — SUMMARY ON THE POINTS OF DIFFERENCE IN THE SEVERAL BREEDS.

I HAVE been led to study domestic pigeons with particular care, because the evidence that all the domestic races have descended from one known source is far clearer than with any other anciently domesticated animal. Secondly, because many treatises in several languages, some of them old, have been written on the pigeon, so that we are enabled to trace the history of several breeds. And lastly, because, from causes which we can partly understand, the amount of variation has been extraordinarily great. The details will often be tediously minute; but no one who really wants to understand the progress of change in domestic animals will regret this; and no one who has kept pigeons and has marked the great difference between the breeds and the trueness with which most of them propagate their kind, will think this care superfluous. Notwithstanding the clear evidence that all the breeds are the descendants of a single species, I could not persuade myself until some years had passed that the whole amount of difference between them had arisen since man first domesticated the wild rock-pigeon.

I have kept alive all the most distinct breeds, which I could procure in England or from the Continent; and have prepared skeletons of all. I have received skins from Persia, and a large number from India and other quarters of the world.[1] Since my admission into two of the London pigeon-clubs, I have received the kindest assistance from many of the most eminent amateurs.[2]

The races of the Pigeon which can be distinguished, and which breed true, are very numerous. MM. Boitard and Corbié[3] describe in detail 122 kinds; and I could add several European kinds not known to them. In India, judging from the skins sent me, there are many breeds unknown here; and Sir W. Elliot informs me that a collection imported by an Indian merchant into Madras from Cairo and Constantinople included several kinds unknown in India. I have no doubt that there exist considerably above 150 kinds which breed true and have been separately named. But of these the far greater number differ from each other only in unimportant characters. Such differences will be here entirely passed over, and I shall confine myself to the more important points of structure. That many important differences exist we shall presently see. I have looked through the magnificent collection of the Columbidæ in the British Museum, and, with the exception of a few forms (such as the Didunculus, Calænas, Goura, &c.), I do not hesitate to

[1] The Hon. C. Murray has sent me some very valuable specimens from Persia; and H.M. Consul, Mr. Keith Abbott, has given me information on the pigeons of the same country. I am deeply indebted to Sir Walter Elliot for an immense collection of skins from Madras, with much information regarding them. Mr. Blyth has freely communicated to me his stores of knowledge on this and all other related subjects. The Rajah Sir James Brooke sent me specimens from Borneo, as has H.M. Consul, Mr. Swinhoe, from Amoy in China, and Dr. Daniell from the west coast of Africa.

[2] Mr. B. P. Brent, well known for his various contributions to poultry literature, has aided me in every way during several years; so has Mr. Tegetmeier, with unwearied kindness. This latter gentleman, who is well known for his works on poultry, and who has largely bred pigeons, has looked over this and the following chapters. Mr. Bult formerly showed me his unrivalled collection of Pouters, and gave me specimens. I had access to Mr. Wicking's collection, which contained a greater assortment of many kinds than could anywhere else be seen; and he has always aided me with specimens and information given in the freest manner. Mr. Haynes and Mr. Corker have given me specimens of their magnificent Carriers. To Mr. Harrison Weir I am likewise indebted. Nor must I by any means pass over the assistance received from Mr. J. M. Eaton, Mr. Baker, Mr. Evans, and Mr. J. Baily, jun., of Mount-street—to the latter gentleman I have been indebted for some valuable specimens. To all these gentlemen I beg permission to return my sincere and cordial thanks.

[3] 'Les Pigeons de Volière et de Colombier,' Paris, 1824. During forty-five years the sole occupation of M. Corbié was the care of the pigeons belonging to the Duchess of Berry.

affirm that some domestic races of the rock-pigeon differ fully as much from each other in external characters as do the most distinct natural genera. We may look in vain through the 288 known species[4] for a beak so small and conical as that of the short-faced tumbler; for one so broad and short as that of the barb; for one so long, straight, and narrow, with its enormous wattles, as that of the English carrier; for an expanded upraised tail like that of the fantail; or for an œsophagus like that of the pouter. I do not for a moment pretend that the domestic races differ from each other in their whole organisation as much as the more distinct natural genera. I refer only to external characters, on which, however, it must be confessed that most genera of birds have been founded. When, in a future chapter, we discuss the principle of selection as followed by man, we shall clearly see why the differences between the domestic races are almost always confined to external, or at least to externally visible, characters.

Owing to the amount and gradations of difference between the several breeds, I have found it indispensable in the following classification to rank them under Groups, Races, and Sub-races; to which varieties and sub-varieties, all strictly inheriting their proper characters, must often be added. Even with the individuals of the same sub-variety, when long kept by different fanciers, different strains can sometimes be recognised. There can be no doubt that, if well-characterized forms of the several Races had been found wild, all would have been ranked as distinct species, and several of them would certainly have been placed by ornithologists in distinct genera. A good classification of the various domestic breeds is extremely difficult, owing to the manner in which many of the forms graduate into each other; but it is curious how exactly the same difficulties are encountered, and the same rules have to be followed, as in the classification of any natural but difficult group of organic beings. An "artificial classification" might be followed which would present fewer difficulties than a "natural classification;" but then it would interrupt many plain affinities. Extreme forms can readily be defined; but intermediate and troublesome forms

[4] 'Coup d'Oeil sur l'Ordre des Pigeons,' par Prince C. L. Bonaparte, Paris, 1855. This author makes 288 species, ranked under 85 genera.

often destroy our definitions. Forms which may be called "aberrant" must sometimes be included within groups to which they do not accurately belong. Characters of all kinds must be used; but as with birds in a state of nature, those afforded by the beak are the best and most readily appreciated. It is not possible to weigh the importance of all the characters which have to be used so as to make the groups and sub-groups of equal value. Lastly, a group may contain only one race, and another and less distinctly defined group may contain several races and sub-races, and in this case it is difficult, as in the classification of natural species, to avoid placing too high a value on characters which are common to a large number of forms.

In my measurements I have never trusted to the eye; and when speaking of a part being large or small, I always refer to the wild rock-pigeon (*Columba livia*) as the standard of comparison. The measurements are given in decimals of an inch.[5]

I will now give a brief description of all the principal breeds. The following diagram may aid the reader in learning their names and seeing their affinities. The rock-pigeon, or *Columba livia* (including under this name two or three closely-allied sub-species or geographical races, hereafter to be described), may be confidently viewed, as we shall see in the next chapter, as the common parent-form. The names in italics on the right-hand side of the table show us the most distinct breeds, or those which have undergone the greatest amount of modification. The lengths of the dotted lines rudely represent the degree of distinctness of each breed from the parent-stock, and the names

[5] As I so often refer to the size of the *C. livia*, or rock-pigeon, it may be convenient to give the mean between the measurements of two wild birds, kindly sent me by Dr. Edmondstone from the Shetland Islands:—

	Inches.
Length from feathered base of beak to end of tail	14·25
" " " " to oil-gland	9·5
" from tip of beak to end of tail	15·02
" of tail-feathers	4·62
" from tip to tip of wing	26·75
" of folded wing	9·25
Beak.—Length from tip of beak to feathered base	·77
" Thickness, measured vertically at further end of nostrils	·23
" Breadth, measured at same place	·16
Feet.—Length from end of middle toe (without claw) to distal end of tibia	2·77
" Length from end of middle toe to end of hind toe (without claws)	2·02

Weight 14½ ounces.

Fig. 17.—The Rock-pigeon, or Columba livia.[6] The parent-form of all domesticated Pigeons.

placed under each other in the columns show the more or less closely connecting links. The distances of the dotted lines from each other approximately represent the amount of difference between the several breeds.

[6] This drawing was made from a dead bird. The six following figures were drawn with great care by Mr. Luke Wells from living birds selected by Mr. Tegetmeier. It may be confidently asserted that the characters of the six breeds which have been figured are not in the least exaggerated.

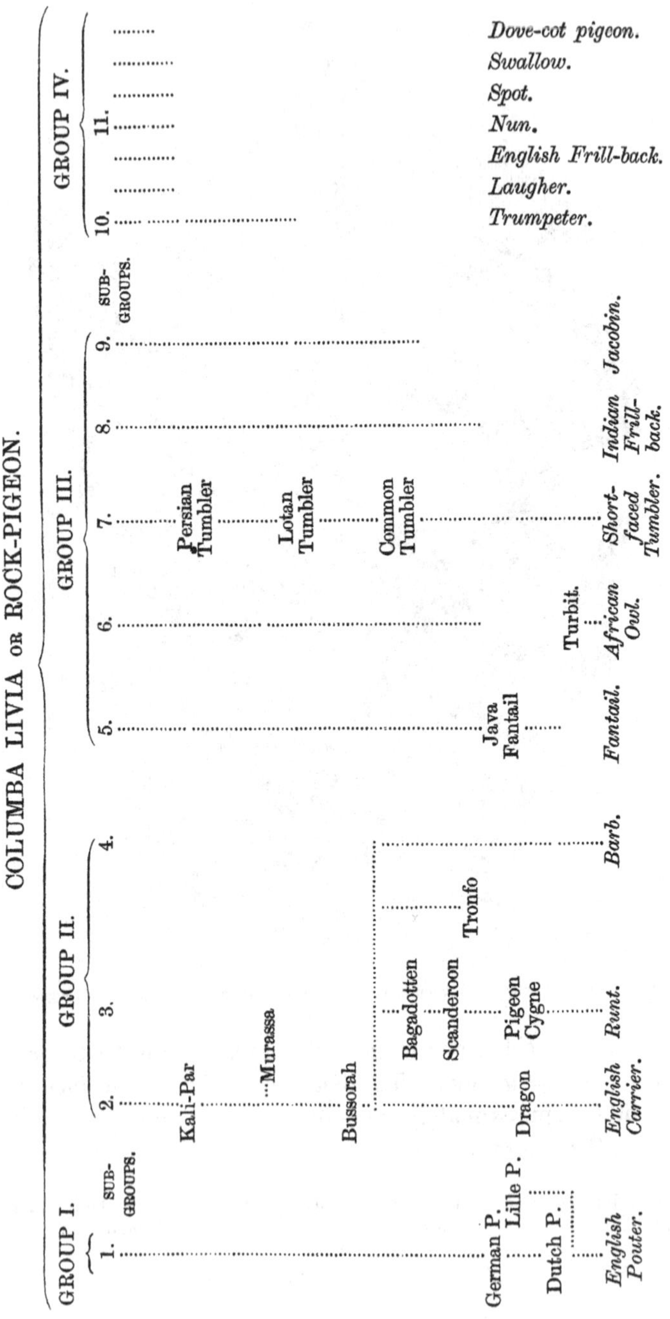

COLUMBA LIVIA OR ROCK-PIGEON.

GROUP I.

SUB-GROUPS.

1. German P. — Lille P. — Dutch P. — *English Pouter.*

GROUP II.

2. Kali-Par — Murassa — Bussorah — Dragon — *English Carrier.*

3. Bagadotten — Scanderoon — Pigeon Cygne — *Runt.*

Tronfo

4. *Barb.*

GROUP III.

5. Java Fantail — *Fantail.*

6. Turbit. — *African Owl.*

7. Persian Tumbler — Lotan Tumbler — Common Tumbler — *Short-faced Tumbler.*

8. *Indian Frill-back.*

9. *Jacobin.*

GROUP IV.

SUB-GROUPS.

10. *Trumpeter.*

11. *Laugher.* — *English Frill-back.* — *Nun.* — *Spot.* — *Swallow.* — *Dove-cot pigeon.*

GROUP I.

This group includes a single race, that of the Pouters. If the most strongly marked sub-race be taken, namely, the Improved English Pouter, this is perhaps the most distinct of all domesticated pigeons.

Fig. 18.—English Pouter.

RACE I.—POUTER PIGEONS. (Kropf-tauben, German. Grosses-gorges, or boulans, French.)

Œsophagus of great size, barely separated from the crop, often inflated. Body and legs elongated. Beak of moderate dimensions.

Sub-race I.—The improved English Pouter, when its crop is fully inflated, presents a truly astonishing appearance. The habit of slightly inflating the crop is common to all domestic pigeons, but is carried to an extreme in the Pouter. The crop does not differ, except in size, from that of other pigeons; but is less plainly separated by an oblique construction from the œsophagus. The diameter of the upper part of the œsophagus is immense, even close up to the head. The beak in one bird which I possessed was almost completely buried when the œsophagus was fully expanded. The males, especially when excited, pout more than the females, and they glory in exercising this power. If a bird will not, to use the technical expression, "play," the fancier, as I have witnessed, by taking the beak into his mouth, blows him up like a balloon; and the bird, then puffed up with wind and pride, struts about, retaining his magnificent size as long as he can. Pouters often take flight with their crops inflated; and after one of my birds had swallowed a good meal of peas and water, as he flew up in order to disgorge them and thus feed his nearly fledged young, I have heard the peas rattling in his inflated crop as if in a bladder. When flying, they often strike the backs of their wings together, and thus make a clapping noise.

Pouters stand remarkably upright, and their bodies are thin and elongated. In connexion with this form of body, the ribs are generally broader and the vertebræ more numerous than in other breeds. From their manner of standing their legs appear longer than they really are, though, in proportion with those of *C. livia*, the legs and feet are actually longer. The wings appear much elongated, but by measurement, in relation to the length of body, this is not the case. The beak likewise appears longer, but it is in fact a little shorter (about ·03 of an inch), proportionally with the size of the body, and relatively to the beak of the rock-pigeon. The Pouter, though not bulky, is a large bird; I measured one which was 34½ inches from tip to tip of wing, and 19 inches from tip of beak to end of tail. In a wild rock-pigeon from the Shetland Islands the same measurements gave only 28¼ and 14¾. There are many sub-varieties of the Pouter of different colours, but these I pass over.

Sub-race II. Dutch Pouter.—This seems to be the parent-form of our improved English Pouters. I kept a pair, but I suspect that they were not pure birds. They are smaller than English pouters, and less well developed in all their characters. Neumeister[7] says that the wings are crossed over the tail, and do not reach to its extremity.

Sub-race III. The Lille Pouter.—I know this breed only from description.[8] It approaches in general form the Dutch Pouter, but the inflated œsophagus assumes a spherical form, as if the pigeon had swallowed a large orange, which had stuck close under the beak. This inflated ball is represented as rising to a level with the crown of the head. The middle toe alone is feathered. A variety of this sub-race, called the claquant, is described by MM. Boitard and Corbié; it pouts but little, and is characterised

[7] 'Das Ganze der Taubenzucht:' Weimar, 1837, pl. 11 and 12.

[8] Boitard and Corbié, 'Les Pigeons,' &c., p. 177, pl. 6.

by the habit of violently hitting its wings together over its back,—a habit which the English Pouter has in a slight degree.

Sub-race IV. Common German Pouter.—I know this bird only from the figures and description given by the accurate Neumeister, one of the few writers on pigeons who, as I have found, may be always trusted. This sub-race seems considerably different. The upper part of the œsophagus is much less distended. The bird stands less upright. The feet are not feathered, and the legs and beak are shorter. In these respects there is an approach in form to the common rock-pigeon. The tail-feathers are very long, yet the tips of the closed wings extend beyond the end of the tail; and the length of the wings, from tip to tip, and of the body, is greater than in the English Pouter.

GROUP II.

This group includes three Races, namely, Carriers, Runts, and Barbs, which are manifestly allied to each other. Indeed, certain carriers and runts pass into each other by such insensible gradations that an arbitrary line has to be drawn between them. Carriers also graduate through foreign breeds into the rock-pigeon. Yet, if well-characterised Carriers and Barbs (see figs. 19 and 20) had existed as wild species, no ornithologist would have placed them in the same genus with each other or with the rock-pigeon. This group may, as a general rule, be recognised by the beak being long, with the skin over the nostrils swollen and often carunculated or wattled, and with that round the eyes bare and likewise carunculated. The mouth is very wide, and the feet are large. Nevertheless the Barb, which must be classed in this same group, has a very short beak, and some runts have very little bare skin round their eyes.

RACE II.—CARRIERS. (Türkische Taube: Pigeons Turcs: Dragons.)

Beak elongated, narrow, pointed; eyes surrounded by much naked, generally carunculated skin; neck and body elongated.

Sub-race I. The English Carrier.—This is a fine bird, of large size, close feathered, generally dark-coloured, with an elongated neck. The beak is attenuated and of wonderful length: in one specimen it was 1·4 inch in length from the feathered base to the tip; therefore nearly twice as long as that of the rock-pigeon, which measured only ·77. Whenever I compare proportionally any part in the carrier and rock-pigeon, I take the length of the body from the base of the beak to the end of the tail as the standard of comparison; and according to this standard, the beak in one

Carrier was nearly half an inch longer than in the rock-pigeon. The upper mandible is often slightly arched. The tongue is very long. The development of the carunculated skin or wattle round the eyes, over the nostrils, and on the lower mandible, is prodigious. The eyelids, measured longitudinally, were in some specimens exactly twice as long as in the

Fig. 19.—English Carrier.

rock-pigeon. The external orifice or furrow of the nostrils was also twice as long. The open mouth in its widest part was in one case ·75 of an inch in width, whereas in the rock-pigeon it is only about ·4 of an inch. This great width of mouth is shown in the skeleton by the reflexed edges of the ramus of the lower jaw. The head is flat on the summit and narrow between the orbits. The feet are large and coarse; the length, as measured from end of hind toe to end of middle toe (without the claws), was in two specimens 2·6 inches; and this, proportionally with the rock-pigeon, is an excess of nearly a quarter of an inch. One very fine Carrier measured 31½ inches from tip to tip of wing. Birds of this sub-race are too valuable to be flown as carriers.

Sub-race II. Dragons; Persian Carriers.—The English Dragon differs from the improved English Carrier in being smaller in all its dimensions, and in having less wattle round the eyes and over the nostrils, and none on the lower mandible. Sir W. Elliot sent me from Madras a Bagdad Carrier (sometimes called khandési), the name of which shows its Persian origin; it would be considered here a very poor Dragon; the body was of the size of the rock-pigeon, with the beak a little longer, namely, 1 inch from the tip to the feathered base. The skin round the eyes was only slightly wattled, whilst that over the nostrils was fairly wattled. The Hon. C. Murray, also, sent me two Carriers direct from Persia; these had nearly the same character as the Madras bird, being about as large as the rock-pigeon, but the beak in one specimen was as much as 1·15 in length; the skin over the nostrils was only moderately, and that round the eyes scarcely at all wattled.

Sub-race III. Bagadotten-Tauben of Neumeister (Pavdotten or Hocker-Tauben).—I owe to the kindness of Mr. Baily, jun., a dead specimen of this singular breed imported from Germany. It is certainly allied to the Runts; nevertheless, from its close affinity with Carriers, it will be convenient here to describe it. The beak is long, and is hooked or bowed downwards in a highly remarkable manner, as will be seen in the woodcut to be hereafter given when I treat of the skeleton. The eyes are surrounded by a wide space of bright red skin, which, as well as that over the nostrils, is moderately wattled. The breast-bone is remarkably protuberant, being abruptly bowed outwards. The feet and tarsi are of great length, larger than in first-rate English Carriers. The whole bird is of large size, but in proportion to the size of the body the feathers of the wing and tail are short; a wild rock-pigeon, of considerably less size, had tail-feathers 4·6 inches in length, whereas in the large Bagadotten these feathers were scarcely over 4·1 inches in length. Riedel[9] remarks that it is a very silent bird.

Sub-race IV. Bussorah Carrier.—Two specimens were sent me by Sir W. Elliot from Madras, one in spirits and the other skinned. The name shows its Persian origin. It is much valued in India, and is considered as a distinct breed from the Bagdad Carrier, which forms my second sub-race. At first I suspected that these two sub-races might have been recently formed by crosses with other breeds, though the estimation in which they are held renders this improbable; but in a Persian treatise,[10] believed to have been written about 100 years ago, the Bagdad and Bussorah breeds are described as distinct. The Bussorah Carrier is of about the same size with the wild rock-pigeon. The shape of the beak, with some little carunculated skin over the nostrils,—the much elongated eyelids,—the

---

[9] 'Die Taubenzucht,' Ulm, 1824, s. 42.

[10] This treatise was written by Sayzid Mohammed Musari, who died in 1770: I owe to the great kindness of Sir W. Elliot a translation of this curious treatise.

broad mouth measured internally,—the narrow head,—the feet proportionally a little longer than in the rock-pigeon,—and the general appearance, all show that this bird is an undoubted Carrier; yet in one specimen the beak was of exactly the same length as in the rock-pigeon. In the other specimen the beak (as well as the opening of the nostrils) was only a very little longer, viz. by ·08 of an inch. Although there was a considerable space of bare and slightly caruneulated skin round the eyes, that over the nostrils was only in a slight degree rugose. Sir W. Elliot informs me that in the living bird the eye seems remarkably large and prominent, and the same fact is noticed in the Persian treatise; but the bony orbit is barely larger than that in the rock-pigeon.

Amongst the several breeds sent to me from Madras by Sir W. Elliot there is a pair of the *Kala Par*, black birds with the beak slightly elongated, with the skin over the nostrils rather full, and with a little naked skin round the eyes. This breed seems more closely allied to the Carrier than to any other breed, being nearly intermediate between the Bussorah Carrier and the rock-pigeon.

The names applied in different parts of Europe and in India to the several kinds of Carriers all point to Persia or the surrounding countries as the source of this Race. And it deserves especial notice that, even if we neglect the Kala Par as of doubtful origin, we get a series broken by very small steps, from the rock-pigeon, through the Bussorah, which sometimes has a beak not at all longer than that of the rock-pigeon and with the naked skin round the eyes and over the nostrils very slightly swollen and carunculated, through the Bagdad sub-race and Dragons, to our improved English Carriers, which present so marvellous a difference from the rock-pigeon or *Columba livia*.

RACE III.—RUNTS. (Scanderoons: Die Florentiner-Taube and Hinkel-Taube of Neumeister: Pigeon Bagadais, Pigeon Romain.)

Beak long, massive; body of great size.

Inextricable confusion reigns in the classification, affinities, and naming of Runts. Several characters which are generally pretty constant in other pigeons, such as the length of the wings, tail, legs, and neck, and the amount of naked skin round the eyes, are excessively variable in Runts. When the naked skin over the nostrils and round the eyes is considerably developed and wattled, and when the size of body is not very great, Runts graduate in so insensible a manner into Carriers, that the distinction is quite arbitrary. This fact is likewise shown by the names given to them in different parts of Europe. Nevertheless, taking the most distinct forms, at least five sub-races (some of them including well-marked varieties) can be distinguished, which differ in such important points of structure, that they would be considered as good species in a state of nature.

Sub-race I. Scanderoon of English writers (Die Florentiner and Hinkel-Taube of Neumeister).—Birds of this sub-race, of which I kept one alive and have since seen two others, differ from the Bagadotten of Neumeister only in not having the beak nearly so much curved downwards, and in the naked skin round the eyes and over the nostrils being hardly at all wattled. Nevertheless I have felt myself compelled to place the Bagadotten in Race II., or that of the Carriers, and the present bird in Race III., or that of the Runts. The Scanderoon has a very short, narrow, and elevated tail; wings extremely short, so that the first primary feathers were not longer than those of a small tumbler pigeon! Neck long, much bowed; breast-bone prominent. Beak long, being 1·15 inch from tip to feathered base; vertically thick; slightly curved downwards. The skin over the nostrils swollen, not wattled; naked skin round the eyes, broad, slightly carunculated. Legs long; feet very large. Skin of neck bright red, often showing a naked medial line, with a naked red patch at the distant end of the radius of the wing. My bird, as measured from the base of the beak to the root of the tail, was fully 2 inches longer than the rock-pigeon; yet the tail itself was only 4 inches in length, whereas in the rock-pigeon, which is a much smaller bird, the tail is $4\frac{5}{8}$ inches in length.

The Hinkel or Florentiner-Taube of Neumeister (Table XIII., fig. 1) agrees with the above description in all the specified characters (for the beak is not mentioned), except that Neumeister expressly says that the neck is short, whereas in my Scanderoon it was remarkably long and bowed; so that the Hinkel forms a well-marked variety.

Sub-race II. Pigeon Cygne and Pigeon Bagadais of Boitard and Corbié (Scanderoon of French writers).—I kept two of these birds alive, imported from France. They differed from the first sub-race or true Scanderoon in the much greater length of the wing and tail, in the beak not being so long, and in the skin about the head being more carunculated. The skin of the neck is red; but the naked patches on the wings are absent. One of my birds measured $38\frac{1}{2}$ inches from tip to tip of wing. By taking the length of the body as the standard of comparison, the two wings were no less than 5 inches longer than those of the rock-pigeon! The tail was $6\frac{1}{4}$ inches in length, and therefore $2\frac{1}{4}$ inches longer than that of the Scanderoon,—a bird of nearly the same size. The beak is longer, thicker, and broader than in the rock-pigeon, proportionally with the size of body. The eyelids, nostrils, and internal gape of mouth are all proportionally very large, as in Carriers. The foot, from the end of the middle to end of hind toe, was actually 2·85 inches in length, which is an excess of ·32 of an inch over the foot of the rock-pigeon, relatively to the size of the two birds.

Sub-race III. Spanish and Roman Runts.—I am not sure that I am right in placing these Runts in a distinct sub-race; yet, if we take well-characterized birds, there can be no doubt of the propriety of the separation. They are heavy, massive birds, with shorter necks, legs, and beaks than in the foregoing races. The skin over the nostrils is swollen, but not carunculated; the naked skin round the eyes is not very wide, and only slightly carunculated; and I have seen a fine so-called Spanish Runt with hardly any naked skin round the eyes. Of the two varieties to be seen in England, one, which is the rarer, has very long wings and tail, and agrees pretty closely with the last sub-race; the other, with shorter wings and tail, is apparently the *Pigeon Romain ordinaire* of Boitard and Corbié. These Runts are apt to tremble like Fantails. They are bad flyers. A few years ago Mr. Gulliver[11] exhibited a Runt which weighed 1 lb. 14 oz.; and, as I am informed by Mr. Tegetmeier, two Runts from the south of France were lately exhibited at the Crystal Palace, each of which weighed 2 lbs. 2½ oz. A very fine rock-pigeon from the Shetland Islands weighed only 14½ oz.

Sub-race IV. Tronfo of Aldrovandi (Leghorn Runt?).—In Aldrovandi's work published in 1600 there is a coarse woodcut of a great Italian pigeon, with an elevated tail, short legs, massive body, and with the beak short and thick. I had imagined that this latter character, so abnormal in the group, was merely a false representation from bad drawing; but Moore, in his work published in 1735, says that he possessed a Leghorn Runt of which "the beak was very short for so large a bird." In other respects Moore's bird resembled the first sub-race or Scanderoon, for it had a long bowed neck, long legs, short beak, and elevated tail, and not much wattle about the head. So that Aldrovandi's and Moore's birds must have formed distinct varieties, both of which seem to be now extinct in Europe. Sir W. Elliot, however, informs me that he has seen in Madras a short-beaked Runt imported from Cairo.

Sub-race V. Murassa (adorned Pigeon) of Madras.—Skins of these handsome chequered birds were sent me from Madras by Sir W. Elliot. They are rather larger than the largest rock-pigeon, with longer and more massive beaks. The skin over the nostrils is rather full and very slightly carunculated, and they have some naked skin round the eyes: feet large. This breed is intermediate between the rock-pigeon and a very poor variety of Runt or Carrier.

From these several descriptions we see that with Runts, as with Carriers, we have a fine gradation from the rock-pigeon (with the Tronfo diverging as a distinct branch) to our largest and most massive Runts. But the chain of affinities, and many points of resemblance, between Runts and Carriers, make me believe that these two races have not descended by independent lines from the rock-pigeon, but from some common parent, as represented in the Table, which had already acquired a moderately long beak, with slightly swollen skin over the nostrils, and with some slightly carunculated naked skin round the eyes.

RACE IV.—BARBS. (Indische-Taube: Pigeons Polonais.)

Beak short, broad, deep; naked skin round the eyes, broad and carunculated; skin over nostrils slightly swollen.

Misled by the extraordinary shortness and form of the beak, I did not at first perceive the near affinity of this Race to that of Carriers until the fact was pointed out to me by Mr. Brent. Subsequently, after examining

---

[11] 'Poultry Chronicle,' vol. ii. p. 573.

the Bussorah Carrier, I saw that no very great amount of modification would be requisite to convert it into a Barb. This view of the affinity of Barbs to Carriers is supported by the analogical difference between the short and long-beaked Runts; and still more strongly by the fact, that young Barbs and Dragons, within 24 hours after being hatched, resemble each

Fig. 20.—English Barb.

other much more closely than do young pigeons of other and equally distinct breeds. At this early age, the length of beak, the swollen skin over the rather open nostrils, the gape of the mouth, and the size of the feet, are the same in both; although these parts afterwards become widely different. We thus see that embryology (as the comparison of very young animals may perhaps be called) comes into play in the classification of domestic varieties, as with species in a state of nature.

Fanciers, with some truth, compare the head and beak of the Barb to that of a bullfinch. The Barb, if found in a state of nature, would certainly have been placed in a new genus formed for its reception. The body is a little larger than that of the rock-pigeon, but the beak is more than ·2 of an inch shorter; although shorter, it is both vertically and horizontally thicker. From the outward flexure of the rami of the lower jaw, the mouth internally is very broad, in the proportion of 6 to ·4 to that of the rock-pigeon. The whole head is broad. The skin over the nostrils is swollen, but not carunculated, except slightly in first-rate birds when old; whilst the naked skin round the eye is broad and much carunculated. It is sometimes so much developed, that a bird belonging to Mr. Harrison Weir could hardly see to pick up food from the ground. The eyelids in one specimen were nearly twice as long as those of the rock-pigeon. The feet are coarse and strong, but proportionally rather shorter than in the rock-pigeon. The plumage is generally dark and uniform. Barbs, in short, may be called short-beaked Carriers, bearing the same relation to Carriers that the Tronfo of Aldrovandi does to the common Runt.

GROUP III.

This group is artificial, and includes a heterogeneous collection of distinct forms. It may be defined by the beak, in well-characterised specimens of the several races, being shorter than in the rock-pigeon, and by the skin round the eyes not being much developed.

RACE V.—FANTAILS.

Sub-race I. European Fantails (Pfauen-Taube; Trembleurs). *Tail expanded, directed upwards, formed of many feathers; oil-gland aborted; body and beak rather short.*

The normal number of tail-feathers in the genus Columba is 12; but Fantails have from only 12 (as has been asserted) up to, according to MM. Boitard and Corbié, 42. I have counted in one of my own birds 33, and at Calcutta Mr. Blyth[12] has counted in an *imperfect* tail 34 feathers. In Madras, as I am informed by Sir W. Elliot, 32 is the standard number; but in England number is much less valued than the position and expansion of the tail. The feathers are arranged in an irregular double row; their permanent expansion, like a fan, and their upward direction, are more remarkable characters than their increased number. The tail is capable of the same movements as in other pigeons, and can be depressed so as to sweep the ground. It arises from a more expanded basis than in

[12] 'Annals and Mag. of Nat. History,' vol. xix., 1847, p. 105.

other pigeons; and in three skeletons there were one or two extra coccygeal vertebræ. I have examined many specimens of various colours

Fig. 21.—English Fantail.

from different countries, and there was no trace of the oil-gland; this is a curious case of abortion.[13] The neck is thin and bowed back-

[13] This gland occurs in most birds; but Nitzsch (in his 'Pterylographie,' 1840, p. 55) states that it is absent in two species of Columba, in several species of Psittacus, in some species of Otis, and in most or all birds of the Ostrich family. It can hardly be an accidental coincidence that the two species of Columba, which are destitute of an oil-gland, have an unusual number of tail-feathers, namely 16, and in this respect resemble Fantails.

wards. The breast is broad and protuberant. The feet are small. The carriage of the bird is very different from that of other pigeons; in good birds the head touches the tail-feathers, which consequently often become crumpled. They habitually tremble much; and their necks have an extraordinary, apparently convulsive, backward and forward movement. Good birds walk in a singular manner, as if their small feet were stiff. Owing to their large tails, they fly badly on a windy day. The dark-coloured varieties are generally larger than white Fantails.

Although between the best and common Fantails, now existing in England, there is a vast difference in the position and size of the tail, in the carriage of the head and neck, in the convulsive movements of the neck, in the manner of walking, and in the breadth of the breast, the differences so graduate away, that it is impossible to make more than one sub-race. Moore, however, an excellent old authority,[14] says, that in 1735 there were two sorts of broad-tailed shakers (*i. e.* fantails), "one having a neck much longer and more slender than the other;" and I am informed by Mr. B. P. Brent that there is an existing German Fantail with a thicker and shorter beak.

Sub-race II. Java Fantail.—Mr. Swinhoe sent me from Amoy, in China, the skin of a Fantail belonging to a breed known to have been imported from Java. It was coloured in a peculiar manner, unlike any European Fantail, and, for a Fantail, had a remarkably short beak. Although a good bird of the kind, it had only 14 tail-feathers; but Mr. Swinhoe has counted in other birds of this breed from 18 to 24 tail-feathers. From a rough sketch sent to me, it is evident that the tail is not so much expanded or so much upraised as in even second-rate European Fantails. The bird shakes its neck like our Fantails. It had a well-developed oil-gland. Fantails were known in India, as we shall hereafter see, before the year 1600; and we may suspect that in the Java Fantail we see the breed in its earlier and less improved condition.

RACE VI.—TURBIT AND OWL. (Möven-Taube: Pigeons à cravate.)

Feathers divergent along the front of the neck and breast; beak very short, vertically rather thick; œsophagus somewhat enlarged.

Turbits and Owls differ from each other slightly in the shape of the head, in the former having a crest, and in the curvature of the beak, but they may be here conveniently grouped together. These pretty birds, some of which are very small, can be recognised at once by the feathers irregularly diverging, like a frill, along the front of the neck, in the same manner, but in a less degree, as along the back of the neck in the Jacobin. This bird has the remarkable habit of continually and momentarily inflating the upper part of the œsophagus, which causes a movement in the frill.

---

[14] *See* the two excellent editions published by Mr. J. M. Eaton in 1852 and 1858, entitled 'A Treatise on Fancy Pigeons.'

When the œsophagus of a dead bird was inflated, it was seen to be larger than in other breeds, and not so distinctly separated from the crop. The Pouter inflates both its true crop and œsophagus; the Turbit inflates in a much less degree the œsophagus alone. The beak of the Turbit

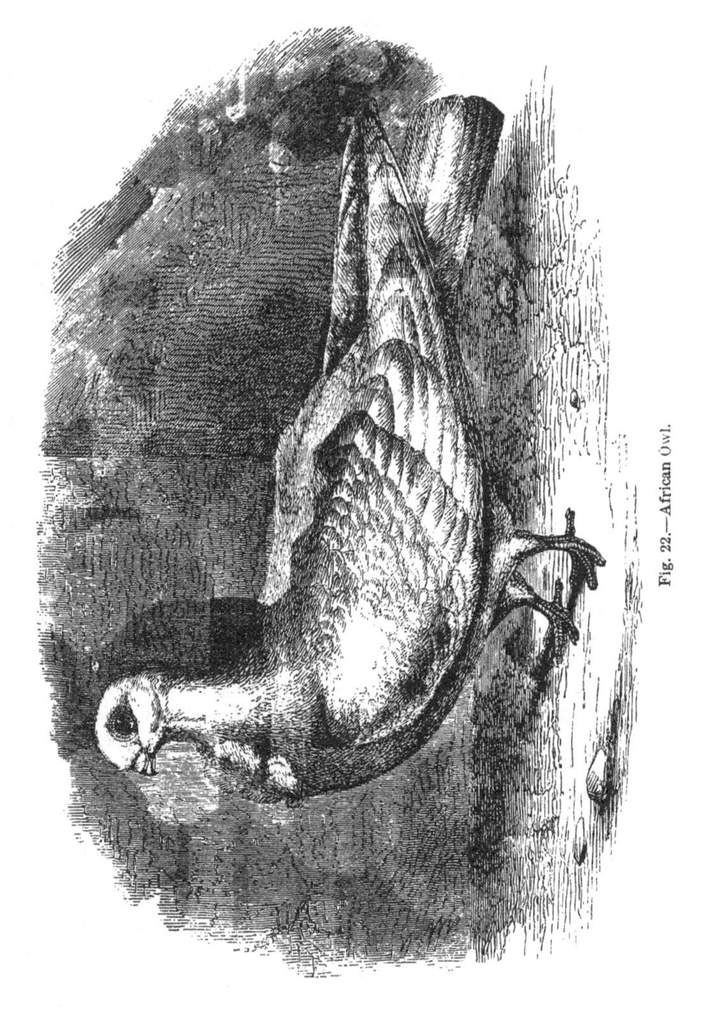

Fig. 22.—African Owl.

is very short, being ·28 of an inch shorter than that of the rock-pigeon, proportionally with the size of their bodies; and in some owls brought by Mr. E. Vernon Harcourt from Tunis, it was even shorter. The beak is vertically thicker, and perhaps a little broader, in proportion to that of the rock-pigeon.

RACE VII.—TUMBLERS. (Tümmler, or Burzel-Tauben: Culbutants.)

During flight, tumble backwards; body generally small; beak generally short, sometimes excessively short and conical.

This Race may be divided into four sub-races, namely, Persian, Lotan, Common, and Short-faced Tumblers. These sub-races include many varieties which breed true. I have examined eight skeletons of various kinds of Tumblers: excepting in one imperfect and doubtful specimen, the ribs are only seven in number, whereas the rock-pigeon has eight ribs.

Sub-race I. Persian Tumblers.—I have received a pair direct from Persia, from the Hon. C. Murray. They were rather smaller birds than the wild rock-pigeon, being about the size of the common dovecot-pigeon, white and mottled, slightly feathered on the feet, with the beak just perceptibly shorter than in the rock-pigeon. H. M. Consul, Mr. Keith Abbott, informs me that the difference in the length of beak is so slight, that only practised Persian fanciers can distinguish these Tumblers from the common pigeon of the country. He informs me that they fly in flocks high up in the air and tumble well. Some of them occasionally appear to become giddy and tumble to the ground, in which respect they resemble some of our Tumblers.

Sub-race II. Lotan, or Lowtun: Indian Ground Tumblers.—These birds present one of the most remarkable inherited habits or instincts which have ever been recorded. The specimens sent to me from Madras by Sir W. Elliot are white, slightly feathered on the feet, with the feathers on the head reversed; and they are rather smaller than the rock or dovecot pigeon. The beak is proportionally only slightly shorter and rather thinner than in the rock-pigeon. These birds when gently shaken and placed on the ground immediately begin tumbling head over heels, and they continue thus to tumble until taken up and soothed,—the ceremony being generally to blow in their faces, as in recovering a person from a state of hypnotism or mesmerism. It is asserted that they will continue to roll over till they die, if not taken up. There is abundant evidence with respect to these remarkable peculiarities; but what makes the case the more worthy of attention is, that the habit has been strictly inherited since before the year 1600, for the breed is distinctly described in the 'Ayeen Akbery.'[15] Mr. Evans kept a pair in London, imported by Captain Vigne; and he assures me that he has seen them tumble in the air, as well as in the manner above described on the ground. Sir W. Elliot, however, writes to me from Madras, that he is informed that they tumble exclusively on the ground, or at a very small height above it. He also

[15] English translation, by F. Gladwin, 4th edition, vol. i. The habit of the Lotan is also described in the Persian treatise before alluded to, published about 100 years ago: at this date the Lotans were generally white and crested as at present. Mr. Blyth describes these birds in 'Annals and Mag. of Nat. Hist.,' vol. xiv., 1847, p. 104: he says that they "may be seen at any of the Calcutta bird-dealers."

mentions another sub-variety, called the Kalmi Lotan, which begins to roll over if only touched on the neck with a rod or wand.

Sub-race III. Common English Tumblers.—These birds have exactly the same habits as the Persian Tumbler, but tumble better. The English bird is rather smaller than the Persian, and the beak is plainly shorter. Compared with the rock-pigeon, and proportionally with the size of body, the beak is from 15 to nearly ·2 of an inch shorter, but it is not thinner. There are several varieties of the common Tumbler, namely, Baldheads, Beards, and Dutch Rollers. I have kept the latter alive; they have differently shaped heads, longer necks, and are feather-footed. They tumble to an extraordinary degree; as Mr. Brent remarks,[16] "Every "few seconds over they go; one, two, or three summersaults at a time. "Here and there a bird gives a very quick and rapid spin, revolving "like a wheel, though they sometimes lose their balance, and make a "rather ungraceful fall, in which they occasionally hurt themselves by "striking some object." From Madras I have received several specimens of the common Tumbler of India, differing slightly from each other in the length of their beaks. Mr. Brent sent me a dead specimen of a "House-tumbler,"[17] which is a Scotch variety, not differing in general appearance and form of beak from the common Tumbler. Mr. Brent states that these birds generally begin to tumble "almost as soon as they can well "fly; at three months old they tumble well, but still fly strong; at five "or six months they tumble excessively; and in the second year they "mostly give up flying, on account of their tumbling so much and so "close to the ground. Some fly round with the flock, throwing a clean "summersault every few yards, till they are obliged to settle from giddiness "and exhaustion. These are called Air Tumblers, and they commonly "throw from twenty to thirty summersaults in a minute, each clear "and clean. I have one red cock that I have on two or three occasions "timed by my watch, and counted forty summersaults in the minute. "Others tumble differently. At first they throw a single summersault, "then it is double, till it becomes a continuous roll, which puts an end "to flying, for if they fly a few yards over they go, and roll till they "reach the ground. Thus I had one kill herself, and another broke his leg. "Many of them turn over only a few inches from the ground, and will "tumble two or three times in flying across their loft. These are called "House-tumblers, from tumbling in the house. The act of tumbling seems "to be one over which they have no control, an involuntary movement "which they seem to try to prevent. I have seen a bird sometimes in his "struggles fly a yard or two straight upwards, the impulse forcing him "backwards while he struggles to go forwards. If suddenly startled, or "in a strange place, they seem less able to fly than if quiet in their accus-"tomed loft." These House-tumblers differ from the Lotan or Ground

[16] 'Journal of Horticulture,' Oct. 22, 1861, p. 76.

[17] *See* the account of the House-tumblers kept at Glasgow, in the 'Cottage Gardener,' 1858, p. 285. Also Mr. Brent's paper, 'Journal of Horticulture,' 1861, p. 76.

Tumbler of India, in not requiring to be shaken in order to begin tumbling. The breed has probably been formed merely by selecting the best common Tumblers, though it is possible that they may have been crossed at some former period with Lotans.

Fig. 23.—Short-faced English Tumbler.

Sub-race IV. Short-faced Tumblers.—These are marvellous birds, and are the glory and pride of many fanciers. In their extremely short, sharp, and conical beaks, with the skin over the nostrils but little developed, they almost depart from the type of the Columbidæ. Their heads are nearly globular and upright in front, so that some fanciers say[18] "the head should resemble a cherry with a barley-corn stuck in it." These are the smallest kind of pigeons. Mr. Esquilant possessed a blue Baldhead, two years old, which when alive weighed, before feeding-time, only 6 oz. 5 drs.; two others, each weighed 7 oz. We have seen that a wild rock-pigeon weighed 14 oz. 2 drs., and a Runt 34 oz. 4 drs. Short-faced Tumblers have a remarkably erect carriage, with prominent breasts, drooping wings, and very small feet. The length of the beak from the tip to the feathered base was in one good bird only ·4 of an inch; in a wild rock-pigeon it was exactly double this length. As these Tumblers have shorter bodies than the wild rock-pigeon, they ought of course to have shorter beaks; but proportionally with the size of body, the beak is ·28 of an inch too short. So, again, the feet of this bird were actually ·45 shorter, and proportionally ·21 of an inch shorter, than the feet of the rock-pigeon. The middle toe has only twelve or thirteen, instead of fourteen or fifteen scutellæ. The primary wing-feathers are not rarely only nine instead of ten in number. The improved short-faced Tumblers have almost lost the power of tumbling; but there are several authentic accounts of their occasionally tumbling. There are several sub-varieties, such as Baldheads, Beards, Mottles, and Almonds; the latter are remarkable from not acquiring their perfectly-coloured plumage until they have moulted three or four times. There is good reason to believe that most of these sub-varieties, some of which breed truly, have arisen since the publication of Moore's treatise in 1735.[19]

Finally, in regard to the whole group of Tumblers, it is impossible to conceive a more perfect gradation than I have now lying before me, from the rock-pigeon, through Persian, Lotan, and Common Tumblers, up to the marvellous short-faced birds; which latter, no ornithologist, judging from mere external structure, would place in the same genus with the rock-pigeon. The differences between the successive steps in this series are not greater than those which may be observed between common dove-cot-pigeons (*C. livia*) brought from different countries.

RACE VIII.—INDIAN FRILL-BACK.

Beak very short; feathers reversed.

A specimen of this bird, in spirits, was sent to me from Madras by Sir W. Elliot. It is wholly different from the Frill-back often exhibited in England. It is a smallish bird, about the size of the common Tumbler, but has a beak in all its proportions like our short-faced Tumblers. The beak, measured from the tip to the feathered base, was only ·46 of an inch in length. The feathers over the whole body are reversed or curl backwards. Had this bird occurred in Europe, I should have thought it only a monstrous variety of our improved Tumbler; but as short-faced Tumblers are not known in India, I think it must rank as a distinct breed. Probably

[18] J. M. Eaton's 'Treatise on Pigeons,' 1852, p. 9.

[19] J. M. Eaton's Treatise, edit. 1858, p. 76.

this is the breed seen by Hasselquist in 1757 at Cairo, and said to have been imported from India.

RACE IX.—JACOBIN. (Zopf or Perücken-Taube: Nonnains.)

Feathers of the neck forming a hood; wings and tail long; beak moderately short.

This pigeon can at once be recognised by its hood, almost enclosing the head and meeting in front of the neck. The hood seems to be merely an exaggeration of the crest of reversed feathers on the back of the head, which is common to many sub-varieties, and which in the Latz-taube[20] is in a nearly intermediate state between a hood and a crest. The feathers of the hood are elongated. Both the wings and tail are likewise much elongated; thus the folded wing of the Jacobin, though a somewhat smaller bird, is fully 1¼ inch longer than in the rock-pigeon. Taking the length of the body without the tail as the standard of comparison, the folded wing, proportionally with the wings of the rock-pigeon, is 2¼ inches too long, and the two wings, from tip to tip, 5¼ inches too long. In disposition this bird is singularly quiet, seldom flying or moving about, as Bechstein and Riedel have likewise remarked in Germany.[21] The latter author also notices the length of the wings and tail. The beak is nearly ·2 of an inch shorter in proportion to the size of the body than in the rock-pigeon; but the internal gape of the mouth is considerably wider.

GROUP IV.

The birds of this group may be characterised by their resemblance in all important points of structure, especially in the beak, to the rock-pigeon. The Trumpeter forms the only well-marked race. Of the numerous other sub-races and varieties I shall specify only a few of the most distinct, which I have myself seen and kept alive.

RACE X.—TRUMPETER. (Trommel-Taube; Pigeon tambour; glougou.)

A tuft of feathers at the base of the beak curling forward; feet much feathered; voice very peculiar; size exceeding that of the rock-pigeon.

This is a well-marked breed, with a peculiar voice, wholly unlike that of any other pigeon. The coo is rapidly repeated, and is continued for

[20] Neumeister, 'Taubenzucht,' Tab. 4, fig. i.

[21] Riedel, 'Die Taubenzucht,' 1824, s. 26. Bechstein, 'Naturgeschichte Deutschlands,' Band iv. s. 36, 1795.

several minutes; hence their name of Trumpeters. They are also characterised by a tuft of elongated feathers, which curls forward over the base of the beak, and which is possessed by no other breed. Their feet are so heavily feathered, that they almost appear like little wings. They are larger birds than the rock-pigeon, but their beak is of very nearly the same proportional size. Their feet are rather small. This breed was perfectly characterised in Moore's time, in 1735. Mr. Brent says that two varieties exist, which differ in size.

RACE XI.—*Scarcely differing in structure from the wild Columba livia.*

Sub-race I. Laughers. Size less than the Rock-pigeon; voice very peculiar.—As this bird agrees in nearly all its proportions with the rock-pigeon, though of smaller size, I should not have thought it worthy of mention, had it not been for its peculiar voice—a character supposed seldom to vary with birds. Although the voice of the Laugher is very different from that of the Trumpeter, yet one of my Trumpeters used to utter a single note like that of the Laugher. I have kept two varieties of Laughers, which differed only in one variety being turn-crowned; the smooth-headed kind, for which I am indebted to the kindness of Mr. Brent, besides its peculiar note, used to coo in a singular and pleasing manner, which, independently, struck both Mr. Brent and myself as resembling that of the turtle-dove. Both varieties come from Arabia. This breed was known by Moore in 1735. A pigeon which seems to say Yak-roo is mentioned in 1600 in the 'Ayeen Akbery,' and is probably the same breed. Sir W. Elliot has also sent me from Madras a pigeon called Yahui, said to have come from Mecca, which does not differ in appearance from the Laugher; it has "a deep melancholy voice, like Yahu, often repeated." Yahu, yahu, means Oh God, Oh God; and Sayzid Mohammed Musari, in the treatise written about 100 years ago, says that these birds "are not flown, because they repeat the name of the Most High God." Mr. Keith Abbott, however, informs me that the common pigeon is called Yahoo in Persia.

Sub-race II. Common Frill-back (Die Strupp-Taube). *Beak rather longer than in the Rock-pigeon; feathers reversed.*—This is a considerably larger bird than the rock-pigeon, and with the beak, proportionally with the size of body, a little (viz. by ·04 of an inch) longer. The feathers, especially on the wing-coverts, have their points curled upwards or backwards.

Sub-race III. Nuns (Pigeons-coquilles).—These elegant birds are smaller than the rock-pigeon. The beak is actually ·17, and proportionally with the size of the body ·1 of an inch shorter than in the rock-pigeons, although of the same thickness. In young birds the scutellæ on the tarsi and toes are generally of a leaden-black colour; and this is a remarkable character (though observed in a lesser degree in some other breeds), as the colour of the legs in the adult state is subject to very little variation in any breed. I have on two or three occasions counted thirteen or fourteen feathers in the tail; this likewise occurs in the barely distinct breed called Helmets.

Nuns are symmetrically coloured, with the head, primary wing-feathers, tail, and tail-coverts of the same colour, namely, black or red, and with the rest of the body white. This breed has retained the same character since Aldrovandi wrote in 1600. I have received from Madras almost similarly coloured birds.

Sub-race IV. Spots (Die Blass-Taube: Pigeons heurtés).—These birds are a very little larger than the rock-pigeon, with the beak a trace smaller in all its dimensions, and with the feet decidedly smaller. They are symmetrically coloured, with a spot on the forehead, with the tail and tail-coverts of the same colour, the rest of the body being white. This breed existed in 1676;[22] and in 1735 Moore remarks that they breed truly, as is the case at the present day.

Sub-race V. Swallows.—These birds, as measured from tip to tip of wing, or from the end of the beak to the end of the tail, exceed in size the rock-pigeon; but their bodies are much less bulky; their feet and legs are likewise smaller. The beak is of about the same length, but rather slighter. Altogether their general appearance is considerably different from that of the rock-pigeon. Their heads and wings are of the same colour, the rest of the body being white. Their flight is said to be peculiar. This seems to be a modern breed, which, however, originated before the year 1795 in Germany, for it is described by Bechstein.

Besides the several breeds now described, three or four other very distinct kinds existed lately, or perhaps still exist, in Germany and France. Firstly, the Karmeliten, or Carme Pigeon, which I have not seen; it is described as of small size, with very short legs, and with an extremely short beak. Secondly, the Finnikin, which is now extinct in England. It had, according to Moore's[23] treatise, published in 1735, a tuft of feathers on the hinder part of the head, which ran down its back not unlike a horse's mane. "When it is salacious it rises over the hen and turns round three or four times, flapping its wings, then reverses and turns as many times the other way." The Turner, on the other hand, when it "plays to the female, turns only one way." Whether these extraordinary statements may be trusted I know not; but the inheritance of any habit may be believed, after what we have seen with respect to the Ground-tumbler of India. MM. Boitard and Corbié describe a pigeon[24] which has the singular habit of sailing for a considerable time through the air, without flapping its wings, like a bird of prey. The confusion is inextricable, from the time of Aldrovandi in 1600 to the present day, in the accounts published of the Draijers, Smiters, Finnikins, Turners, Claquers, &c., which are all remarkable from their manner of flight. Mr. Brent informs me that he has seen one of these breeds in Germany with its wing-feathers injured from having been so often struck together; but he did not see it flying. An old stuffed specimen of a Finnikin in the British Museum presents no well-marked character. Thirdly, a singular pigeon

[22] Willoughby's 'Ornithology,' edited by Ray.

[23] J. M. Eaton's edition (1858) of Moore, p. 98.

[24] Pigeon Patu Plongeur. 'Les Pigeons,' &c., p. 165.

with a forked tail is mentioned in some treatises; and as Bechstein[25] briefly describes and figures this bird, with a tail "having completely the structure of that of the house-swallow," it must once have existed, for Bechstein was far too good a naturalist to have confounded any distinct species with the domestic pigeon. Lastly, an extraordinary pigeon imported from Belgium has lately been exhibited at the Philoperisteron Society in London,[26] which "conjoins the colour of an archangel with the head of an owl or barb, its most striking peculiarity being the extraordinary length of the tail and wing-feathers, the latter crossing beyond the tail, and giving to the bird the appearance of a gigantic swift (Cypselus), or long-winged hawk." Mr. Tegetmeier informs me that this bird weighed only 10 ounces, but in length was 15½ inches from tip of beak to end of tail, and 32½ inches from tip to tip of wing; now the wild rock-pigeon weighs 14½ ounces, and measures from tip of beak to end of tail 15 inches, and from tip to tip of wing only 26¾ inches.

I have now described all the domestic pigeons known to me, and have added a few others on reliable authority. I have classed them under four Groups, in order to mark their affinities and degrees of difference; but the third group is artificial. The kinds examined by me form eleven races, which include several sub-races; and even these latter present differences that would certainly have been thought of specific value if observed in a state of nature. The sub-races likewise include many strictly inherited varieties; so that altogether there must exist, as previously stated, above 150 kinds which can be distinguished, though generally by characters of extremely slight importance. Many of the genera of the Columbidæ, which are admitted by ornithologists, do not differ in any great degree from each other; taking this into consideration, there can be no doubt that several of the most strongly characterised domestic forms, if found wild, would have been placed in at least five new genera. Thus, a new genus would have been formed for the reception of the improved English Pouter: a second genus for Carriers and Runts; and this would have been a wide or comprehensive genus, for it would have admitted common Spanish Runts without any wattle, short-beaked Runts like the Tronfo, and the improved English Carrier: a third genus would have been formed for the Barb: a fourth for the Fantail: and lastly, a fifth for the short-beaked, not-wattled pigeons, such as Turbits

[25] 'Naturgesch. Deutschlands,' Band iv. s. 47.

[26] Mr. W. B. Tegetmeier, 'Journal of Horticulture,' Jan. 20th, 1863, p. 58.

and short-faced Tumblers. The remaining domestic forms might have been included in the same genus with the wild rock-pigeon.

Individual Variability; Variations of a remarkable nature.

The differences which we have as yet considered are characteristic of distinct breeds; but there are other differences, either confined to individual birds, or often observed in certain breeds but not characteristic of them. These individual differences are of importance, as they might in most cases be secured and accumulated by man's power of selection; and thus an existing breed might be greatly modified or a new one formed. Fanciers notice and select only those slight differences which are externally visible; but the whole organisation is so tied together by correlation of growth, that a change in one part is frequently accompanied by other changes. For our purpose, modifications of all kinds are equally important, and, if affecting a part which does not commonly vary, are of more importance than a modification in some conspicuous part. At the present day any visible deviation of character in a well-established breed is rejected as a blemish; but it by no means follows that at an early period, before well-marked breeds had been formed, such deviations would have been rejected; on the contrary, they would have been eagerly preserved as presenting a novelty, and would then have been slowly augmented, as we shall hereafter more clearly see, by the process of unconscious selection.

I have made numerous measurements of the various parts of the body in the several breeds, and have hardly ever found them quite the same in birds of the same breed,—the differences being greater than we commonly meet with in wild species. To begin with the primary feathers of the wing and tail; but I may first mention, as some readers may not be aware of the fact, that the number of the primary wing and tail feathers in wild birds is generally constant, and characterises, not only whole genera, but even whole families. When the tail-feathers are unusually numerous, as for instance in the swan, they are apt to be variable in number; but this does not apply to the several species and genera of the Columbidæ, which never (as far as I can hear) have less than twelve or more than sixteen tail-feathers; and these numbers characterise, with rare exception, whole sub-families.[27] The wild rock-pigeon has twelve tail-feathers. With Fan-

---

[27] 'Coup-d'œil sur l'Ordre des Pigeons,' par C. L. Bonaparte (Comptes Rendus), 1854-55. Mr. Blyth, in 'Annals of Nat. Hist.,' vol. xix., 1847,

tails, as we have seen, the number varies from fourteen to forty-two. In two young birds in the same nest I counted twenty-two and twenty-seven feathers. Pouters are very liable to have additional tail-feathers, and I have seen on several occasions fourteen or fifteen in my own birds. Mr. Bult had a specimen, examined by Mr. Yarrell, with seventeen tail-feathers. I had a Nun with thirteen, and another with fourteen tail-feathers; and in a Helmet, a breed barely distinguishable from the Nun, I have counted fifteen, and have heard of other such instances. On the other hand, Mr. Brent possessed a Dragon, which during its whole life never had more than ten tail-feathers; and one of my Dragons, descended from Mr. Brent's, had only eleven. I have seen a Baldhead-Tumbler with only ten; and Mr. Brent had an Air-Tumbler with the same number, but another with fourteen tail-feathers. Two of these latter Tumblers, bred by Mr. Brent, were remarkable,—one from having the two central tail-feathers a little divergent, and the other from having the two outer feathers longer by three-eighths of an inch than the others; so that in both cases the tail exhibited a tendency, but in different ways, to become forked. And this shows us how a swallow-tailed breed, like that described by Bechstein, might have been formed by careful selection.

With respect to the primary wing-feathers, the number in the Columbidæ, as far as I can find out, is always nine or ten. In the rock-pigeon it is ten; but I have seen no less than eight short-faced Tumblers with only nine primaries, and the occurrence of this number has been noticed by fanciers, owing to ten flight-feathers of a white colour being one of the points in Short-faced Baldhead-Tumblers. Mr. Brent, however, had an Air-Tumbler (not short-faced) which had in both wings eleven primaries. Mr. Corker, the eminent breeder of prize Carriers, assures me that some of his birds had eleven primaries in both wings. I have seen eleven in one wing in two Pouters. I have been assured by three fanciers that they have seen twelve in Scanderoons; but as Neumeister asserts that in the allied Florence Runt the middle flight-feather is often double, the number twelve may have been caused by two of the ten primaries having each two shafts to a single feather. The secondary wing-feathers are difficult to count, but the number seems to vary from twelve to fifteen. The length of the wing and tail relatively to the body, and of the wings to the tail, certainly varies; I have especially noticed this in Jacobins. In Mr. Bult's magnificent collection of Pouters, the wings and tail varied greatly in length; and were sometimes so much elongated that the birds could hardly play upright. In the relative length of the few first primaries I have observed only a slight degree of variability. Mr. Brent informs me that he has observed the shape of the first feather to vary very slightly. But the variation in these latter points is extremely slight compared with what may often be observed in the natural species of the Columbidæ.

In the beak I have observed very considerable differences in birds of the

---

p. 41, mentions, as a very singular fact, "that of the two species of Ectopistes, which are nearly allied to each other, one should have fourteen tail-feathers, while the other, the passenger pigeon of North America, should possess but the usual number—twelve."

same breed, as in carefully bred Jacobins and Trumpeters. In Carriers there is often a conspicuous difference in the degree of attenuation and curvature of the beak. So it is indeed in many breeds: thus I had two strains of black Barbs, which evidently differed in the curvature of the upper mandible. In width of mouth I have found a great difference in two Swallows. In Fantails of first-rate merit I have seen some birds with much longer and thinner necks than in others. Other analogous facts could be given. We have seen that the oil-gland is aborted in all Fantails (with the exception of the sub-race from Java), and, I may add, so hereditary is this tendency to abortion, that some, although not all, of the mongrels from the Fantail and Pouter had no oil-gland; in one Swallow out of many which I have examined, and in two Nuns, there was no oil-gland.

The number of the scutellæ on the toes often varies in the same breed, and sometimes even differs on the two feet of the same individual; the Shetland rock-pigeon has fifteen on the middle, and six on the hinder toe; whereas I have seen a Runt with sixteen on the middle and eight on the hind toe; and a short-faced Tumbler with only twelve and five on these same toes. The rock-pigeon has no sensible amount of skin between its toes; but I possessed a Spot and a Nun with the skin extending for a space of a quarter of an inch from the fork, between the two *inner* toes. On the other hand, as will hereafter be more fully shown, pigeons with feathered feet very generally have the bases of their *outer* toes connected by skin. I had a red Tumbler, which had a coo unlike that of its fellows, approaching in tone to that of the Laugher: this bird had the habit, to a degree which I never saw equalled in any other pigeon, of often walking with its wings raised and arched in an elegant manner. I need say nothing on the great variability, in almost every breed, in size of body, in colour, in the feathering of the feet, and in the feathers on the back of the head being reversed. But I may mention a remarkable Tumbler[28] exhibited at the Crystal Palace, which had an irregular crest of feathers on its head, somewhat like the tuft on the head of the Polish fowl. Mr. Bult reared by accident a hen Jacobin with the feathers on the thigh so long as to reach the ground, and a cock having, but in a lesser degree, the same peculiarity: from these two birds he bred others similarly characterised, which were exhibited at the Philoperisteron Club. I bred a mongrel pigeon which had fibrous feathers, and the wing and tail-feathers so short and imperfect that the bird could not fly even a foot in height.

There are many singular and inherited peculiarities in the plumage of pigeons: thus Almond-Tumblers do not acquire their perfect mottled feathers until they have moulted three or four times: the Kite-Tumbler is at first brindled black and red with a barred appearance, but when "it throws its nest feathers it becomes almost black, generally with a bluish tail, and a reddish colour on the inner webs of the primary wing feathers."[29] Neu-

[28] Described and figured in the 'Poultry Chronicle,' vol. iii., 1855, p. 82.

[29] 'The Pigeon Book,' by Mr. B. P. Brent, 1859, p. 41.

meister describes a breed of a black colour with white bars on the wing and a white crescent-shaped mark on the breast; these marks are generally rusty-red before the first moult, but after the third or fourth moult they undergo a change; the wing-feathers and the crown of the head likewise then become white or grey.[30]

It is an important fact, and I believe there is hardly an exception to the rule, that the especial characters for which each breed is valued are eminently variable: thus, in the Fantail, the number and direction of the tail-feathers, the carriage of the body, and the degree of trembling are all highly variable points; in Pouters, the degree to which they pout, and the shape of their inflated crops; in the Carrier, the length, narrowness, and curvature of the beak, and the amount of wattle; in Short-faced Tumblers, the shortness of the beak, the prominence of the forehead, and general carriage,[31] and in the Almond Tumbler the colour of the plumage; in common Tumblers, the manner of tumbling; in the Barb, the breadth and shortness of the beak and the amount of eye-wattle; in Runts, the size of body; in Turbits, the frill; and lastly in Trumpeters, the cooing, as well as the size of the tuft of feathers over the nostrils. These, which are the distinctive and selected characters of the several breeds, are all eminently variable.

There is another interesting fact with respect to the character of the different breeds, namely, that they are often most strongly displayed in the male bird. In Carriers, when the males and females are exhibited in separate pens, the wattle is plainly seen to be much more developed in the males, though I have seen a hen Carrier belonging to Mr. Haynes heavily wattled. Mr. Tegetmeier informs me that, in twenty Barbs in Mr. P. H. Jones's possession, the males had generally the largest eye-wattles; Mr. Esquilant also believes in this rule, but Mr. H. Weir, a first-rate judge, entertains some doubt on the subject. Male Pouters distend their crops to a much greater size than do the females; I have, however, seen a hen in the possession of Mr. Evans which pouted excellently; but this is an unusual circumstance. Mr. Harrison Weir, a successful breeder of prize

[30] 'Die Staarhälsige Taube, Das Ganze, &c.,' s. 21, tab. i. fig. 4.

[31] 'A Treatise on the Almond Tumbler,' by J. M. Eaton, 1852, p. 8, et passim.

Fantails, informs me that his cock birds often have a greater number of tail-feathers than the hens. Mr. Eaton asserts[32] that, if a cock and hen Tumbler were of equal merit, the hen would be worth double the money; and as pigeons always pair, so that an equal number of both sexes is necessary for reproduction, this seems to show that high merit is rarer in the female than in the male. In the development of the frill in Turbits, of the hood in Jacobins, of the tuft in Trumpeters, of tumbling in Tumblers, there is no difference between the males and females. I may here add a rather different case, namely, the existence in France[33] of a wine-coloured variety of the Pouter, in which the male is generally chequered with black, whilst the female is never so chequered. Dr. Chapuis also remarks[34] that in certain light-coloured pigeons the males have their feathers striated with black, and these striæ increase in size at each moult, so that the male ultimately becomes spotted with black. With Carriers, the wattle, both on the beak and round the eyes, and with Barbs that round the eyes, goes on increasing with age. This augmentation of character with advancing age, and more especially the difference between the males and females in the above-mentioned several respects, are highly remarkable facts, for there is no sensible difference at any age between the two sexes in the aboriginal rock-pigeon; and rarely any such difference throughout the whole family of the Columbidæ.[35]

Osteological Characters.

In the skeletons of the various breeds there is much variability; and though certain differences occur frequently, and others rarely, in certain breeds, yet none can be said to be absolutely characteristic of any breed. Considering that strongly-marked domestic races have been formed chiefly by man's power

[32] A Treatise, &c., p. 10.

[33] Boitard and Corbié, 'Les Pigeons,' &c., 1824, p. 173.

[34] 'Le Pigeon Voyageur Belge,' 1865, p. 87.

[35] Prof. A. Newton ('Proc. Zoolog. Soc.,' 1865, p. 716) remarks that he knows no species which presents any remarkable sexual distinction; but it is stated ('Naturalist's Library, Birds,' vol. ix. p. 117) that the excrescence at the base of the beak in the *Carpophaga oceanica* is sexual. this, if correct, is an interesting point of analogy with the male Carrier, which has the wattle at the base of its beak so much more developed than in the female. Mr. Wallace informs me that in the sub-family of the Treronidæ the sexes often differ in vividness of colour.

of selection, we ought not to expect to find great and constant differences in the skeleton; for fanciers can neither see, nor do they care for, modifications of structure in the internal framework. Nor ought we to expect changes in the skeletons from changed habits of life; as every facility is given to the most distinct breeds to follow the same habits, and the much modified races are never allowed to wander abroad and procure their own food in various ways. Moreover, I find, on comparing the skeletons of *Columba livia*, *œnas*, *palumbus*, and *turtur*, which are ranked by all systematists in two or three distinct though allied genera, that the differences are extremely slight, certainly less than between the skeletons of some of the most distinct domestic breeds. How far the skeleton of the wild rock-pigeon is constant I have no means of judging, as I have examined only two.

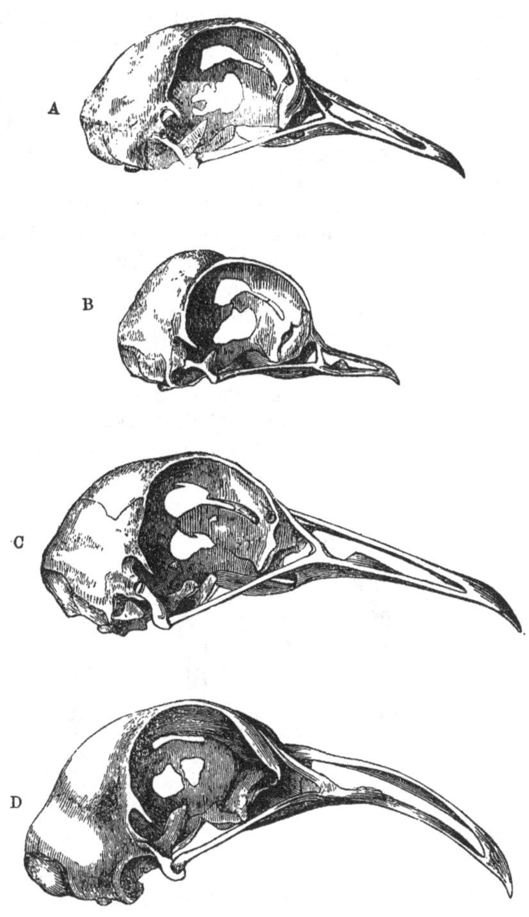

Fig. 24.—Skulls of Pigeons, viewed laterally, of natural size. A. Wild Rock-pigeon, *Columba livia*. B. Short-faced Tumbler. C. English Carrier. D. Bagadotten Carrier.

Skull.—The individual bones, especially those at the base, do not differ in shape. But the whole skull, in its proportions, outline, and relative direction of the bones, differs greatly in some of the breeds, as may be seen by comparing the figures of (A) the wild rock-pigeon, (B) the short-faced tumbler, (C) the English carrier, and (D) the Bagadotten carrier (of Neumeister), all drawn of the natural size and viewed laterally. In the carrier, besides the elongation of the bones of the face, the space between the orbits is proportionally a little narrower than in the rock-pigeon. In the Bagadotten the upper mandible is remarkably arched, and the premaxillary bones are proportionally broader. In the short-faced tumbler the skull is more globular; all the bones of the face are much shortened, and the front of the skull and descending nasal bones are almost perpendicular; the maxillo-jugal arch and premaxillary bones form an almost straight line; the space between the prominent edges of the eye-orbits is depressed. In the barb the premaxillary bones are much shortened, and their anterior portion is thicker than in the rock-pigeon, as is the lower part of the nasal bone. In two nuns the ascending branches of the premaxillaries, near their tips, were somewhat attenuated, and in these birds, as well as in some others, for instance in the spot, the occipital crest over the foramen was considerably more prominent than in the rock-pigeon.

In the lower jaw, the articular surface is proportionally smaller in many breeds than in the rock-pigeon; and the vertical diameter more especially of the outer part of the articular surface is considerably shorter. May not this be accounted for by the lessened use of the jaws, owing to nutritious food having been given during a long period to all highly improved pigeons? In runts, carriers, and barbs (and in a lesser degree in several breeds), the whole side of the jaw near the articular end is bent inwards in a highly remarkable manner; and the superior margin of the ramus, beyond the middle, is reflexed in an equally remarkable manner, as may be seen in the accompanying figures, in comparison with the jaw of the rock-pigeon.

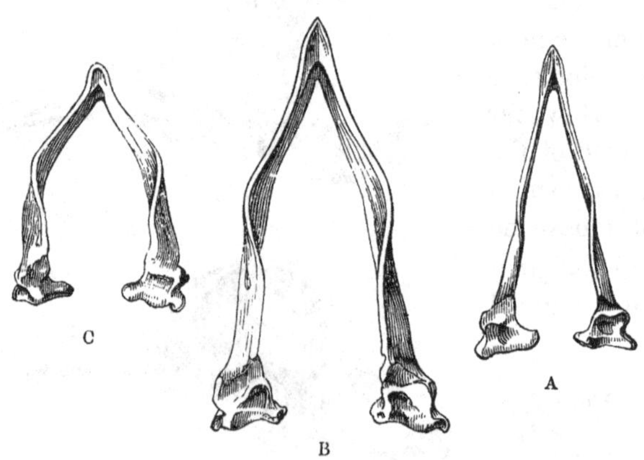

Fig. 25.—Lower jaws, seen from above, of natural size. A. Rock-pigeon. B. Runt. C. Barb.

This reflexion of the upper margin of the lower jaw is plainly connected with the singularly wide gape of the mouth, as has been described in runts, carriers, and barbs. The reflexion is well shown in fig. 26 of the head of a runt seen from above; here a wide open space may be observed on each side, between the edges of the lower jaw and of the premaxillary bones. In the rock-pigeon, and in several domestic breeds, the edges of the lower jaw on each side come close up to the premaxillary bones, so

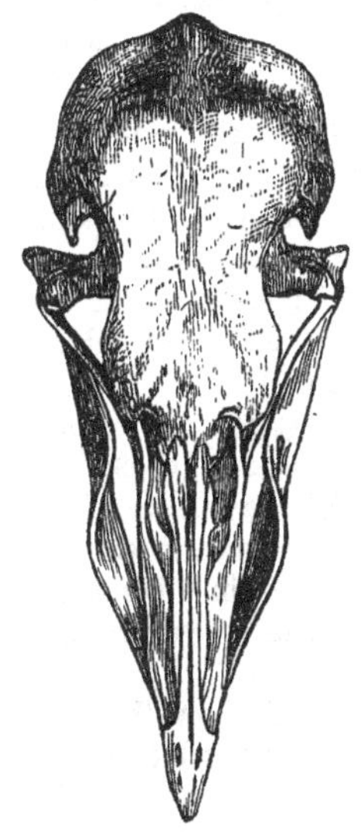

Fig. 26.—Skull of Runt, seen from above, of natural size, showing the reflexed margin of the distal portion of the lower jaw.

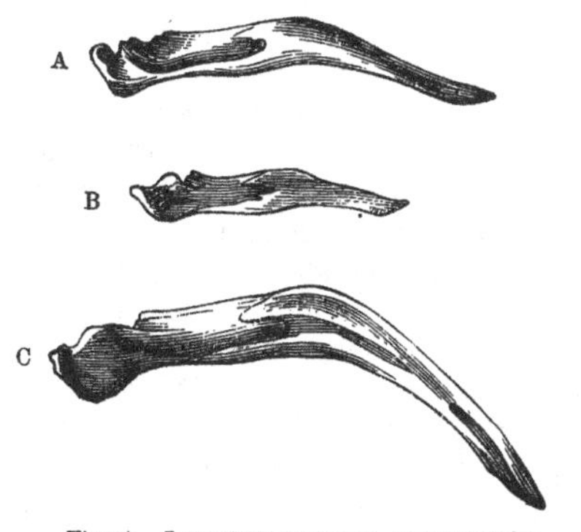

Fig. 27.—Lateral view of jaws, of natural size. A. Rock-pigeon. B. Short-faced Tumbler. C. Bagadotten Carrier.

that no open space is left. The degree of downward curvature of the distal half of the lower jaw also differs to an extraordinary degree in some breeds, as may be seen in the drawings (fig. A) of the rock-pigeon, (B) of the short-faced tumbler, and (C) of the Bagadotten carrier of Neumeister. In some runts the symphysis of the lower jaw is remarkably solid. No one would readily have believed that jaws differing so greatly in the several above-specified points could have belonged to the same species.

Vertebræ.—All the breeds have twelve cervical vertebræ.[36] But in a Bussorah carrier from India, the twelfth vertebra carried a small rib, a quarter of an inch in length, with a perfect double articulation.

The *dorsal vertebræ* are always eight. In the rock-pigeon all eight bear ribs; the eighth rib being very thin, and the seventh having no process. In pouters all the ribs are extremely broad, and, in three out of four skeletons examined by me, the eighth rib was twice or even thrice as broad as in the rock-pigeon; and the seventh pair had distinct processes. In many breeds there are only seven ribs, as in seven out of eight skeletons of various tumblers, and in several skeletons of fantails, turbits, and nuns. In all these breeds the seventh pair was very small, and was destitute of processes, in which respect it differed from the same rib in the rock-pigeon. In one tumbler, and in the Bussorah carrier, even the sixth pair had no process. The hypapophysis of the second dorsal vertebra varies much in development; being sometimes (as in several, but

[36] I am not sure that I have designated the different kinds of vertebræ correctly: but I observe that different anatomists follow in this respect different rules, and, as I use the same terms in the comparison of all the skeletons, this, I hope, will not signify.

not all tumblers) nearly as prominent as that of the third dorsal vertebra; and the two hypapophyses together tend to form an ossified arch. The development of the arch, formed by the hypapophyses of the third and fourth dorsal vertebræ, also varies considerably, as does the size of the hypapophysis of the fifth vertebra.

The rock-pigeon has twelve *sacral vertebræ*; but these vary in number, relative size, and distinctness in the different breeds. In pouters, with their elongated bodies, there are thirteen or even fourteen, and, as we shall immediately see, an additional number of caudal vertebræ. In runts and carriers there is generally the proper number, namely twelve; but in one runt, and in the Bussorah carrier, there were only eleven. In tumblers there are either eleven, twelve, or thirteen sacral vertebræ.

The *caudal vertebræ* are seven in number in the rock-pigeon. In fantails, which have their tails so largely developed, there are either eight or nine, and apparently in one case ten, and they are a little longer than in the rock-pigeon, and their shape varies considerably. Pouters, also, have eight or nine caudal vertebræ. I have seen eight in a nun and jacobin. Tumblers, though such small birds, always have the normal number seven; as have carriers, with one exception, in which there were only six.

The following table will serve as a summary, and will show the most remarkable deviations in the number of the vertebræ and ribs which I have observed:—

	Rock Pigeon.	Pouter, from Mr. Bult.	Tumbler, Dutch Roller.	Bussorah Carrier.
Cervical Vertebræ	12	12	12	12 The 12th bore a small rib.
Dorsal Vertebræ ..	8	8	8	8
„ Ribs	8 The 6th pair with processes, the 7th pair without a process.	8 The 6th and 7th pair with processes.	7 The 6th and 7th pair without processes.	7 The 6th and 7th pair without processes.
Sacral Vertebræ ..	12	14	11	11
Caudal Vertebræ ..	7	8 or 9	7	7
Total Vertebræ ..	39	42 or 43	38	38

The *pelvis* differs very little in any breed. The anterior margin of the ilium, however, is sometimes a little more equally rounded on both sides than in the rock-pigeon, The ischium is also frequently rather more elongated. The obturator-notch is sometimes, as in many tumblers, less developed than in the rock-pigeon. The ridges on the ilium are very prominent in most runts.

In the bones of the extremities I could detect no difference, except in their proportional lengths; for instance, the metatarsus in a pouter was 1·65 inch, and in a short-faced tumbler only ·95 in length; and this is a greater difference than would naturally follow from their differently-sized bodies; but long legs in the pouter, and small feet in the tumbler, are selected points. In some pouters the *scapula* is rather straighter, and in some tumblers it is straighter, with the apex less elongated, than in the rock-pigeon: in the woodcut, fig. 28, the scapulæ of the rock-pigeon (A), and of a short-faced tumbler (B), are given. The processes at the summit of the *coracoid*, which receive the extremities of the furcula, form a more perfect cavity in some tumblers than in the rock-pigeon: in pouters these processes are larger and differently shaped, and the exterior angle of the extremity of the coracoid, which is articulated to the sternum, is squarer.

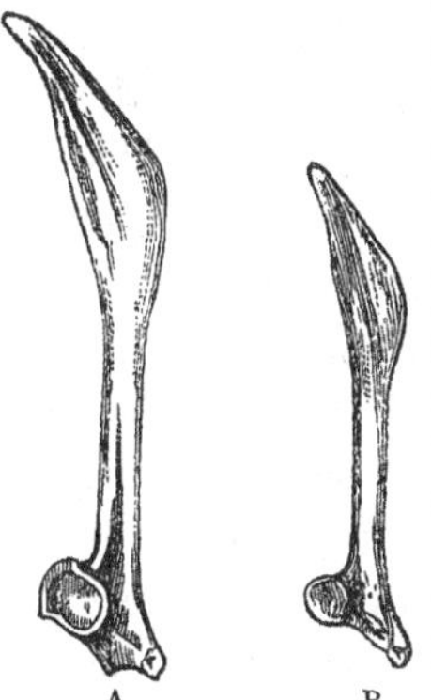

Fig. 28.—Scapulæ, of natural size. A. Rock-pigeon. B. Short-faced Tumbler.

The two arms of the *furcula* in pouters diverge less, proportionally to their length, than in the rock-pigeon; and the symphysis is more solid and pointed. In fantails the degree of divergence of the two arms varies in a remarkable manner. In fig. 29, B and C represent the furculæ of two fantails; and it will be seen that the divergence in B is rather less even than in the furcula of the short-faced, small-sized tumbler (A); whereas the divergence in C equals that in a rock-pigeon, or in the pouter (D), though the latter is a much larger bird. The extremities of the furcula, where articulated to the coracoids, vary considerably in outline.

In the *sternum* the differences in form are slight, except in the size and outline of the perforations, which, both in the larger and lesser sized breeds, are sometimes small. These perforations, also, are sometimes either nearly circular, or elongated, as is often the case with carriers. The posterior perforations occasionally are not complete, being left open posteriorly. The marginal apophyses forming the anterior perforations vary greatly in development. The degree of convexity of the posterior part of the sternum differs much, being sometimes almost perfectly flat. The manubrium is rather more prominent in some individuals than in others, and the pore immediately under it varies greatly in size.

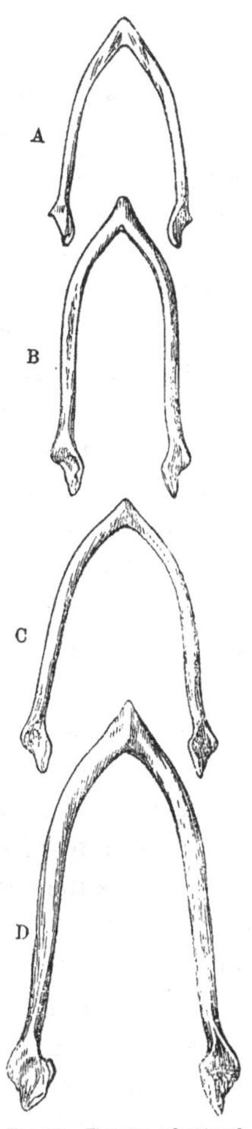

Fig. 29.—Furculæ, of natural size. A. Short-faced Tumbler. B and C. Fantails. D. Pouter.

Correlation of Growth.—By this term I mean that the whole organisation is so connected, that when one part varies, other parts vary; but which of two correlated variations ought to be looked at as the cause and which as the effect, or whether both result from some common cause, we can seldom or never tell. The point of interest for us is that, when fanciers, by the continued selection of slight variations, have largely modified one part, they often unintentionally produce other modifications. For instance, the beak is readily acted on by selection, and, with its increased or diminished length, the tongue increases or diminishes, but not in due proportion; for, in a barb and short-faced tumbler, both of which have very short beaks, the tongue, taking the rock-pigeon as the standard of comparison, was proportionally not shortened enough, whilst in two carriers and in a runt the tongue, proportionally with the beak, was not lengthened enough. Thus, in a first-rate English carrier, in which the beak from the tip to the feathered base was exactly thrice as long as in a first-rate short-faced tumbler, the tongue was only a little more than twice as long. But the tongue varies in length independently of the beak: thus, in a carrier with a beak 1·2 inch in length, the tongue was ·67 in length; whilst in a runt which equalled the carrier in length of body and in stretch of wings from tip to tip, the beak was ·92 whilst the tongue was ·73 of an inch in length, so that the tongue was actually longer than in the carrier with its long beak. The tongue of the runt was also very broad at the root. Of two runts, one had its beak longer by ·23 of an inch, whilst its tongue was shorter by ·14 than in the other.

With the increased or diminished length of the beak the length of the slit forming the external orifice of the nostrils varies, but not in due proportion, for, taking the rock-pigeon as the standard, the orifice in a short-faced tumbler was not shortened in due proportion with its very short beak. On the other hand (and this could not have been anticipated), the orifice in three English carriers, in the Bagadotten carrier, and in a runt (*pigeon cygne*), was longer by above the tenth of an inch than would follow from the length of the beak proportionally with that of the rock-pigeon. In one carrier the orifice of the nostrils was thrice as long as in the rock-pigeon, though in body and length of beak this bird was not nearly double the size of the rock-pigeon. This greatly increased length of the orifice of the nostrils seems to stand partly in correlation with the enlargement of the wattled skin on the upper mandible and over the nostrils; and this is a character which is selected by fanciers. So again, the broad, naked, and wattled skin round the eyes of carriers and barbs is a selected character; and in obvious correlation with this, the eyelids, measured longitudinally, are proportionally more than double the length of those of the rock-pigeon.

The great difference (see woodcut No. 27) in the curvature of the lower jaw in the rock-pigeon, the tumbler, and Bagadotten carrier, stands in obvious relation to the curvature of the upper jaw, and more especially to the angle formed by the maxillo-jugal arch with the premaxillary bones. But in carriers, runts, and barbs the singular reflexion of the upper margin of the middle part of the lower jaw (see woodcut No. 25) is not strictly correlated with the width or divergence (as may be clearly seen in woodcut No. 26) of the premaxillary bones, but with the breadth of the horny and soft parts of the upper mandible, which are always overlapped by the edges of the lower mandible.

In pouters, the elongation of the body is a selected character, and the ribs, as we have seen, have generally become very broad, with the seventh pair furnished with processes; the sacral and caudal vertebræ have been augmented in number; the sternum has likewise increased in length (but not in the depth of the crest) by ·4 of an inch more than would follow from the greater bulk of the body in comparison with that of the rock-pigeon. In fantails, the length and number of the caudal vertebræ have increased. Hence, during the gradual progress of variation and selection, the internal bony frame-work and the external shape of the body have been, to a certain extent, modified in a correlated manner.

Although the wings and tail often vary in length independently of each other, it is scarcely possible to doubt that they generally tend to become elongated or shortened in correlation. This is well seen in jacobins, and still more plainly in runts, some varieties of which have their wings and tail of great length, whilst others have both very short. With jacobins, the remarkable length of the tail and wing-feathers is not a character which is intentionally selected by fanciers; but fanciers have been trying for centuries, at least since the year 1600, to increase the length of the reversed feathers on the neck, so that the hood may more completely enclose the head; and it may be suspected that the increased length of the wing and tail-feathers stands in correlation with the increased length of the neck-feathers. Short-faced tumblers have short wings in nearly due proportion with the reduced size of their bodies; but it is remarkable, seeing that the number of the primary wing-feathers is a constant character in most birds, that these tumblers generally have only nine instead of ten primaries. I have myself observed this in eight birds; and the Original Columbarian Society[37] reduced the standard for bald-head tumblers from ten to nine white flight-feathers, thinking it unfair that a bird which had only nine feathers should be disqualified for a prize because it had not ten *white* flight-feathers. On the other hand, in carriers and runts, which have large bodies and long wings, eleven primary feathers have occasionally been observed.

Mr. Tegetmeier has informed me of a curious and inexplicable case of correlation, namely, that young pigeons of all breeds, which when mature become white, yellow, silver (*i.e.* extremely pale blue), or dun-coloured, are born almost naked; whereas other coloured pigeons are born well clothed with down. Mr. Esquilant, however, has observed that young dun carriers are not so bare as young dun barbs and tumblers. Mr. Tegetmeier has seen two young birds in the same nest, produced from differently coloured parents, which differed greatly in the degree to which they were at first clothed with down.

I have observed another case of correlation which at first sight appears quite inexplicable, but on which, as we shall see in a future chapter, some light can be thrown by the law of homologous parts varying in the same manner. The case is, that, when the feet are much feathered, the roots of the feathers are connected by a web of skin, and apparently in correlation with this the two outer toes become connected for a considerable space by skin. I have observed this in very many

[37] J. M. Eaton's Treatise, edit. 1858, p. 78.

specimens of pouters, trumpeters, swallows, roller-tumblers (likewise observed in this breed by Mr. Brent), and in a lesser degree in other feather-footed pigeons.

The feet of the smaller and larger breeds are of course much smaller or larger than those of the rock-pigeon; but the scutellæ or scales covering the toes and tarsi have not only decreased or increased in size, but likewise in number. To give a single instance, I have counted eight scutellæ on the hind toe of a runt, and only five on that of a short-faced tumbler. With birds in a state of nature the number of the scutellæ on the feet is usually a constant character. The length of the feet and the length of the beak apparently stand in correlation; but as disuse apparently has affected the size of the feet, this case may come under the following discussion.

On the Effects of Disuse.—In the following discussion on the relative proportions of the feet, sternum, furcula, scapulæ, and wings, I may premise, in order to give some confidence to the reader, that my measurements were all made in the same manner, and that all the measurements of the external parts were made without the least intention of applying them to the following purpose.

I measured most of the birds which came into my possession, from the feathered *base* of the beak (the length of beak itself being so variable) to the end of the tail, and to the oil-gland, but unfortunately (except in a few cases) not to the root of the tail; I measured each bird from the extreme tip to tip of wing; and the length of the terminal folded part of the wing, from the extremity of the primaries to the joint of the radius. I measured the feet without the claws, from the end of the middle toe to the end of the hind toe; and the tarsus together with the middle toe. I have taken in every case the mean measurement of two wild rock-pigeons from the Shetland Islands, as the standard of comparison. The following table shows the actual length of the feet in each bird; and the difference between the length which the feet ought to have had according to the size of body of each, in comparison with the size of body and length of feet of the rock-pigeon, calculated (with a few specified exceptions) by the standard of the length of the body from the base of the beak to the oil-gland. I have preferred this standard, owing to the variability of the length of tail. But I have made similar calculations, taking as the standard the length from tip to tip of wing, and likewise in most cases from the base of the beak to the end of the tail; and the result has always been closely similar. To give an example: the first bird in the table, being a short-faced tumbler, is much smaller than the rock-pigeon, and would naturally have shorter feet; but it is found on calculation to have feet too short by ·11 of an inch, in comparison with the feet of the rock-pigeon, relatively to the size of the body in these two birds, as measured from the base of beak to the oil-gland. So again, when this same tumbler and the rock-pigeon were compared by the length of their wings, or by the extreme length of their bodies, the feet of the tumbler were likewise found to be too short in very nearly the same proportion. I am well aware that the measurements pretend to greater accuracy than is possible, but it was less trouble to write down the actual measurements given by the compasses in each case than an approximation.

TABLE I.

Pigeons with their beaks generally shorter than that of the Rock-pigeon, proportionally with the size of their bodies.

Name of Breed.	Actual length of Feet.	Difference between actual and calculated length of feet, in proportion to length of feet and size of body in the Rock-pigeon.	
		Too short by	Too long by
Wild rock-pigeon (mean measurement)	2·02		
Short-faced Tumbler, bald-head	1·57	0·11	..
" " almond	1·60	0·16	..
Tumbler, red magpie	1·75	0·19	..
" red common (by standard to end of tail)	1·85	0·07	..
" common bald-head	1·85	0·18	..
" roller	1·80	0·06	..
Turbit	1·75	0·17	..
"	1·80	0·01	..
"	1·84	0·15	..
Jacobin	1·90	0·02	..
Trumpeter, white	2·02	0·06	..
" mottled	1·95	0·18	..
Fantail (by standard to end of tail)	1·85	0·15	..
" " "	1·95	0·15	..
" crested var. "	1·95	0·0	0·0
Indian Frill-back "	1·80	0·19	..
English Frill-back	2·10	0·03	..
Nun	1·82	0·02	..
Laugher	1·65	0·16	..
Barb	2·00	0·03	..
"	2·00	..	0·03
Spot	1·90	0·02	..
"	1·90	0·07	..
Swallow, red	1·85	0·18	..
" blue	2·00	..	0·03
Pouter	2·42	..	0·11
" German	2·30	..	0·09
Bussorah Carrier	2·17	..	0·09
Number of specimens	28	22	5

TABLE II.

Pigeons with their beaks longer than that of the Rock-pigeon, proportionally with the size of their bodies.

Name of Breed.	Actual length of Feet.	Difference between actual and calculated length of feet, in proportion to length of feet and size of body in the Rock-pigeon.	
		Too short by	Too long by
Wild rock-pigeon (mean measurement)	2·02		
Carrier	2·60	..	0·31
"	2·60	..	0·25
"	2·40	..	0·21
" Dragon	2·25	..	0·06
Bagadotten Carrier	2·80	..	0·56
Scanderoon, white	2·80	..	0·37
" Pigeon cygne	2·85	..	0·29
Runt	2·75	..	0·27
Number of specimens	8	..	8

In these two tables we see in the first column the actual length of the feet in thirty-six birds belonging to various breeds, and in the two other columns we see by how much the feet are too short or too long, according to the size of bird, in comparison with the rock-pigeon. In the first table twenty-two specimens have their feet too short, on an average by a little above the tenth of an inch (viz. ·107); and five specimens have their feet on an average a very little too long, namely, by ·07 of an inch. But some of these latter and exceptional cases can be explained; for instance, with pouters the legs and feet are selected for length, and thus any natural tendency to a diminution in the length of the feet will have been counteracted. In the swallow and barb, when the calculation was made on any standard of comparison excepting the one above used (viz. length of body from base of beak to oil-gland), the feet were found to be too small.

In the second table we have eight birds, with their beaks much longer than in the rock-pigeon, both actually and proportionally with the size of body, and their feet are in an equally marked manner longer, namely, in proportion, on an average by ·29 of an inch. I should here state that in Table I. there are a few partial exceptions to the beak being proportionally shorter than in the rock-pigeon: thus the beak of the English frill-back is just perceptibly longer, and that of the Bussorah carrier of the same length or slightly longer, than in the rock-pigeon. The beaks of spots, swallows, and laughers are only a very little shorter, or of the same proportional length, but slenderer. Nevertheless, these two tables, taken conjointly, indicate pretty plainly some kind of correlation between the length of the beak and the size of the feet. Breeders of cattle and horses believe that there is an analogous connection between the length of the limbs and head; they assert that a race-horse with the head of a dray-horse, or a greyhound with the head of a bulldog, would be a monstrous production. As fancy pigeons are generally kept in small aviaries, and are abundantly supplied with food, they must walk about much less than the wild rock-pigeon; and it may be admitted as highly probable that the reduction in the size of the feet in the twenty-two birds in the first table has been caused by disuse,[38] and that this reduction has acted by correlation on the beaks of the great majority of the birds in Table I. When, on the other hand, the beak has been much elongated by the continued selection of successive slight increments of length, the feet by correlation have likewise become much elongated in comparison with those of the wild rock-pigeon, notwithstanding their lessened use.

As I had taken measures from the end of the middle toe to the heel of the tarsus in the rock-pigeon and in the above thirty-six birds, I have made calculations analogous with those above given, and the result is the same,—namely, that in the short-beaked breeds, with equally few exceptions as in the former case, the middle toe conjointly with the tarsus has decreased in length; whereas in the long-beaked breeds it has increased in length, though not quite so uniformly as in the former case, for the leg in some varieties of the runt varies much in length.

As fancy pigeons are generally confined in aviaries of moderate size, and as even when not confined they do not search for their own food, they must during many generations have used their wings incomparably less than the wild rock-pigeon. Hence it seemed to me probable that all the parts of the skeleton subservient to flight would be found to be reduced in size. With respect to the sternum, I have carefully measured its extreme length in twelve birds of different breeds, and in two wild rock-pigeons from the Shetland Islands. For the proportional comparison I have tried with all twelve birds three standards of measurement, namely, the length from the base of the beak to the oil-gland, to the end of the tail, and from the extreme tip to tip of wings. The result has been in each case nearly the same, the sternum being invariably found to be shorter than in the wild rock-pigeon. I will give only a single table, as calculated by the standard from the base of the beak to the oil-gland; for the result in this case is nearly the mean between the results obtained by the two other standards.

Length of Sternum.

Name of Breed.	Actual Length. Inches.	Too Short by	Name of Breed.	Actual Length. Inches.	Too Short by
Wild Rock-pigeon...	2·55	..	Barb	2·35	0·34
Pied Scanderoon ..	2·80	0·60	Nun	2·27	0·15
Bagadotten Carrier	2·80	0·17	German Pouter ..	2·36	0·54
Dragon	2·45	0·41	Jacobin	2·33	0·22
Carrier	2·75	0·35	English Frill-back ..	2·40	0·43
Short-faced Tumbler	2·05	0·28	Swallow	2·45	0·17

[38] In an analogous, but converse, manner, certain natural groups of the Columbidæ, from being more terrestrial in their habits than other allied groups, have larger feet. *See* Prince Bonaparte's 'Coup-d'œil sur l'Ordre des Pigeons.'

This table shows that in these twelve breeds the sternum is on an average one-third of an inch (exactly ·332) shorter than in the rock-pigeon, proportionally with the size of their bodies; so that the sternum has been reduced by between one-seventh and one-eighth of its entire length; and this is a considerable reduction.

I have also measured in twenty-one birds, including the above dozen, the prominence of the crest of the sternum relatively to its length, independently of the size of the body. In two of the twenty-one birds the crest was prominent in the same relative degree as in the rock-pigeon; in seven it was more prominent; but in five out of these seven, namely, in a fantail, two scanderoons, and two English carriers, this greater prominence may to a certain extent be explained, as a prominent breast is admired and selected by fanciers; in the remaining twelve birds the prominence was less. Hence it follows that the crest exhibits a slight, though uncertain, tendency to become reduced in prominence in a greater degree than does the length of the sternum relatively to the size of body, in comparison with the rock-pigeon.

I have measured the length of the scapula in nine different large and small-sized breeds, and in all the scapula is proportionally shorter (taking the same standard as before) than in the wild rock-pigeon. The reduction in length on an average is very nearly one-fifth of an inch, or about one-ninth of the length of the scapula in the rock-pigeon.

The arms of the furcula in all the specimens which I compared, diverged less, proportionally with the size of body, than in the rock-pigeon; and the whole furcula was proportionally shorter. Thus in a runt, which measured from tip to tip of wings 38½ inches, the furcula was only a very little longer (with the arms hardly more divergent) than in a rock-pigeon which measured from tip to tip 26½ inches. In a barb, which in all its measurements was a little larger than the same rock-pigeon, the furcula was a quarter of an inch shorter. In a pouter, the furcula had not been lengthened proportionally with the increased length of the body. In a short-faced tumbler, which measured from tip to tip of wings 24 inches, therefore only 2½ inches less than the rock-pigeon, the furcula was barely two-thirds of the length of that of the rock-pigeon.

We thus clearly see that the sternum, scapulæ, and furcula are all reduced in proportional length; but when we turn to the wings we find what at first appears a wholly different and unexpected result. I may here remark that I have not picked out specimens, but have used every measurement made by me. Taking the length from the base of beak to the end of the tail as the standard of comparison, I find that, out of thirty-five birds of various breeds, twenty-five have wings of greater, and ten have them of less proportional length, than in the rock-pigeon. But from the frequently correlated length of the tail and wing-feathers, it is better to take as the standard of comparison the length from the base of the beak to the oil-gland; and by this standard, out of twenty-six of the same birds which had been thus measured, twenty-one had wings too long, and only five had them too short. In the twenty-one birds the wings exceeded in length those of the rock-pigeon, on an average, by $1\frac{1}{3}$ inch; whilst in the five birds they were less in length by only ·8 of an inch. As I was much surprised that the wings of closely confined birds should thus so frequently have been increased in length, it occurred to me that it might be solely due to the greater length of the wing-feathers; for this certainly is the case with the jacobin, which has wings of unusual length. As in almost every case I had measured the folded wings, I subtracted the length of this terminal part from that of the expanded wings, and thus I obtained, with a moderate degree of accuracy, the length of the wings from the ends of the two radii, answering from wrist to wrist in our arms. The wings, thus measured in the same twenty-five birds, now gave a widely different result; for they were proportionally with those of the rock-pigeon too short in seventeen birds, and in only eight too long. Of these eight birds, five were long-beaked,[39] and this fact perhaps indicates that there is some correlation between the length of the beak and the length of the bones of the wings, in the same manner as with the feet and tarsi. The shortening of the humerus and radius in the seventeen birds may probably be attributed to disuse, as in the case of the scapulæ and furcula to which the wing-bones are attached;—the lengthening of the wing-feathers, and consequently the expansion of the wings from tip to tip, being, on the other hand, as completely independent of use and disuse as is the growth of the hair or wool on our long-haired dogs or long-woolled sheep.

To sum up: we may confidently admit that the length of the sternum, and frequently the prominence of its crest, the length of the scapulæ and furcula, have all been reduced in size in comparison with the same parts in the rock-pigeon. And I

[39] It perhaps deserves notice that besides these five birds two of the eight were barbs, which, as I have shown, must be classed in the same group with the long-beaked carriers and runts. Barbs may properly be called short-beaked carriers. It would, therefore, appear as if, during the reduction of their beaks, their wings had retained a little of that excess of length which is characteristic of their nearest relations and progenitors.

presume that this may be safely attributed to disuse or lessened exercise. The wings, as measured from the ends of the radii, have likewise been generally reduced in length; but, owing to the increased growth of the wing-feathers, the wings, from tip to tip, are commonly longer than in the rock-pigeon. The feet, as well as the tarsi conjointly with the middle toe, have likewise in most cases become reduced; and this it is probable has been caused by their lessened use; but the existence of some sort of correlation between the feet and beak is shown more plainly than the effects of disuse. We have also some faint indication of a similar correlation between the main bones of the wing and the beak.

Summary on the Points of Difference between the several Domestic Races, and between the individual Birds.—The beak, together with the bones of the face, differ remarkably in length, breadth, shape, and curvature. The skull differs in shape, and greatly in the angle formed by the union of the premaxillary, nasal, and maxillo-jugal bones. The curvature of the lower jaw and the reflexion of its upper margin, as well as the gape of the mouth, differ in a highly remarkable manner. The tongue varies much in length, both independently and in correlation with the length of the beak. The development of the naked, wattled skin over the nostrils and round the eyes varies in an extreme degree. The eyelids and the external orifices of the nostrils vary in length, and are to a certain extent correlated with the degree of development of the wattle. The size and form of the œsophagus and crop, and their capacity for inflation, differ immensely. The length of the neck varies. With the varying shape of the body, the breadth and number of the ribs, the presence of processes, the number of the sacral vertebræ, and the length of the sternum, all vary. The number and size of the coccygeal vertebræ vary, apparently in correlation with the increased size of the tail. The size and shape of the perforations in the sternum, and the size and divergence of the arms of the furcula, differ. The oil-gland varies in development, and is sometimes quite aborted. The direction and length of certain feathers have been much modified, as in the hood of the Jacobin and the frill of the Turbit. The wing and tail feathers generally vary in length together, but sometimes independently of each other and of the size of the body. The number and position of the tail-feathers vary to an unparalleled degree. The primary and secondary wing-feathers occasionally vary in number, apparently in correlation with the length of the wing. The length of the leg and the size of the feet, and, in connection with the latter, the number of the scutellæ, all vary. A web of skin sometimes connects the bases of the two inner toes, and almost invariably the two outer toes when the feet are feathered.

The size of the body differs greatly: a runt has been known to weigh more than five times as much as a short-faced tumbler. The eggs differ in size and shape. According to Parmentier,[40] some races use much straw in building their nests, and others use little; but I cannot hear of any recent corroboration of this statement. The length of time required for hatching the eggs is uniform in all the breeds. The period at which the characteristic plumage of some breeds is acquired, and at which certain changes of colour supervene, differs. The degree to which the young birds are clothed with down when first hatched is different, and is correlated in a singular manner with the future colour of the plumage. The manner of flight, and certain inherited movements, such as clapping the wings, tumbling either in the air or on the ground, and the manner of courting the female, present the most singular differences. In disposition the several races differ. Some races are very silent; others coo in a highly peculiar manner.

Although many different races have kept true in character during several centuries, as we shall hereafter more fully see, yet there is far more individual variability in the truest breeds than in birds in a state of nature. There is hardly any exception to the rule that those characters vary most which are now most valued and attended to by fanciers, and which consequently are now being improved by continued selection. This is indirectly admitted by fanciers when they complain that it is much more difficult to breed high fancy pigeons up to the proper standard of excellence than the so-called toy pigeons, which differ from

[40] Temminck, 'Hist. Nat. Gén. des Pigeons et des Gallinacés,' tom. i., 1813, p. 170.

each other merely in colour; for particular colours when once acquired are not liable to continued improvement or augmentation. Some characters become attached, from quite unknown causes, more strongly to the male than to the female sex; so that we have, in certain races, a tendency towards the appearance of secondary sexual characters,[41] of which the aboriginal rock-pigeon displays not a trace.

[41] This term was used by John Hunter for such differences in structure between the males and females, as are not directly connected with the act of reproduction, as the tail of the peacock, the horns of deer, &c.

CHAPTER VI.

PIGEONS—*continued.*

ON THE ABORIGINAL PARENT-STOCK OF THE SEVERAL DOMESTIC RACES — HABITS OF LIFE — WILD RACES OF THE ROCK-PIGEON — DOVECOT-PIGEONS — PROOFS OF THE DESCENT OF THE SEVERAL RACES FROM COLUMBA LIVIA — FERTILITY OF THE RACES WHEN CROSSED — REVERSION TO THE PLUMAGE OF THE WILD ROCK-PIGEON — CIRCUMSTANCES FAVOURABLE TO THE FORMATION OF THE RACES — ANTIQUITY AND HISTORY OF THE PRINCIPAL RACES — MANNER OF THEIR FORMATION — SELECTION — UNCONSCIOUS SELECTION — CARE TAKEN BY FANCIERS IN SELECTING THEIR BIRDS — SLIGHTLY DIFFERENT STRAINS GRADUALLY CHANGE INTO WELL-MARKED BREEDS — EXTINCTION OF INTERMEDIATE FORMS — CERTAIN BREEDS REMAIN PERMANENT, WHILST OTHERS CHANGE — SUMMARY.

THE differences described in the last chapter between the eleven chief domestic races and between individual birds of the same race, would be of little significance, if they had not all descended from a single wild stock. The question of their origin is therefore of fundamental importance, and must be discussed at considerable length. No one will think this superfluous who considers the great amount of difference between the races, who knows how ancient many of them are, and how truly they breed at the present day. Fanciers almost unanimously believe that the different races are descended from several wild stocks, whereas most naturalists believe that all are descended from the *Columba livia* or rock-pigeon.

Temminck[1] has well observed, and Mr. Gould has made the same remark to me, that the aboriginal parent must have been a species which roosted and built its nest on rocks; and I may add that it must have been a social bird. For all the domestic races are highly social, and none are known to build or habitually to roost on trees. The awkward manner in which some pigeons, kept by me in a summer-house near an old walnut-tree, occasionally alighted on the barer branches, was

[1] Temminck, 'Hist. Nat. Gén. des Pigeons,' &c., tom. i. p. 191.

evident.[2] Nevertheless, Mr. R. Scot Skirving informs me that he often saw crowds of pigeons in Upper Egypt settling on the low trees, but not on the palms, in preference to the mud hovels of the natives. In India Mr. Blyth[3] has been assured that the wild *C. livia*, var. *intermedia*, sometimes roosts in trees. I may here give a curious instance of compulsion leading to changed habits: the banks of the Nile above lat. 28° 30′ are perpendicular for a long distance, so that when the river is full the pigeons cannot alight on the shore to drink, and Mr. Skirving repeatedly saw whole flocks settle on the water, and drink whilst they floated down the stream. These flocks seen from a distance resembled flocks of gulls on the surface of the sea.

If any domestic race had descended from a species which was not social, or which built its nest or roosted in trees,[4] the sharp eyes of fanciers would assuredly have detected some vestige of so different an aboriginal habit. For we have reason to believe that aboriginal habits are long retained under domestication. Thus with the common ass we see signs of its original desert life in its strong dislike to cross the smallest stream of water, and in its pleasure in rolling in the dust. The same strong dislike to cross a stream is common to the camel, which has been domesticated from a very ancient period. Young pigs, though so tame, sometimes squat when frightened, and thus try to conceal themselves even on an open and bare place. Young turkeys, and occasionally even young fowls, when the hen gives the danger-cry, run away and try to hide themselves, like young partridges or pheasants, in order that their mother may take flight, of which she has lost the power. The musk-duck (*Dendrocygna viduata*) in its native

[2] I have heard through Sir C. Lyell from Miss Buckley, that some half-bred carriers kept during many years near London regularly settled by day on some adjoining trees, and, after being disturbed in their loft by their young being taken, roosted on them at night.

[3] 'Annals and Mag. of Nat. Hist.,' 2nd ser., vol. xx., 1857, p. 509; and in a late volume of the Journal of the Asiatic Society.

[4] In works written on the pigeon by fanciers I have sometimes observed the mistaken belief expressed that the species which naturalists call ground-pigeons (in contradistinction to arboreal pigeons) do not perch and build on trees. In these same works wild species resembling the chief domestic races are often said to exist in various parts of the world, but such species are quite unknown to naturalists.

country often perches and roosts on trees,[5] and our domesticated musk-ducks, though such sluggish birds, "are fond of perching on the tops of barns, walls, &c., and, if allowed to spend the night in the hen-house, the female will generally go to roost by the side of the hens, but the drake is too heavy to mount thither with ease."[6] We know that the dog, however well and regularly fed, often buries, like the fox, any superfluous food; and we see him turning round and round on a carpet, as if to trample down grass to form a bed; we see him on bare pavements scratching backwards as if to throw earth over his excrement, although, as I believe, this is never effected even where there is earth. In the delight with which lambs and kids crowd together and frisk on the smallest hillock, we see a vestige of their former alpine habits.

We have therefore good reason to believe that all the domestic races of the pigeon are descended either from some one or from several species which both roosted and built their nests on rocks, and were social in disposition. As only five or six wild species with these habits and making any near approach in structure to the domesticated pigeon are known to exist, I will enumerate them.

Firstly, the *Columba leuconota* resembles certain domestic varieties in its plumage, with the one marked and never-failing difference of a white band which crosses the tail at some distance from the extremity. This species, moreover, inhabits the Himalaya, close to the limit of perpetual snow; and therefore, as Mr. Blyth has remarked, is not likely to have been the parent of our domestic breeds, which thrive in the hottest countries. Secondly, the *C. rupestris*, of Central Asia, which is intermediate[7] between the *C. leuconota* and *livia*; but has nearly the same coloured tail with the former species. Thirdly, the *Columba littoralis* builds and roosts, according to Temminck, on rocks in the Malayan archipelago; it is white, excepting parts of the wing and the tip of the tail, which are black; its legs are livid-coloured, and this is a character not observed in any adult domestic pigeon; but I need not have mentioned this species or the closely-allied *C. luctuosa*, as they in fact belong to the genus Carpophaga. Fourthly, *Columba Guinea*, which ranges from Guinea[8] to the Cape of Good Hope,

---

[5] Sir R. Schomburgk, in 'Journal R. Geograph. Soc.,' vol. xiii., 1844, p. 32.

[6] Rev. E. S. Dixon, 'Ornamental Poultry,' 1848, pp. 63, 66.

[7] Proc. Zoolog. Soc., 1859, p. 400.

[8] Temminck, 'Hist. Nat. Gén. des Pigeons,' tom. i.; also 'Les Pigeons,' par Mad. Knip and Temminck. Bonaparte however, in his 'Coup-d'œil,' believes that two closely allied species are confounded together under this name.

and roosts either on trees or rocks, according to the nature of the country. This species belongs to the genus Strictœnas of Reichenbach, but is closely allied to true Columba; it is to some extent coloured like certain domestic races, and has been said to be domesticated in Abyssinia; but Mr. Mansfield Parkyns, who collected the birds of that country and knows the species, informs me that this is a mistake. Moreover the *C. Guinea* is characterized by the feathers of the neck having peculiar notched tips,—a character not observed in any domestic race. Fifthly, the *Columba œnas* of Europe, which roosts on trees, and builds its nest in holes, either in trees or the ground; this species, as far as external characters go, might be the parent of several domestic races; but, though it crosses readily with the true rock-pigeon, the offspring, as we shall presently see, are sterile hybrids, and of such sterility there is not a trace when the domestic races are intercrossed. It should also be observed that if we were to admit, against all probability, that any of the foregoing five or six species were the parents of some of our domestic pigeons, not the least light would be thrown on the chief differences between the eleven most strongly-marked races.

We now come to the best known rock-pigeon, the *Columba livia*, which is often designated in Europe pre-eminently as the Rock-pigeon, and which naturalists believe to be the parent of all the domesticated breeds. This bird agrees in every essential character with the breeds which have been only slightly modified. It differs from all other species in being of a slaty-blue colour, with two black bars on the wings, and with the croup (or loins) white. Occasionally birds are seen in Faroe and the Hebrides with the black bars replaced by two or three black spots; this form has been named by Brehm[9] *C. amaliæ*, but this species has not been admitted as distinct by other ornithologists. Graba[10] even found a difference between the wing-bars of the same bird in Faroe. Another and rather more distinct form is either truly wild or has become feral on the cliffs of England, and was doubtfully named by Mr. Blyth[11] as *C. affinis*, but is now no longer considered by him as a distinct species. *C. affinis* is rather smaller than the rock-pigeon of the Scottish islands, and has a very different appearance owing to the wing-coverts being chequered with black, with similar marks often extending over the back. The chequering consists of a large black spot on the two sides, but chiefly on the outer side, of each feather. The wing-bars in the true rock-pigeon and in the chequered variety are, in fact, due to similar though larger spots symmetrically crossing the secondary wing-feather and the larger coverts. Hence the chequering arises merely from an extension of these marks to other parts of the plumage. Chequered birds are not confined to the coasts of England; for

---

The *C. leucocephala* of the West Indies is stated by Temminck to be a rock-pigeon; but I am informed by Mr. Gosse that this is an error.

[9] 'Handbuch der Naturgesch. Vogel Deutschlands.'

[10] 'Tagebuch Reise nach Färo,' 1830, s. 62.

[11] 'Annals and Mag. of Nat. Hist.,' vol. xix., 1847, p. 102. This excellent paper on pigeons is well worth consulting.

they were found by Graba at Faroe; and W. Thompson[12] says that at Islay fully half the wild rock-pigeons were chequered. Colonel King, of Hythe, stocked his dovecot with young wild birds which he himself procured from nests at the Orkney Islands; and several specimens, kindly sent to me by him, were all plainly chequered. As we thus see that chequered birds occur mingled with the true rock-pigeon at three distinct sites, namely, Faroe, the Orkney Islands, and Islay, no importance can be attached to this natural variation in the plumage.

Prince C. L. Bonaparte,[13] a great divider of species, enumerates, with a mark of interrogation, as distinct from *C. livia*, the *C. turricola* of Italy, the *C. rupestris* of Daouria, and the *C. Schimperi* of Abyssinia; but these birds differ from *C. livia* in characters of the most trifling value. In the British Museum there is a chequered pigeon, probably the *C. Schimperi* of Bonaparte, from Abyssinia. To these may be added the *C. gymnocyclus* of G. R. Gray from W. Africa, which is slightly more distinct, and has rather more naked skin round the eyes than the rock-pigeon; but from information given me by Dr. Daniell, it is doubtful whether this is a wild bird, for dovecot-pigeons (which I have examined) are kept on the coast of Guinea.

The wild rock-pigeon of India (*C. intermedia* of Strickland) has been more generally accepted as a distinct species. It chiefly differs in the croup being blue instead of snow-white; but as Mr. Blyth informs me, the tint varies, being sometimes albescent. When this form is domesticated chequered birds appear, just as occurs in Europe with the truly wild *C. livia*. Moreover we shall immediately have proof that the blue and white croup is a highly variable character; and Bechstein[14] asserts that with dovecot-pigeons in Germany this is the most variable of all the characters of the plumage. Hence it may be concluded that *C. intermedia* cannot be ranked as specifically distinct from *C. livia*.

In Madeira there is a rock-pigeon which a few ornithologists have suspected to be distinct from *C. livia*. I have examined numerous specimens collected by Mr. E. V. Harcourt and Mr. Mason. They are rather smaller than the rock-pigeon from the Shetland Islands, and their beaks are plainly thinner; but the thickness of the beak varied in the several specimens. In plumage there is remarkable diversity; some specimens are identical in every feather (I speak after actual comparison) with the rock-pigeon of the Shetland Islands; others are chequered, like *C. affinis* from the cliffs of England, but generally to a greater degree, being almost black over the whole back; others are identical with the so-called *C. intermedia* of India in the degree of blueness of the croup; whilst others have this part very pale or very dark blue, and are likewise chequered. So much variability raises a strong suspicion that these birds are domestic pigeons which have become feral.

---

12 'Natural History of Ireland,' Birds, vol. ii. (1850), p. 11. For Graba, *see* previous reference.

13 'Coup-d'œil sur l'Ordre des Pigeons,' Comptes Rendus, 1854-55.

14 'Naturgesch. Deutschlands,' Band iv., 1795, s. 14.

From these facts it can hardly be doubted that *C. livia*, *affinis*, *intermedia*, and the forms marked with an interrogation by Bonaparte, ought all to be included under a single species. But it is quite immaterial whether or not they are thus ranked, and whether some one of these forms or all are the progenitors of the various domestic kinds, as far as any light is thus thrown on the differences between the more strongly-marked races. That common dovecot-pigeons, which are kept in various parts of the world, are descended from one or from several of the above-mentioned wild varieties of *C. livia*, no one who compares them will doubt. But before making a few remarks on dovecot-pigeons, it should be stated that the wild rock-pigeon has been found easy to tame in several countries. We have seen that Colonel King at Hythe stocked his dovecot more than twenty years ago with young wild birds taken at the Orkney Islands, and since this time they have greatly multiplied. The accurate Macgillivray[15] asserts that he completely tamed a wild rock-pigeon in the Hebrides; and several accounts are on record of these pigeons having bred in dovecots in the Shetland Islands. In India, as Captain Hutton informs me, the wild rock-pigeon is easily tamed, and breeds readily with the domestic kind; and Mr. Blyth[16] asserts that wild birds come frequently to the dovecots and mingle freely with their inhabitants. In the ancient 'Ayeen Akbery' it is written that, if a few wild pigeons be taken, "they are speedily joined by a thousand others of their kind."

Dovecot-pigeons are those which are kept in dovecots in a semi-domesticated state; for no special care is taken of them, and they procure their own food, except during the severest weather. In England, and, judging from MM. Boitard and Corbié's work, in France, the common dovecot-pigeon exactly resembles the chequered variety of *C. livia*; but I have seen dovecots brought from Yorkshire, without any trace of chequering, like the wild rock-pigeon of the Shetland Islands. The chequered dovecots from the Orkney Islands, after having been domesticated by Colonel King for more than twenty years, differed slightly from each other in the darkness of their plumage, and in the thickness of their beaks; the thinnest beak being rather thicker than the thickest one in the Madeira birds. In Germany, according to Bechstein, the common dovecot-pigeon is not chequered. In India they often become chequered, and sometimes pied with white; the croup also, as I am informed by Mr. Blyth, becomes nearly white. I have received from Sir J. Brooke some dovecot-pigeons,

---

[15] 'History of British Birds,' vol. i. pp. 275-284. Mr. Andrew Duncan tamed a rock-pigeon in the Shetland Islands. Mr. James Barclay, and Mr. Smith of Uyea Sound, both say that the wild rock-pigeon can be easily tamed; and the former gentleman asserts that the tamed birds breed four times a year. Dr. Lawrence Edmondstone informs me that a wild rock-pigeon came and settled in his dovecot in Balta Sound in the Shetland Islands, and bred with his pigeons; he has also given me other instances of the wild rock-pigeon having been taken young and breeding in captivity.

[16] 'Annals and Mag. of Nat. History,' vol. xix., 1847, p. 103, and vol. for 1857, p. 512.

which originally came from the S. Natunas Islands in the Malay archipelago, and which had been crossed with the Singapore dovecots; they were small, and the darkest variety was extremely like the dark chequered variety with a blue croup from Madeira; but the beak was not so thin, though decidedly thinner than in the rock-pigeon from the Shetland Islands. A dovecot-pigeon sent to me by Mr. Swinhoe from Foochow, in China, was likewise rather small, but differed in no other respect. I have also received, through the kindness of Dr. Daniell, four living dovecot-pigeons from Sierra Leone;[17] these were fully as large as the Shetland rock-pigeon, with even bulkier bodies. In plumage some of them were identical with the Shetland rock-pigeon, but with the metallic tints apparently rather more brilliant; others had a blue croup and resembled the chequered variety of *C. intermedia* of India; and some were so much chequered as to be nearly black. In these four birds the beak differed slightly in length, but in all it was decidedly shorter, more massive, and stronger than in the wild rock-pigeon from the Shetland Islands, or in the English dovecot. When the beaks of these African pigeons were compared with the thinnest beaks of the wild Madeira specimens, the contrast was great; the former being fully one-third thicker in a vertical direction than the latter; so that any one at first would have felt inclined to rank these birds as specifically distinct; yet so perfectly graduated a series could be formed between the above-mentioned varieties, that it was obviously impossible to separate them.

To sum up: the wild *Columba livia*, including under this name *C. affinis*, *intermedia*, and the other still more closely-affined geographical races, has a vast range from the southern coast of Norway and the Faroe Islands to the shores of the Mediterranean, to Madeira and the Canary Islands, to Abyssinia, India, and Japan. It varies greatly in plumage, being in many places chequered with black, and having either a white or blue croup or loins; it varies also slightly in the size of the beak and body. Dovecot-pigeons, which no one disputes are descended from one or more of the above wild forms, present a similar but greater range of variation in plumage, in the size of body, and in the length and thickness of the beak. There seems to be some relation between the croup being blue or white, and the temperature of the country inhabited by both wild and dovecot pigeons; for nearly all the dovecot-pigeons in the northern parts of Europe have a white croup, like that of the wild European

[17] Domestic pigeons of the common kind are mentioned as being pretty numerous in John Barbut's 'Description of the Coast of Guinea' (p. 215), published in 1746; they are said, in accordance with the name which they bear, to have been imported.

rock-pigeon; and nearly all the dovecot-pigeons of India have a blue croup like that of the wild *C. intermedia* of India. As in various countries the wild rock-pigeon has been found easy to tame, it seems extremely probable that the dovecot-pigeons throughout the world are the descendants of at least two and perhaps more wild stocks, but these, as we have just seen, cannot be ranked as specifically distinct.

With respect to the variation of *C. livia,* we may without fear of contradiction go one step further. Those pigeon-fanciers who believe that all the chief races, such as Carriers, Pouters, Fantails, &c., are descended from distinct aboriginal stocks, yet admit that the so-called toy-pigeons, which differ from the rock-pigeon in little except in colour, are descended from this bird. By toy-pigeons are meant such birds as Spots, Nuns, Helmets, Swallows, Priests, Monks, Porcelains, Swabians, Archangels, Breasts, Shields, and others in Europe, and many others in India. It would indeed be as puerile to suppose that all these birds are descended from so many distinct wild stocks as to suppose this to be the case with the many varieties of the gooseberry, heartsease, or dahlia. Yet these pigeons all breed true, and many of them present sub-varieties which likewise truly transmit their character. They differ greatly from each other and from the rock-pigeon in plumage, slightly in size and proportions of body, in size of feet, and in the length and thickness of their beaks. They differ from each other in these respects more than do dovecot-pigeons. Although we may safely admit that the latter, which vary slightly, and that the toy-pigeons, which vary in a greater degree in accordance with their more highly-domesticated condition, are descended from *C. livia,* including under this name the above-enumerated wild geographical races; yet the question becomes far more difficult when we consider the eleven principal races, most of which have been so profoundly modified. It can, however, be shown, by indirect evidence of a perfectly conclusive nature, that these principal races are not descended from so many wild stocks; and if this be once admitted, few will dispute that they are the descendants of *C. livia,* which agrees with them so closely in habits and in most characters, which varies in a state of nature, and which has certainly undergone a considerable amount of variation, as in the toy-pigeons. We shall moreover presently see how eminently favourable circumstances have been for a great amount of modification in the more carefully tended breeds.

The reasons for concluding that the several principal races have not descended from so many aboriginal and unknown stocks may be grouped under the following six heads:—*Firstly*, if the eleven chief races have not arisen from the variation of some one species, together with its geographical races, they must be descended from several extremely distinct aboriginal species; for no amount of crossing between only six or seven wild forms could produce races so distinct as pouters, carriers, runts, fantails, turbits, short-faced tumblers, jacobins, and trumpeters. How could crossing produce, for instance, a pouter or a fantail, unless the two supposed aboriginal parents possessed the remarkable characters of these breeds? I am aware that some naturalists, following Pallas, believe that crossing gives a strong tendency to variation, independently of the characters inherited from either parent. They believe that it would be easier to raise a pouter or fantail pigeon from crossing two distinct species, neither of which possessed the characters of these races, than from any single species. I can find few facts in support of this doctrine, and believe in it only to a limited degree; but in a future chapter I shall have to recur to this subject. For our present purpose the point is not material. The question which concerns us is, whether or not many new and important characters have arisen since man first domesticated the pigeon. On the ordinary view, variability is due to changed conditions of life; on the Pallasian doctrine, variability, or the appearance of new characters, is due to some mysterious effect from the crossing of two species, neither of which possess the characters in question. In some few instances it is credible, though for several reasons not probable, that well-marked races have been formed by crossing; for instance, a barb might perhaps have been formed by a cross between a long-beaked carrier, having large eye-wattles, and some short-beaked pigeon. That many races have been in some degree modified by crossing, and that certain varieties which are distinguished only by peculiar tints have arisen from crosses between differently-coloured varieties, may be admitted as almost certain. On the doctrine, therefore, that the chief races owe their differences to their descent from distinct species, we must admit that at least eight or nine, or more probably a dozen species, all having the same habit of breeding and roosting on rocks and living in society, either now exist somewhere, or formerly existed but have become extinct as wild birds. Considering how carefully wild pigeons have been collected throughout the world, and what conspicuous birds they are, especially when frequenting rocks, it is extremely improbable that eight or nine species, which were long ago domesticated and therefore must have inhabited some anciently known country, should still exist in the wild state and be unknown to ornithologists.

The hypothesis that such species formerly existed, but have become extinct, is in some slight degree more probable. But the extinction of so many species within the historical period is a bold hypothesis, seeing how little influence man has had in exterminating the common rock-pigeon, which agrees in all its habits of life with the domestic races. The *C. livia* now exists and flourishes on the small northern islands of Faroe, on many islands off the coast of Scotland, on Sardinia and the shores of the Mediterranean, and in the centre of India. Fanciers have sometimes imagined that the several supposed parent-species were originally confined to small islands, and thus might readily have been exterminated; but the facts just given do not favour the probability of their extinction, even on small islands. Nor is it probable, from what is known of the distribution of birds, that the islands near Europe should have been inhabited by peculiar species of pigeons; and if we assume that distant oceanic islands were the homes of the supposed parent-species, we must remember that ancient voyages were tediously slow, and that ships were then ill-provided with fresh food, so that it would not have been easy to bring home living birds. I have said ancient voyages, for nearly all the races of the pigeon were known before the year 1600, so that the supposed wild species must have been captured and domesticated before that date.

Secondly.—The doctrine that the chief domestic races have descended from several aboriginal species, implies that several species were formerly so thoroughly domesticated as to breed readily when confined. Although it is easy to tame most wild birds, experience shows us that it is difficult to get them to breed freely under confinement; although it must be owned that this is less difficult with pigeons than with most other birds. During the last two or three hundred years, many birds have been kept in aviaries, but hardly one has been added to our list of thoroughly reclaimed species; yet on the above doctrine we must admit that in ancient times nearly a dozen kinds of pigeons, now unknown in the wild state, were thoroughly domesticated.

Thirdly.—Most of our domesticated animals have run wild in various parts of the world; but birds, owing apparently to their partial loss of the power of flight, less often than quadrupeds. Nevertheless I have met with accounts showing that the common fowl has become feral in South America and perhaps in West Africa, and on several islands: the turkey was at one time almost feral on the banks of the Parana; and the Guinea-fowl has become perfectly wild at Ascension and in Jamaica. In this latter island the peacock, also, "has become a maroon bird." The common duck wanders from its home and becomes almost wild in Norfolk. Hybrids between the common and musk-duck which have become wild have been shot in North America, Belgium, and near the Caspian Sea. The goose is said to have run wild in La Plata. The common dovecot-pigeon has become wild at Juan Fernandez, Norfolk Island, Ascension, probably at Madeira, on the shores of Scotland, and, as is asserted, on the banks of the Hudson in North America.[18] But how different is the case, when we turn

[18] With respect to feral pigeons—for Juan Fernandez, *see* Bertero in 'Annal. des Sc. Nat.,' tom. xxi. p. 351. For Norfolk Island, *see* Rev. E. S. Dixon in the 'Dovecote,' 1851, p. 14, on the authority of Mr. Gould. For Ascension I rely on MS. information given me by Mr. Layard. For the banks of the Hudson, *see* Blyth in 'Annals of Nat. Hist.,' vol. xx., 1857, p. 511. For Scotland, *see* Macgillivray, 'British Birds,' vol. i. p. 275; also Thompson's 'Nat. History of Ireland, Birds,' vol. ii. p. 11. For ducks, *see* Rev. E. S. Dixon, 'Ornamental Poultry,' 1847, p. 122. For the feral hybrids of the common and musk-ducks, *see* Audubon's 'American Ornithology,' and Selys-Longchamp's 'Hybrides dans la Famille des Anatides.' For the goose, Isidore Geoffroy St. Hilaire, 'Hist. Nat. Gén.,' tom. iii. p. 498. For guinea-fowls, *see* Gosse's 'Naturalist's Sojourn in Jamaica,' p. 124; and his 'Birds of Jamaica' for fuller particulars. I saw the wild guinea-fowl in Ascension. For the peacock, *see* 'A Week at

to the eleven chief domestic races of the pigeon, which are supposed by some authors to be descended from so many distinct species! no one has ever pretended that any one of these races has been found wild in any quarter of the world; yet they have been transported to all countries, and some of them must have been carried back to their native homes. On the view that all the races are the product of variation, we can understand why they have not become feral, for the great amount of modification which they have undergone shows how long and how thoroughly they have been domesticated; and this would unfit them for a wild life.

Fourthly.—If it be assumed that the characteristic differences between the various domestic races are due to descent from several aboriginal species, we must conclude that man chose for domestication in ancient times, either intentionally or by chance, a most abnormal set of pigeons; for that species resembling such birds as pouters, fantails, carriers, barbs, short-faced tumblers, turbits, &c., would be in the highest degree abnormal, as compared with all the existing members of the great pigeon-family, cannot be doubted. Thus we should have to believe that man not only formerly succeeded in thoroughly domesticating several highly abnormal species, but that these same species have since all become extinct, or are at least now unknown. This double accident is so extremely improbable that the assumed existence of so many abnormal species would require to be supported by the strongest evidence. On the other hand, if all the races are descended from *C. livia*, we can understand, as will hereafter be more fully explained, how any slight deviation in structure which first appeared would continually be augmented by the preservation of the most strongly marked individuals; and as the power of selection would be applied according to man's fancy, and not for the bird's own good, the accumulated amount of deviation would certainly be of an abnormal nature in comparison with the structure of pigeons living in a state of nature.

I have already alluded to the remarkable fact, that the cha-

Port Royal,' by a competent authority, Mr. R. Hill, p. 42. For the turkey I rely on oral information; I ascertained that they were not Curassows. With respect to fowls I will give the references in the next chapter.

racteristic differences between the chief domestic races are eminently variable: we see this plainly in the great difference in the number of the tail-feathers in the fantail, in the development of the crop in pouters, in the length of the beak in tumblers, in the state of the wattle in carriers, &c. If these characters are the result of successive variations added together by selection, we can understand why they should be so variable: for these are the very parts which have varied since the domestication of the pigeon, and therefore would be likely still to vary; these variations moreover have been recently, and are still being accumulated by man's selection; therefore they have not as yet become firmly fixed.

Fifthly.—All the domestic races pair readily together, and, what is equally important, their mongrel offspring are perfectly fertile. To ascertain this fact I made many experiments, which are given in the note below; and recently Mr. Tegetmeier has made similar experiments with the same result.[19] The accurate Neumeister[20] asserts that when dovecots

[19] I have drawn out a long table of the various crosses made by fanciers between the several domestic breeds, but I do not think it worth publishing. I have myself made for this special purpose many crosses, and all were perfectly fertile. I have united in one bird five of the most distinct races, and with patience I might undoubtedly have thus united all. The case of five distinct breeds being blended together with unimpaired fertility is important, because Gärtner has shown that it is a very general, though not, as he thought, universal rule, that complex crosses between several species are excessively sterile. I have met with only two or three cases of reported sterility in the offspring of certain races when crossed. Von Pistor ('Das Ganze der Feld-taubenzucht,' 1831, s. 15) asserts that the mongrels from barbs and fantails are sterile: I have proved this to be erroneous, not only by crossing these hybrids with several other hybrids of the same parentage, but by the more severe test of pairing brother and sister hybrids *inter se*, and they were *perfectly* fertile. Temminck has stated ('Hist. Nat. Gén. des Pigeons,' tom. i. p. 197) that the turbit or owl will not cross readily with other breeds: but my turbits crossed, when left free, with almond tumblers and with trumpeters; the same thing has occurred (Rev. E. S. Dixon, 'The Dovecot,' p. 107) between turbits and dovecots and nuns. I have crossed turbits with barbs, as has M. Boitard (p. 34), who says the hybrids were very fertile. Hybrids from a turbit and fantail have been known to breed *inter se* (Riedel, Taubenzucht, s. 25, and Bechstein, 'Naturgesch. Deutsch.' B. iv. s. 44). Turbits (Riedel, s. 26) have been crossed with pouters and with jacobins, and with a hybrid jacobin-trumpeter (Riedel, s. 27). The latter author has, however, made some vague statements (s. 22) on the sterility of turbits when crossed with certain other crossed breeds. But I have little doubt that the Rev. E. S. Dixon's explanation of such statements is correct, viz. that individual birds both with turbits and other breeds are occasionally sterile.

[20] 'Das Ganze der Taubenzucht,' s. 18.

are crossed with pigeons of any other breed, the mongrels are extremely fertile and hardy. MM. Boitard and Corbié[21] affirm, after their great experience, that with crossed pigeons the more distinct the breeds, the more productive are their mongrel offspring. I admit that the doctrine first broached by Pallas is highly probable, if not actually proved, namely, that closely allied species, which in a state of nature or when first captured would have been in some degree sterile when crossed, lose this sterility after a long course of domestication; yet when we consider the great difference between such races as pouters, carriers, runts, fantails, turbits, tumblers, &c., the fact of their perfect, or even increased, fertility when intercrossed in the most complicated manner becomes a strong argument in favour of their having all descended from a single species. This argument is rendered much stronger when we hear (I append in a note[22]

[21] 'Les Pigeons,' &c., p. 35.

[22] Domestic pigeons pair readily with the allied *C. oenas* (Bechstein, 'Naturgesch. Deutschlands,' B. iv. s. 3); and Mr. Brent has made the same cross several times in England, but the young were very apt to die at about ten days old; one hybrid which he reared (from *C. oenas* and a male Antwerp carrier) paired with a dragon, but never laid eggs. Bechstein further states (s. 26) that the domestic pigeon will cross with *C. palumbus*, *Turtur risoria*, and *T. vulgaris*, but nothing is said of the fertility of the hybrids, and this would have been mentioned had the fact been ascertained. In the Zoological Gardens (MS. report to me from Mr. James Hunt) a male hybrid from *Turtur vulgaris* and a domestic pigeon "paired with several different species of pigeons and doves, but none of the eggs were good." Hybrids from *C. oenas* and *gymnophthalmos* were sterile. In Loudon's 'Mag. of Nat. Hist.' vol. vii. 1834, p. 154, it is said that a male hybrid (from *Turtur vulgaris* male, and the cream-coloured *T. risoria* female) paired during two years with a female *T. risoria*, and the latter laid many eggs, but all were sterile. MM. Boitard and Corbié ('Les Pigeons,' p. 235) state that the hybrids from these two turtle-doves are invariably sterile both *inter se* and with either pure parent. The experiment was tried by M. Corbié "avec une espèce d'obstination;" and likewise by M. Manduyt, and by M. Vieillot. Temminck also found the hybrids from these two species quite barren. Therefore, when Bechstein ('Naturgesch. Vogel. Deutschlands,' B. 4, s. 101) asserts that the hybrids from these two turtle-doves propagate *inter se* equally well with pure species, and when a writer in the 'Field' newspaper (in a letter dated Nov. 10th, 1858) makes a similar assertion, it would appear that there must be some mistake; though what the mistake is I know not, as Bechstein at least must have known the white *variety* of *T. risoria*: it would be an unparalleled fact if the same two species sometimes produced *extremely* fertile, and sometimes *extremely* barren, offspring. In the MS. report from the Zoological Gardens it is said that hybrids from *Turtur vulgaris* and *suratensis*, and from *T. vulgaris* and *Ectopistes migratorius*, were sterile. Two of the latter male hybrids paired with their pure parents, viz. *Turtur vulgaris* and the Ectopistes, and likewise with *T. risoria* and with *Columba oenas*, and

all the cases which I have collected) that hardly a single well-ascertained instance is known of hybrids between two true species of pigeons being fertile, *inter se*, or even when crossed with one of their pure parents.

Sixthly.—Excluding certain important characteristic differences, the chief races agree most closely both with each other and with *C. livia* in all other respects. As previously observed, all are eminently sociable; all dislike to perch or roost, and refuse to build in trees; all lay two eggs, and this is not a universal rule with the Columbidæ; all, as far as I can hear, require the same time for hatching their eggs; all can endure the same great range of climate; all prefer the same food, and are passionately fond of salt; all exhibit (with the asserted exception of the finnikin and turner, which do not differ much in any other character) the same peculiar gestures when courting the females; and all (with the exception of trumpeters and laughers, which likewise do not differ much in any other character) coo in the same peculiar manner, unlike the voice of any other wild pigeon. All the coloured breeds display the same peculiar metallic tints on the breast, a character far from general with pigeons. Each race presents nearly the same range of variation in colour; and in most of the races we have the same singular correlation between the development of down in the young and the future colour of plumage. All have the proportional length of their toes, and of their primary wing-feathers, nearly the same,—characters which are apt to differ in the several members of the Columbidæ. In those races which present some remarkable deviation of structure, such as in the tail of fantails, crop of pouters, beak of carriers and tumblers, &c., the other parts remain nearly unaltered. Now every naturalist will admit that it would be scarcely possible to pick out a dozen natural species in any Family, which should agree closely in habits and in general structure, and yet should differ greatly in a few cha-

many eggs were produced, but all were barren. At Paris, hybrids have been raised (Isid. Geoffroy Saint Hilaire, 'Hist. Nat. Générale,' tom. iii. p. 180) from *Turtur auritus* with *T. cambayensis* and with *T. suratensis*; but nothing is said of their fertility. At the Zoological Gardens of London the *Goura coronata* and *victoriæ* produced a hybrid, which paired with the pure *G. coronata*, and laid several eggs, but these proved barren. In 1860 *Columba gymnophthalmos* and *maculosa* produced hybrids in these same gardens.

racters alone. This fact is explicable through the doctrine of natural selection; for each successive modification of structure in each natural species is preserved, solely because it is of service; and such modifications when largely accumulated imply a great change in the habits of life, and this will almost certainly lead to other changes of structure throughout the whole organisation. On the other hand, if the several races of the pigeon have been produced by man through selection and variation, we can readily understand how it is that they should still all resemble each other in habits and in those many characters which man has not cared to modify, whilst they differ to so prodigious a degree in those parts which have struck his eye or pleased his fancy.

Besides the points above enumerated, in which all the domestic races resemble *C. livia* and each other, there is one which deserves special notice. The wild rock-pigeon is of a slaty-blue colour; the wings are crossed by two black bars; the croup varies in colour, being generally white in the pigeon of Europe, and blue in that of India; the tail has a black bar close to the end, and the outer webs of the outer tail-feathers are edged with white, except near the tips. These combined characters are not found in any wild pigeon besides *C. livia.* I have looked carefully through the great collection of pigeons in the British Museum, and I find that a dark bar at the end of the tail is common; that the white edging to the outer tail-feathers is not rare; but that the white croup is extremely rare, and the two black bars on the wings occur in no other pigeon, excepting the alpine *C. leuconota* and *C. rupestris* of Asia. Now if we turn to the domestic races, it is highly remarkable, as an eminent fancier, Mr. Wicking, observed to me, that, whenever a blue bird appears in any race, the wings almost invariably show the double black bars.[23] The primary wing-feathers may be white or black, and the whole body may be

[23] There is one exception to the rule, namely in a sub-variety of the swallow of German origin, which is figured by Neumeister, and was shown to me by Mr. Wicking. This bird is blue, but has not the black wing-bars; for our object, however, in tracing the descent of the chief races, this exception signifies the less as the swallow approaches closely in structure to *C. livia.* In many sub-varieties, the black bars are replaced by bars of various colours. The figures given by Neumeister are sufficient to show that, if the wings alone are blue, the black wing-bars appear.

of any colour, but if the wing-coverts alone are blue, the two black bars surely appear. I have myself seen, or acquired trustworthy evidence, as given below,[24] of blue birds with black bars on the wing, with the croup either white or very pale or dark blue, with the tail having a terminal black bar, and with the outer feathers externally edged with white or very pale coloured, in the following races, which, as I carefully observed in each case, appeared to be perfectly pure: namely, in Pouters, Fantails, Tumblers, Jacobins, Turbits, Barbs, Carriers, Runts of three distinct varieties, Trumpeters, Swallows, and in many other toy-pigeons, which, as being closely allied to *C. livia,* are not worth enumerating. Thus we see that, in purely-bred races of every kind known in Europe, blue birds occasionally appear having all the marks which characterise *C. livia,* and which concur in no other wild species. Mr. Blyth, also, has made the same observation with respect to the various domestic races known in India.

Certain variations in the plumage are equally common in the wild *C. livia,* in dovecot-pigeons, and in all the most highly modified races. Thus, in all, the croup varies from white to

[24] I have observed blue birds with all the above-mentioned marks in the following races, which seemed to be perfectly pure, and were shown at various exhibitions. Pouters, with the double black wing-bars, with white croup, dark bar to end of tail, and white edging to outer tail feathers. Turbits, with all these same characters. Fantails, with the same; but the croup in some was bluish or pure blue: Mr. Wicking bred blue fantails from two black birds. Carriers (including the Bagadotten of Neumeister), with all the marks: two birds which I examined had white, and two had blue croups; the white edging to the outer tail-feathers was not present in all. Mr. Corker, a great breeder, assures me that, if black carriers are matched for many successive generations, the offspring become first ash-coloured, and then blue with black wing-bars. Runts of the elongated breed had the same marks, but the croup was pale blue; the outer tail-feathers had white edges. Neumeister figures the great Florence Runt of a blue colour with black bars. Jacobins are very rarely blue, but I have received authentic accounts of at least two instances of the blue variety with black bars having appeared in England: blue jacobins were bred by Mr. Brent from two black birds. I have seen common tumblers, both Indian and English, and short-faced tumblers, of a blue colour, with black wing-bars, with the black bar at the end of the tail, and with the outer tail-feathers edged with white; the croup in all was blue, or extremely pale blue, never absolutely white. Blue barbs and trumpeters seem to be excessively rare; but Neumeister, who may be implicitly trusted, figures blue varieties of both, with black wing-bars. Mr. Brent informs me that he has seen a blue barb; and Mr. H. Weir, as I am informed by Mr. Tegetmeier, once bred a silver (which means very pale blue) barb from two yellow birds.

blue, being most frequently white in Europe, and very generally blue in India.[25] We have seen that the wild *C. livia* in Europe, and dovecots in all parts of the world, often have the upper wing-coverts chequered with black; and all the most distinct races, when blue, are occasionally chequered in precisely the same manner. Thus I have seen Pouters, Fantails, Carriers, Turbits, Tumblers (Indian and English), Swallows, Bald-pates, and other toy-pigeons blue and chequered; and Mr. Esquilant has seen a chequered Runt. I bred from two pure blue Tumblers a chequered bird.

The facts hitherto given refer to the occasional appearance in pure races of blue birds with black wing-bars, and likewise of blue and chequered birds; but it will now be seen that when two birds belonging to distinct races are crossed, neither of which have, nor probably have had during many generations, a trace of blue in their plumage, or a trace of wing-bars and the other characteristic marks, they very frequently produce mongrel offspring of a blue colour, sometimes chequered, with black wing-bars, &c.; or if not of a blue colour, yet with the several characteristic marks more or less plainly developed. I was led to investigate this subject from MM. Boitard and Corbié[26] having asserted that from crosses between certain breeds it is rare to get anything but bisets or dovecot-pigeons, which, as we know, are blue birds with the usual characteristic marks. We shall hereafter see that this subject possesses, independently of our present object, considerable interest, so that I will give the results of my own trials in full. I selected for experiment races which, when pure, very seldom produce birds of a blue colour, or have bars on their wings and tail.

The nun is white, with the head, tail, and primary wing-feathers black; it is a breed which was established as long ago

[25] Mr. Blyth informs me that all the domestic races in India have the croup blue; but this is not invariable, for I possess a very pale blue Simmali pigeon with the croup perfectly white, sent to me by Sir W. Elliot from Madras. A slaty-blue and chequered Nakshi pigeon has some white feathers on the croup alone. In some other Indian pigeons there were a few white feathers confined to the croup, and I have noticed the same fact in a carrier from Persia. The Java fantail (imported into Amoy, and thence sent me) has a perfectly white croup.

[26] 'Les Pigeons,' &c., p. 37.

as the year 1600. I crossed a male nun with a female red common tumbler, which latter variety generally breeds true. Thus neither parent had a trace of blue in the plumage, or of bars on the wing and tail. I should premise that common tumblers are rarely blue in England. From the above cross I reared several young: one was red over the whole back, but with the tail as blue as that of the rock-pigeon; the terminal bar, however, was absent, but the outer feathers were edged with white: a second and third nearly resembled the first, but the tail in both presented a trace of the bar at the end: a fourth was brownish, and the wings showed a trace of the double bar: a fifth was pale blue over the whole breast, back, croup, and tail, but the neck and primary wing-feathers were reddish; the wings presented two distinct bars of a red colour; the tail was not barred, but the outer feathers were edged with white. I crossed this last curiously coloured bird with a black mongrel of complicated descent, namely, from a black barb, a spot, and almond tumbler, so that the two young birds produced from this cross included the blood of five varieties, none of which had a trace of blue or of wing and tail bars: one of the two young birds was brownish-black, with black wing-bars; the other was reddish-dun, with reddish wing-bars, paler than the rest of the body, with the croup pale blue, the tail bluish, with a trace of the terminal bar.

Mr. Eaton[27] matched two short-faced tumblers, namely, a splash cock and kite hen (neither of which are blue or barred), and from the first nest he got a perfect blue bird, and from the second a silver or pale blue bird, both of which, in accordance with all analogy, no doubt presented the usual characteristic marks.

I crossed two male black barbs with two female red spots. These latter have the whole body and wings white, with a spot on the forehead, the tail and tail-coverts red; the race existed at least as long ago as 1676, and now breeds perfectly true, as was known to be the case in the year 1735.[28] Barbs are uniformly-coloured birds, with rarely even a trace of bars on the wing or tail; they are known to breed very true. The mongrels thus raised were black or nearly black, or dark or pale brown,

[27] 'Treatise on Pigeons,' 1858, p. 145.

[28] J. Moore's 'Columbarium,' 1735, in J. M. Eaton's edition, 1852, p. 71.

sometimes slightly piebald with white: of these birds no less than six presented double wing-bars; in two the bars were conspicuous and quite black; in seven some white feathers appeared on the croup; and in two or three there was a trace of the terminal bar to the tail, but in none were the outer tail-feathers edged with white.

I crossed black barbs (of two excellent strains) with purely-bred, snow-white fantails. The mongrels were generally quite black, with a few of the primary wing and tail-feathers white: others were dark reddish-brown, and others snow-white: none had a trace of wing-bars or of the white croup. I then paired together two of these mongrels, namely, a brown and black bird, and their offspring displayed wing-bars, faint, but of a darker brown than the rest of body. In a second brood from the same parents a brown bird was produced, with several white feathers confined to the croup.

I crossed a male dun dragon belonging to a family which had been dun-coloured without wing-bars during several generations, with a uniform red barb (bred from two black barbs); and the offspring presented decided but faint traces of wing-bars. I crossed a uniform red male runt with a white trumpeter; and the offspring had a slaty-blue tail, with a bar at the end, and with the outer feathers edged with white. I also crossed a female black and white chequered trumpeter (of a different strain from the last) with a male almond-tumbler, neither of which exhibited a trace of blue, or of the white croup, or of the bar at end of tail: nor is it probable that the progenitors of these two birds had for many generations exhibited any of these characters, for I have never even heard of a blue trumpeter in this country, and my almond-tumbler was purely bred; yet the tail of this mongrel was bluish, with a broad black bar at the end, and the croup was perfectly white. It may be observed in several of these cases, that the tail first shows a tendency to become by reversion blue; and this fact of the persistency of colour in the tail and tail-coverts [29] will surprise no one who has attended to the crossing of pigeons.

[29] I could give numerous examples; two will suffice. A mongrel, whose four grandparents were a white turbit, white trumpeter, white fantail, and blue pouter, was white all over, except a very few feathers about the head and on the

The last case which I will give is the most curious. I paired a mongrel female barb-fantail with a mongrel male barb-spot; neither of which mongrels had the least blue about them. Let it be remembered that blue barbs are excessively rare; that spots, as has been already stated, were perfectly characterized in the year 1676, and breed perfectly true; this likewise is the case with white fantails, so much so that I have never heard of white fantails throwing any other colour. Nevertheless the offspring from the above two mongrels was of exactly the same blue tint as that of the wild rock-pigeon from the Shetland Islands over the whole back and wings; the double black wing-bars were equally conspicuous; the tail was exactly alike in all its characters, and the croup was pure white; the head, however, was tinted with a shade of red, evidently derived from the spot, and was of a paler blue than in the rock-pigeon, as was the stomach. So that two black barbs, a red spot, and a white fantail, as the four purely-bred grandparents, produced a bird of the same general blue colour, together with every characteristic mark, as in the wild *Columba livia*.

With respect to crossed breeds frequently producing blue birds chequered with black, and resembling in all respects both the dovecot-pigeon and the chequered wild variety of the rock-pigeon, the statement before referred to by MM. Boitard and Corbié would almost suffice; but I will give three instances of the appearance of such birds from crosses in which one alone of the parents or great-grandparents was blue, but not chequered. I crossed a male blue turbit with a snow-white trumpeter, and the following year with a dark, leaden-brown, short faced tumbler; the offspring from the first cross were as perfectly chequered as any dovecot-pigeon; and from the second, so much so as to be nearly as black as the most darkly chequered rock-pigeon from Madeira. Another bird, whose great-grandparents were a white trumpeter, a white fantail, a white red-spot, a red runt, and a blue pouter, was slaty-blue and chequered exactly like a dovecot-pigeon. I may here

wings, but the whole tail and tail-coverts were dark bluish-grey. Another mongrel, whose four grandparents were a red runt, white trumpeter, white fantail, and the same blue pouter, was pure white all over, except the tail and upper tail-coverts, which were pale fawn, and except the faintest trace of double wing-bars of the same pale fawn tint.

add a remark made to me by Mr. Wicking, who has had more experience than any other person in England in breeding pigeons of various colours: namely, that when a blue, or a blue and chequered bird, having black wing-bars, once appears in any race and is allowed to breed, these characters are so strongly transmitted that it is extremely difficult to eradicate them.

What, then, are we to conclude from this tendency in all the chief domestic races, both when purely bred and more especially when intercrossed, to produce offspring of a blue colour, with the same characteristic marks, varying in the same manner, as in *Columba livia?* If we admit that these races have all descended from *C. livia*, no breeder will doubt that the occasional appearance of blue birds thus characterised is accounted for on the well-known principle of "throwing back" or reversion. Why crossing should give so strong a tendency to reversion, we do not with certainty know; but abundant evidence of this fact will be given in the following chapters. It is probable that I might have bred even for a century pure black barbs, spots, nuns, white fantails, trumpeters, &c., without obtaining a single blue or barred bird; yet by crossing these breeds I reared in the first and second generation, during the course of only three or four years, a considerable number of young birds, more or less plainly coloured blue, and with most of the characteristic marks. When black and white, or black and red birds, are crossed, it would appear that a slight tendency exists in both parents to produce blue offspring, and that this, when combined, overpowers the separate tendency in either parent to produce black, or white, or red offspring.

If we reject the belief that all the races of the pigeon are the modified descendants of *C. livia*, and suppose that they are descended from several aboriginal stocks, then we must choose between the three following assumptions: firstly, that at least eight or nine species formerly existed which were aboriginally coloured in various ways, but have since varied in so exactly the same manner as to assume the colouring of *C. livia*; but this assumption throws not the least light on the appearance of such colours and marks when the races are crossed. Or secondly, we may assume that the aboriginal species were all coloured blue, and had the wing-bars and other characteristic marks of *C. livia*,—a supposition which is highly improbable, as besides this one species no existing member of the Columbidæ presents these combined characters; and it would not be possible to find any other instance of several species identical in plumage, yet as different in important points of structure as are pouters, fantails, carriers, tumblers, &c. Or lastly, we may assume that all the races, whether descended from *C. livia* or from several aboriginal species, although they have been bred with so much care and are so highly valued by fanciers, have all been crossed within a dozen or score of generations with *C. livia*, and have thus acquired their tendency to produce blue birds with the several characteristic marks. I have said that it must be assumed that each race has been crossed with *C. livia* within a dozen, or, at the utmost, within a score of generations; for there is no reason to believe that crossed offspring ever revert to one of their ancestors when removed by a greater number of generations. In a breed which has been crossed only once, the tendency to reversion will naturally become less and less in the succeeding generations, as in each there will be less and less of the blood of the foreign breed; but when there has been no cross with a distinct breed, and there is a tendency in both parents to revert to some long-lost character, this tendency, for all that we can see to the contrary, may be transmitted undiminished for an indefinite number of generations. These two distinct cases of reversion are often confounded together by those who have written on inheritance.

Considering, on the one hand, the improbability of the three assumptions which have just been discussed, and, on the other hand, how simply the facts are explained on the principle of reversion, we may conclude that the occasional appearance in all the races, both when purely bred and more especially when crossed, of blue birds, sometimes chequered, with double wing-bars, with white or blue croups, with a bar at the end of the tail, and with the outer tail-feathers edged with white, affords an argument of the greatest weight in favour of the view that all are descended from *Columba livia*, including under this name the three or four wild varieties or sub-species before enumerated.

To sum up the six foregoing arguments, which are opposed to the belief that the chief domestic races are the descendants of at least eight or nine or perhaps a dozen species; for the crossing of any less number would not yield the characteristic differences between the several races. *Firstly*, the improbability that so many species should still exist somewhere, but be unknown to ornithologists, or that they should have become within the historical period extinct, although man has had so little influence in exterminating the wild *C. livia*. *Secondly*, the improbability of man in former times having thoroughly domesticated and rendered fertile under confinement so many species. *Thirdly*, these supposed species having nowhere become feral. *Fourthly*, the extraordinary fact that man should, intentionally or by chance, have chosen for domestication several species, extremely abnormal in character; and furthermore, the points of structure which render these supposed species so abnormal being now highly variable. *Fifthly*, the fact of all the races, though differing in many important points of structure, producing perfectly fertile mongrels; whilst all the hybrids which have been produced between even closely allied species in the pigeon-family are sterile. *Sixthly*, the remarkable statements just given on the tendency in all the races, both when purely bred and when crossed, to revert in numerous minute details of colouring to the character of the wild rock-pigeon, and to vary in a similar manner. To these arguments may be added the extreme improbability that a number of species formerly existed, which differed greatly from each other in some few points, but which resembled each other as closely as do the domestic races in other points of structure, in voice, and in all their habits of life. When these several facts and arguments are fairly taken into consideration, it would require an overwhelming amount of evidence to make us admit that the chief domestic races are descended from several aboriginal stocks; and of such evidence there is absolutely none.

The belief that the chief domestic races are descended from several wild stocks no doubt has arisen from the apparent improbability of such great modifications of structure having been effected since man first domesticated the rock-pigeon. Nor am I surprised at any degree of hesitation in admitting their common origin: formerly, when I went into my aviaries and watched such birds as pouters, carriers, barbs, fantails, and short-faced tumblers, &c., I could not persuade myself that they had all descended from the same wild stock, and that man had consequently in one sense created these remarkable modifications. Therefore I have argued the question of their origin at great, and, as some will think, superfluous length.

Finally, in favour of the belief that all the races are descended from a single stock, we have in *Columba livia* a still existing and widely distributed species, which can be and has been domesticated in various countries. This species agrees in most points of structure and in all its habits of life, as well as occasionally in every detail of plumage, with the several domestic races. It breeds freely with them, and produces fertile offspring. It varies in a state of nature,[30] and still more so when semi-domesticated, as shown by comparing the Sierra Leone pigeons with those of India, or with those which apparently have run wild in Madeira. It has undergone a still greater amount of variation in the case of the numerous toy-pigeons, which no one supposes to be descended from distinct species; yet some of these toy-pigeons have transmitted their character truly for centuries. Why, then, should we hesitate to believe in that greater amount of variation which is necessary for the production of the eleven chief races? It should be borne in mind that in two of the most strongly-marked races, namely, carriers and short-faced tumblers, the extreme forms can be connected with the parent-species by graduated differences not greater than those which may be observed between the dovecot-pigeons inhabiting different countries, or between the various kinds of toy-pigeons,—gradations which must certainly be attributed to variation.

That circumstances have been eminently favourable for the modification of the pigeon through variation and selection will now be shown. The earliest record, as has been pointed out to me by Professor Lepsius, of pigeons in a domesticated condition, occurs in the fifth Egyptian dynasty, about

[30] It deserves notice, as bearing on the general subject of variation, that not only *C. livia* presents several wild forms, regarded by some naturalists as species and by others as sub-species or as mere varieties, but that the species of several allied genera are in the same predicament. This is the case, as Mr. Blyth has remarked to me, with Treron, Palumbus, and Turtur.

3000 B.C.;[31] but Mr. Birch, of the British Museum, informs me that the pigeon appears in a bill of fare in the previous dynasty. Domestic pigeons are mentioned in Genesis, Leviticus, and Isaiah.[32] In the time of the Romans, as we hear from Pliny,[33] immense prices were given for pigeons; "nay, they are come to this pass, that they can reckon up their pedigree and race." In India, about the year 1600, pigeons were much valued by Akber Khan: 20,000 birds were carried about with the court, and the merchants brought valuable collections. "The monarchs of Iran and Turan sent him some very rare breeds. His Majesty," says the courtly historian, "by crossing the breeds, which method was never practised before, has improved them astonishingly."[34] Akber Khan possessed seventeen distinct kinds, eight of which were valuable for beauty alone. At about this same period of 1600 the Dutch, according to Aldrovandi, were as eager about pigeons as the Romans had formerly been. The breeds which were kept during the fifteenth century in Europe and in India apparently differed from each other. Tavernier, in his Travels in 1677, speaks, as does Chardin in 1735, of the vast number of pigeon-houses in Persia; and the former remarks that, as Christians were not permitted to keep pigeons, some of the vulgar actually turned Mahometans for this sole purpose. The Emperor of Morocco had his favourite keeper of pigeons, as is mentioned in Moore's treatise, published 1737. In England, from the time of Willughby in 1678 to the present day, as well as in Germany and in France, numerous treatises have been published on the pigeon. In India, about a hundred years ago, a Persian treatise was written; and the writer thought it no light affair, for he begins with a solemn invocation, "in the name of God, the gracious and merciful." Many large towns, in Europe and the United States, now have their societies of devoted pigeon-fanciers: at present there are three such societies in London. In India, as I hear from

[31] 'Denkmaler,' Abth. ii. Bl. 70.

[32] The 'Dovecote,' by the Rev. E. S. Dixon, 1851, pp. 11-13. Adolphe Pictet (in his 'Les Origines Indo-Européennes,' 1859, p. 399) states that there are in the ancient Sanscrit language between 25 and 30 names for the pigeon, and other 15 or 16 Persian names; none of these are common to the European languages. This fact indicates the antiquity of the domestication in the East of the pigeon.

[33] English translation, 1601, book x. ch. xxxvii.

[34] 'Ayeen Akbery,' translated by F. Gladvin, 4to. edit., vol. i. p. 270.

Mr. Blyth, the inhabitants of Delhi and of some other great cities are eager fanciers. Mr. Layard informs me that most of the known breeds are kept in Ceylon. In China, according to Mr. Swinhoe of Amoy, and Dr. Lockhart of Shangai, carriers, fantails, tumblers, and other varieties are reared with care, especially by the bonzes or priests. The Chinese fasten a kind of whistle to the tail-feathers of their pigeons, and as the flock wheels through the air they produce a sweet sound. In Egypt the late Abbas Pacha was a great fancier of fantails. Many pigeons are kept at Cairo and Constantinople, and these have lately been imported by native merchants, as I hear from Sir W. Elliot, into Southern India, and sold at high prices.

The foregoing statements show in how many countries, and during how long a period, many men have been passionately devoted to the breeding of pigeons. Hear how an enthusiastic fancier at the present day writes: "If it were possible for noblemen and gentlemen to know the amazing amount of solace and pleasure derived from Almond Tumblers, when they begin to understand their properties, I should think that scarce any nobleman or gentleman would be without their aviaries of Almond Tumblers."[35] The pleasure thus taken is of paramount importance, as it leads amateurs carefully to note and preserve each slight deviation of structure which strikes their fancy. Pigeons are often closely confined during their whole lives; they do not partake of their naturally varied diet; they have often been transported from one climate to another; and all these changes in their conditions of life would be likely to cause variability. Pigeons have been domesticated for nearly 5000 years, and have been kept in many places, so that the numbers reared under domestication must have been enormous; and this is another circumstance of high importance, for it obviously favours the chance of rare modifications of structure occasionally appearing. Slight variations of all kinds would almost certainly be observed, and, if valued, would, owing to the following circumstances, be preserved and propagated with unusual facility. Pigeons, differently from any other domesticated animal, can easily be mated for life, and, though kept with other pigeons, they rarely prove unfaithful to each other. Even when the

[35] J. M. Eaton, 'Treatise on the Almond Tumbler,' 1851; Preface, p. vi.

male does break his marriage-vow, he does not permanently desert his mate. I have bred in the same aviaries many pigeons of different kinds, and never reared a single bird of an impure strain. Hence a fancier can with the greatest ease select and match his birds. He will also soon see the good results of his care; for pigeons breed with extraordinary rapidity. He may freely reject inferior birds, as they serve at an early age as excellent food. To sum up, pigeons are easily kept, paired, and selected; vast numbers have been reared; great zeal in breeding them has been shown by many men in various countries; and this would lead to their close discrimination, and to a strong desire to exhibit some novelty, or to surpass other fanciers in the excellence of already established breeds.

History of the principal Races of the Pigeon.[36]

Before discussing the means and steps by which the chief races have been formed, it will be advisable to give some historical details, for more is known of the history of the pigeon, little though this be, than of any other domesticated animal. Some of the cases are interesting as proving how long domestic varieties may be propagated with exactly the same or nearly the same characters; and other cases are still more interesting as showing how slowly but steadily races have been greatly modified during successive generations. In the last chapter I stated that Trumpeters and Laughers, both so remarkable for their voices, seem to have been perfectly characterized in 1735; and Laughers were apparently known in India before the year 1600. Spots in 1676, and Nuns in the time of Aldrovandi, before 1600, were coloured exactly as they now are. Common Tumblers and Ground Tumblers exhibited in India, before the year 1600, the same extraordinary peculiarities of flight as at the present day, for they are well described in the 'Ayeen Akbery.' These breeds may all have existed for a much longer period; we know only that they were perfectly characterized at the dates above given. The *average* length of life of the domestic pigeon is probably about five or six years; if so, some of these races have retained their character perfectly for at least forty or fifty generations.

Pouters.—These birds, as far as a very short description serves for comparison, appear to have been well characterized in Aldrovandi's time,[37] before the year 1600. Length of body and length of leg are at the present time the two chief points of excellence. In 1735 Moore said (see Mr. J. M. Eaton's edition)—and Moore was a first-rate fancier—that he once saw a bird with

---

[36] As in the following discussion I often speak of the present time, I should state that this chapter was completed in the year 1858.

[37] 'Ornithologie,' 1600, vol. ii. p. 360.

a body 20 inches in length, "though 17 or 18 inches is reckoned a very good length;" and he has seen the legs very nearly 7 inches in length, yet a leg 6½ or 6¾ long "must be allowed to be a very good one." Mr. Bult, the most successful breeder of Pouters in the world, informs me that at present (1858) the standard length of the body is not less than 18 inches; but he has measured one bird 19 inches in length, and has heard of 20 and 22 inches, but doubts the truth of these latter statements. The standard length of the leg is now 7 inches, but Mr. Bult has recently measured two of his own birds with legs 7½ long. So that in the 123 years which have elapsed since 1735 there has been hardly any increase in the standard length of the body; 17 or 18 inches was formerly reckoned a very good length, and now 18 inches is the minimum standard; but the length of leg seems to have increased, as Moore never saw one quite 7 inches long; now the standard is 7, and two of Mr. Bult's birds measured 7½ inches in length. The extremely slight improvement in Pouters, except in the length of the leg, during the last 123 years, may be partly accounted for by the neglect which they suffered, as I am informed by Mr. Bult, until within the last 20 or 30 years. About 1765[38] there was a change of fashion, stouter and more feathered legs being preferred to thin and nearly naked legs.

Fantails.—The first notice of the existence of this breed is in India, before the year 1600, as given in the 'Ayeen Akbery;'[39] at this date, judging from Aldrovandi, the breed was unknown in Europe. In 1677 Willughby speaks of a Fantail with 26 tail-feathers; in 1735 Moore saw one with 36 feathers; and in 1824 MM. Boitard and Corbié assert that in France birds can easily be found with 42 tail-feathers. In England, the number of the tail-feathers is not at present so much regarded as their upward direction and expansion. The general carriage of the bird is likewise now much regarded. The old descriptions do not suffice to show whether in these latter respects there has been much improvement; but if fantails had formerly existed with their heads and tails touching each other, as at the present time, the fact would almost certainly have been noticed. The Fantails which are now found in India probably show the state of the race, as far as carriage is concerned, at the date of their introduction into Europe; and some, said to have been brought from Calcutta, which I kept alive, were in a marked manner inferior to our exhibition birds. The Java Fantail shows the same difference in carriage; and although Mr. Swinhoe has counted 18 and 24 tail-feathers in his birds, a first-rate specimen sent to me had only 14 tail-feathers.

Jacobins.—This breed existed before 1600, but the hood, judging from the figure given by Aldrovandi, did not enclose the head nearly so perfectly as at present: nor was the head then white; nor were the wings and tail so long, but this last character might have been overlooked by the rude artist. In Moore's time, in 1735, the Jacobin was considered the

[38] 'A Treatise on Domestic Pigeons,' dedicated to Mr. Mayor, 1765. Preface, p. xiv.

[39] Mr. Blyth has given a translation of part of the 'Ayeen Akbery' in 'Annals and Mag. of Nat. Hist.,' vol. xix., 1847, p. 104.

smallest kind of pigeon, and the bill is said to be very short. Hence either the Jacobin, or the other kinds with which it was then compared, must have been since considerably modified; for Moore's description (and it must be remembered that he was a first-rate judge) is clearly not applicable, as far as size of body and length of beak are concerned, to our present Jacobins. In 1795, judging from Bechstein, the breed had assumed its present character.

Turbits.—It has generally been supposed by the older writers on pigeons, that the Turbit is the Cortbeck of Aldrovandi; but if this be the case, it is an extraordinary fact that the characteristic frill should not have been noticed. The beak, moreover, of the Cortbeck is described as closely resembling that of the Jacobin, which shows a change in the one or the other race. The Turbit, with its characteristic frill and bearing its present name, is described by Willughby in 1677; and the bill is said to be like that of the bullfinch,—a good comparison, but now more strictly applicable to the beak of the Barb. The sub-breed called the Owl was well known in Moore's time, in 1735.

Tumblers.—Common Tumblers, as well as Ground Tumblers, perfect as far as tumbling is concerned, existed in India before the year 1600; and at this period diversified modes of flight, such as flying at night, the ascent to a great height, and manner of descent, seem to have been much attended to, as at the present time, in India. Belon[40] in 1555 saw in Paphlagonia what he describes as "a very new thing, viz. pigeons which flew so high in the air that they were lost to view, but returned to their pigeon-house without separating." This manner of flight is characteristic of our present Tumblers, but it is clear that Belon would have mentioned the act of tumbling if the pigeons described by him had tumbled. Tumblers were not known in Europe in 1600, as they are not mentioned by Aldrovandi, who discusses the flight of pigeons. They are briefly alluded to by Willughby, in 1687, as small pigeons "which show like footballs in the air." The short-faced race did not exist at this period, as Willughby could not have overlooked birds so remarkable for their small size and short beaks. We can even trace some of the steps by which this race has been produced. Moore in 1735 enumerates correctly the chief points of excellence, but does not give any description of the several sub-breeds; and from this fact Mr. Eaton infers[41] that the short-faced Tumbler had not then come to full perfection. Moore even speaks of the Jacobin as being the smallest pigeon. Thirty years afterwards, in 1765, in the Treatise dedicated to Mayor, short-faced Almond Tumblers are fully described, but the author, an excellent fancier, expressly states in his Preface (p. xiv.) that, "from great care and expense in breeding them, they have arrived to so great perfection and are so different from what they were 20 or 30 years past, that an old fancier would have condemned them for no other reason than because they are not like what used to be thought good when he was in the fancy before."

---

[40] 'L'Hist de la Nature des Oiseaux,' p. 314.

[41] 'Treatise on Pigeons,' 1852, p. 64.

Hence it would appear that there was a rather sudden change in the character of the short-faced Tumbler at about this period; and there is reason to suspect that a dwarfed and half-monstrous bird, the parent-form of the several short-faced sub-breeds, then appeared. I suspect this because short-faced Tumblers are born with their beaks (ascertained by careful measurement) as short, proportionally with the size of their bodies, as in the adult bird; and in this respect they differ greatly from all other breeds, which slowly acquire during growth their various characteristic qualities.

Since the year 1765 there has been some change in one of the chief characters of the short-faced Tumbler, namely, in the length of the beak. Fanciers measure the "head and beak" from the tip of the beak to the front corner of the eyeball. About the year 1765 a "head and beak" was considered good,[42] which, measured in the usual manner, was $\frac{7}{8}$ of an inch in length; now it ought not to exceed $\frac{5}{8}$ of an inch; "it is however possible," as Mr. Eaton candidly confesses, "for a bird to be considered as pleasant or neat even at $\frac{6}{8}$ of an inch, but exceeding that length it must be looked upon as unworthy of attention." Mr. Eaton states that he has never seen in the course of his life more than two or three birds with the "head and beak" not exceeding half an inch in length; "still I believe in the course of a few years that the head and beak will be shortened, and that half-inch birds will not be considered so great a curiosity as at the present time." That Mr. Eaton's opinion deserves attention cannot be doubted, considering his success in winning prizes at our exhibitions. Finally in regard to the Tumbler it may be concluded from the facts above given that it was originally introduced into Europe, probably first into England, from the East; and that it then resembled our common English Tumbler, or more probably the Persian or Indian Tumbler, with a beak only just perceptibly shorter than that of the common dovecot-pigeon. With respect to the short-faced Tumbler, which is not known to exist in the East, there can hardly be a doubt that the whole wonderful change in the size of the head, beak, body, and feet, and in general carriage, has been produced during the last two centuries by continued selection, aided probably by the birth of a semi-monstrous bird somewhere about the year 1750.

Runts.—Of their history little can be said. In the time of Pliny the pigeons of Campania were the largest known; and from this fact alone some authors assert that they were Runts. In Aldrovandi's time, in 1600, two sub-breeds existed; but one of them, the short-beaked, is now extinct in Europe.

Barbs.—Notwithstanding statements to the contrary, it seems to me impossible to recognise the barb in Aldrovandi's descriptions and figures; four breeds, however, existed in the year 1600 which were evidently allied both to Barbs and Carriers. To show how difficult it is to recognise some of the breeds described by Aldrovandi, I will give the different opinions in regard to the above four kinds, named by him *C. Indica*, *Cretensis*, *Gutturosa*, and *Persica*. Willughby thought that the *Columba Indica* was a

[42] J. M. Eaton's 'Treatise on the Breeding and Managing of the Almond Tumbler,' 1851. Compare p. v. of Preface, p. 9, and p. 32.

Turbit, but the eminent fancier Mr. Brent believes that it was an inferior Barb: *C. Cretensis*, with a short beak and a swelling on the upper mandible, cannot be recognised: *C.* (falsely called) *gutturosa*, which from its *rostrum, breve, crassum, et tuberosum* seems to me to come nearest to the Barb, Mr. Brent believes to be a Carrier; and lastly, the *C. Persica et Turcica*, Mr. Brent thinks, and I quite concur with him, was a short-beaked Carrier with very little wattle. In 1687 the Barb was known in England, and Willughby describes the beak as like that of the Turbit; but it is not credible that his Barb should have had a beak like that of our present birds, for so accurate an observer could not have overlooked its great breadth.

English Carrier.—We may look in vain in Aldrovandi's work for any bird resembling our prize Carriers; the *C. Persica et Turcica* of this author comes the nearest, but is said to have had a short thick beak; therefore it must have approached in character a Barb, and have differed greatly from our Carriers. In Willughby's time, in 1677, we can clearly recognise the Carrier, but he adds, "the bill is not short, but of a moderate length," a description which no one would apply to our present Carriers, so conspicuous for the extraordinary length of their beaks. The old names given in Europe to the Carrier, and the several names now in use in India, indicate that Carriers originally came from Persia; and Willughby's description would perfectly apply to the Bussorah Carrier as it now exists in Madras. In later times we can partially trace the progress of change in our English Carriers: Moore in 1735 says "an inch and a half is reckoned a long beak, though there are very good Carriers that are found not to exceed an inch and a quarter." These birds must have resembled, or perhaps been a little superior to, the Carriers, previously described, which are now found in Persia. In England at the present day "there are," as Mr. Eaton[43] states, "beaks that would measure (from edge of eye to tip of beak) one inch and three-quarters, and some few even two inches in length."

From theseish torical details we see that nearly all the chief domestic races existed before the year 1600. Some remarkable only for colour appear to have been identical with our present breeds, some were nearly the same, some considerably different, and some have since become extinct. Several breeds, such as Finnikins and Turners, the swallow-tailed pigeon of Bechstein and the Carmelite, seem both to have originated and to have disappeared within this same period. Any one now visiting a well-stocked English aviary would certainly pick out as the most distinct kinds, the massive Runt, the Carrier with its wonderfully elongated beak and great wattles, the Barb with its short broad beak and eye-wattles, the short-faced Tumbler

[43] 'Treatise on Pigeons,' 1852, p. 41.

with its small conical beak, the Pouter with its great crop, long legs and body, the Fantail with its upraised, widely-expanded, well-feathered tail, the Turbit with its frill and short blunt beak, and the Jacobin with its hood. Now, if this same person could have viewed the pigeons kept before 1600 by Akber Khan in India and by Aldrovandi in Europe, he would have seen the Jacobin with a less perfect hood; the Turbit apparently without its frill; the Pouter with shorter legs, and in every way less remarkable—that is, if Aldrovandi's Pouter resembled the old German kind; the Fantail would have been far less singular in appearance, and would have had much fewer feathers in its tail; he would have seen excellent flying Tumblers, but he would in vain have looked for the marvellous short-faced breeds; he would have seen birds allied to barbs, but it is extremely doubtful whether he would have met with our actual Barbs; and lastly, he would have found Carriers with beaks and wattle incomparably less developed than in our English Carriers. He might have classed most of the breeds in the same groups as at present; but the differences between the groups were then far less strongly pronounced than at present. In short, the several breeds had at this early period not diverged in so great a degree from their aboriginal common parent, the wild rock-pigeon.

Manner of Formation of the chief Races.

We will now consider more closely the probable steps by which the chief races have been formed. As long as pigeons are kept semi-domesticated in dovecots in their native country, without any care in selecting and matching them, they are liable to little more variation than the wild *C. livia*, namely, in the wings becoming chequered with black, in the croup being blue or white, and in the size of the body. When, however, dovecot-pigeons are transported into diversified countries, such as Sierra Leone, the Malay archipelago, and Madeira (where the wild *C. livia* is not known to exist), they are exposed to new conditions of life; and apparently in consequence they vary in a somewhat greater degree. When closely confined, either for the pleasure of watching them, or to prevent their straying, they must be exposed, even under their native climate, to considerably different conditions; for they cannot obtain their natural diversity of food; and, what is probably more important, they are abundantly fed, whilst debarred from taking much exercise. Under these circumstances we might expect to find, from the analogy of all other domesticated animals, a greater amount of individual variability than with the wild pigeon; and this is the case. The want of exercise apparently tends to reduce the size of the feet and organs of flight; and then, from the law of correlation of growth, the beak apparently becomes affected. From what we now see occasionlly taking place in our aviaries, we may conclude that sudden variations or sports, such as the appearance of a crest of feathers on the head, of feathered feet, of a new shade of colour, of an additional feather in the tail or wing, would occur at rare intervals during the many centuries which have elapsed since the pigeon was first domesticated. At the present day such "sports" are generally rejected as blemishes; and there is so much mystery in the breeding of pigeons that, if a valuable sport did occur, its history would often be concealed. Before the last hundred and fifty years, there is hardly a chance of the history of any such sport having been recorded. But it by no means follows from this that such sports in former times, when the pigeon had undergone much less variation, would have been rejected. We are profoundly ignorant of the cause of each sudden and apparently spontaneous variation, as well as of the infinitely numerous shades of difference between the birds of the same family. But in a future chapter we shall see that all such variations appear to be the indirect result of changes of some kind in the conditions of life.

Hence, after a long course of domestication, we might expect to see in the pigeon much individual variability, and occasional sudden variations, as well as slight modifications from the lessened use of certain parts, together with the effects of correlation of growth. But without selection all this would produce only a trifling or no result; for without such aid differences of all kinds would, from the two following causes, soon disappear. In a healthy and vigorous lot of pigeons many more young birds are killed for food or die than are reared to maturity; so that an individual having any peculiar character, if not selected, would run a good chance of being destroyed; and if not destroyed, the peculiarity in question would almost certainly be obliterated by free intercrossing. It might, however, occasionally happen that the same variation repeatedly occurred, owing to the action of peculiar and uniform conditions of life, and in this case it would prevail independently of selection. But when selection is brought into play all is changed; for this is the foundation-stone in the formation of new races; and with the pigeon, circumstances, as we have already seen, are eminently favourable for selection. When a bird presenting some conspicuous variation has been preserved, and its offspring have been selected, carefully matched, and again propagated, and so onwards during successive generations, the principle is so obvious that nothing more need be said about it. This may be called *methodical selection*, for the breeder has a distinct object in view, namely, to preserve some character which has actually appeared; or to create some improvement already pictured in his mind.

Another form of selection has hardly been noticed by those authors who have discussed this subject, but is even more important. This form may be called *unconscious selection*, for the breeder selects his birds unconsciously, unintentionally, and without method, yet he surely though slowly produces a great result. I refer to the effects which follow from each fancier at first procuring and afterwards rearing as good birds as he can, according to his skill, and according to the standard of excellence at each successive period. He does not wish permanently to modify the breed; he does not look to the distant future, or speculate on the final result of the slow accumulation during many generations of successive slight changes: he is content if he possesses a good stock, and more than content if he can beat his rivals. The fancier in the time of Aldrovandi, when in the year 1600 he admired his own jacobins, pouters, or carriers, never reflected what their descendants in the year 1860 would become; he would have been astonished could he have seen our jacobins, our improved English carriers, and our pouters; he would probably have denied that they were the descendants of his own once admired stock, and he would perhaps not have valued them, for no other reason, as was written in 1765, "than because they were not like what used to be thought good when he was in the fancy." No one will attribute the lengthened beak of the carrier, the shortened beak of the short-faced tumbler, the lengthened leg of the pouter, the more perfectly-enclosed hood of the jacobin, &c.,—changes effected since the time of Aldrovandi, or even since a much later period,—to the direct and immediate action of the conditions of life. For these several races have been modified in various and even in directly opposite ways, though kept under the same climate and treated in all respects in as nearly uniform a manner as possible. Each slight change in the length or shortness of the beak, in the length of leg, &c., has no doubt been indirectly and remotely caused by some change in the conditions to which the bird has been subjected, but we must attribute the final result, as is manifest in those cases of which we have any historical record, to the continued selection and accumulation of many slight successive variations.

The action of unconscious selection, as far as pigeons are concerned, depends on a universal principle in human nature, namely, on our rivalry, and desire to outdo our neighbours. We see this in every fleeting fashion, even in our dress, and it leads the fancier to endeavour to exaggerate every peculiarity in his breeds. A great authority on pigeons[44] says, "Fanciers do not and will not admire a medium standard, that is, half and half, which is neither here nor there, but admire extremes." After remarking that the fancier of short-faced beard tumblers wishes for a very short beak, and that the fancier of long-faced beard tumblers wishes for a very long beak, he says, with respect to one of intermediate length, "Don't deceive yourself. Do you suppose for a moment the short or the long-faced fancier would accept such a bird as a gift? Certainly not; the short-faced fancier could see no beauty in it; the long-faced fancier would swear thére was no use in it, &c." In these comical passages, written seriously, we see the principle which has ever guided fanciers, and has led to such great modifications in all the domestic races which are valued solely for their beauty or curiosity.

Fashions in pigeon-breeding endure for long periods; we cannot change the structure of a bird as quickly as we can the fashion of our dress. In the time of Aldrovandi, no doubt the more the pouter inflated his crop, the more he was valued. Nevertheless, fashions do to a certain extent change; first one

[44] Eaton's 'Treatise on Pigeons,' 1858, p. 86.

point of structure and then another is attended to; or different breeds are admired at different times and in different countries. As the author just quoted remarks, "the fancy ebbs and flows; a thorough fancier now-a-days never stoops to breed toy-birds;" yet these very "toys" are now most carefully bred in Germany. Breeds which at the present time are highly valued in India are considered worthless in England. No doubt, when breeds are neglected, they degenerate; still we may believe that, as long as they are kept under the same conditions of life, characters once gained will be partially retained for a long time, and may form the starting-point for a future course of selection.

Let it not be objected to this view of the action of unconscious selection that fanciers would not observe or care for extremely slight differences. Those alone who have associated with fanciers can be thoroughly aware of their accurate powers of discrimination acquired by long practice, and of the care and labour which they bestow on their birds. I have known a fancier deliberately study his birds day after day to settle which to match together and which to reject. Observe how difficult the subject appears to one of the most eminent and experienced fanciers. Mr. Eaton, the winner of many prizes, says, "I would here particularly guard you against keeping too great a variety of pigeons, otherwise you will know a little about all the kinds, but nothing about one as it ought to be known." "It is possible there may be a few fanciers that have a good general knowledge of the several fancy pigeons, but there are many who labour under the delusion of supposing they know what they do not." Speaking exclusively of one sub-variety of one race, namely, the short-faced almond tumbler, and after saying that some fanciers sacrifice every property to obtain a good head and beak, and that other fanciers sacrifice everything for plumage, he remarks: "Some young fanciers who are over covetous go in for all the five properties at once, and they have their reward by getting nothing." In India, as I hear from Mr. Blyth, pigeons are likewise selected and matched with the greatest care. But we must not judge of the slight differences which would have been valued in ancient days, by those which are now valued after the formation of many races, each with its own standard of perfection, kept uniform by our numerous

Exhibitions. The ambition of the most energetic fancier may be fully satisfied by the difficulty of excelling other fanciers in the breeds already established, without trying to form a new one.

A difficulty with respect to the power of selection will perhaps already have occurred to the reader, namely, what could have led fanciers first to attempt to make such singular breeds as pouters, fantails, carriers, &c.? But it is this very difficulty which the principle of unconscious selection removes. Undoubtedly no fancier ever did intentionally make such an attempt. All that we need suppose is that a variation occurred sufficiently marked to catch the discriminating eye of some ancient fancier, and then unconscious selection carried on for many generations, that is, the wish of succeeding fanciers to excel their rivals, would do the rest. In the case of the fantail we may suppose that the first progenitor of the breed had a tail only slightly erected, as may now be seen in certain runts,[45] with some increase in the number of the tail-feathers, as now occasionally occurs with nuns. In the case of the pouter we may suppose that some bird inflated its crop a little more than other pigeons, as is now the case in a slight degree with the œsophagus of the turbit. We do not in the least know the origin of the common tumbler, but we may suppose that a bird was born with some affection of the brain, leading it to make somersaults in the air; and the difficulty in this case is lessened, as we know that, before the year 1600, in India, pigeons remarkable for their diversified manner of flight were much valued, and by the order of the Emperor Akber Khan were sedulously trained and carefully matched.

In the foregoing cases we have supposed that a sudden variation, conspicuous enough to catch a fancier's eye, first appeared; but even this degree of abruptness in the process of variation is not necessary for the formation of a new breed. When the same kind of pigeon has been kept pure, and has been bred during a long period by two or more fanciers, slight differences in the strain can often be recognised. Thus I have seen first-rate jacobins in one man's possession which certainly

[45] *See* Neumeister's figure of the Florence runt, tab. 13, in 'Das Ganze der Taubenzucht.'

differed slightly in several characters from those kept by another. I possessed some excellent barbs descended from a pair which had won a prize, and another lot descended from a stock formerly kept by that famous fancier Sir John Sebright, and these plainly differed in the form of the beak; but the differences were so slight, that they could hardly be described by words. Again, the common English and Dutch tumbler differ in a somewhat greater degree, both in length of beak and shape of head. What first caused these slight differences cannot be explained any more than why one man has a long nose and another a short one. In the strains long kept distinct by different fanciers, such differences are so common that they cannot be accounted for by the accident of the birds first chosen for breeding having been originally as different as they now are. The explanation no doubt lies in selection of a slightly different nature having been applied in each case; for no two fanciers have exactly the same taste, and consequently no two, in choosing and carefully matching their birds, prefer or select exactly the same. As each man naturally admires his own birds, he goes on continually exaggerating by selection whatever slight peculiarities they may possess. This will more especially happen with fanciers living in different countries, who do not compare their stocks and aim at a common standard of perfection. Thus, when a mere strain has once been formed, unconscious selection steadily tends to augment the amount of difference, and thus converts the strain into a sub-breed, and this ultimately into a well-marked breed or race.

The principle of correlation of growth should never be lost sight of. Most pigeons have small feet, apparently caused by their lessened use, and from correlation, as it would appear, their beaks have likewise become reduced in length. The beak is a conspicuous organ, and, as soon as it had thus become perceptibly shortened, fanciers would almost certainly strive to reduce it still more by the continued selection of birds with the shortest beaks; whilst at the same time other fanciers, as we know has actually been the case, would, in other sub-breeds, strive to increase its length. With the increased length of the beak, the tongue would become greatly lengthened, as would the eyelids with the increased development of the eye-wattles; with the reduced or increased size of the feet the number of the scutellæ would vary; with the length of the wing the number of the primary wing-feathers would differ; and with the increased length of the body in the pouter the number of the sacral vertebræ would be augmented. These important and correlated differences of structure do not invariably characterise any breed; but if they had been attended to and selected with as much care as the more conspicuous external differences, there can hardly be a doubt that they would have been rendered constant. Fanciers could assuredly have made a race of tumblers with nine instead of ten primary wing-feathers, seeing how often the number nine appears without any wish on their part, and indeed in the case of the white-winged varieties in opposition to their wish. In a similar manner, if the vertebræ had been visible and had been attended to by fanciers, assuredly an additional number might easily have been fixed in the pouter. If these latter characters had once been rendered constant we should never have suspected that they had at first been highly variable, or that they had arisen from correlation, in the one case with the shortness of the wings, and in the other case with the length of the body.

In order to understand how the chief domestic races have become distinctly separated from each other, it is important to bear in mind, that fanciers constantly try to breed from the best birds, and consequently that those which are inferior in the requisite qualities are in each generation neglected; so that after a time the less improved parent-stocks and many subsequently formed intermediate grades become extinct. This has occurred in the case of the pouter, turbit, and trumpeter, for these highly improved breeds are now left without any links closely connecting them either with each other or with the aboriginal rock-pigeon. In other countries, indeed, where the same care has not been applied, or where the same fashion has not prevailed, the earlier forms may long remain unaltered or altered only in a slight degree, and we are thus sometimes enabled to recover the connecting links. This is the case in Persia and India with the tumbler and carrier, which there differ but slightly from the rock-pigeon in the proportions of their beaks. So again in Java, the fantail sometimes has only fourteen caudal feathers, and the tail is much less elevated and expanded than in our improved birds; so that the Java bird forms a link between a first-rate fantail and the rock-pigeon.

Occasionally a breed may be retained for some particular quality in a nearly unaltered condition in the same country, together with highly modified offshoots or sub-breeds, which are valued for some distinct property. We see this exemplified in England, where the common tumbler, which is valued only for its flight, does not differ much from its parent-form, the Eastern tumbler; whereas the short-faced tumbler has been prodigiously modified, from being valued, not for its flight, but for other qualities. But the common-flying tumbler of Europe has already begun to branch out into slightly different sub-breeds, such as the common English tumbler, the Dutch roller, the Glasgow house-tumbler, and the long-faced beard tumbler, &c.; and in the course of centuries, unless fashions greatly change, these sub-breeds will diverge through the slow and insensible process of unconscious selection, and become modified, in a greater and greater degree. After a time the perfectly graduated links, which now connect all these sub-breeds together, will be lost, for there would be no object and much difficulty in retaining such a host of intermediate sub-varieties.

The principle of divergence, together with the extinction of the many previously existing intermediate forms, is so important for understanding the origin of domestic races, as well as of species in a state of nature, that I will enlarge a little more on this subject. Our third main group includes carriers, barbs, and runts, which are plainly related to each other, yet wonderfully distinct in several important characters. According to the view given in the last chapter, these three races have probably descended from an unknown race having an intermediate character, and this from the rock-pigeon. Their characteristic differences are believed to be due to different breeders having at an early period admired different points of structure; and then, on the acknowledged principle of admiring extremes, having gone on breeding, without any thought of the future, as good birds as they could,—carrier-fanciers preferring long beaks with much wattle,—barb-fanciers preferring short thick beaks with much eye-wattle,—and runt-fanciers not caring about the beak or wattle, but only for the size and weight of the body. This process will have led to the neglect and final extinction of the earlier, inferior, and intermediate birds; and thus it has come to pass, that in Europe these three races are now so extraordinarily distinct from each other. But in the East, whence they were originally brought, the fashion has been different, and we there see breeds which connect the highly modified English carrier with the rock-pigeon, and others which to a certain extent connect carriers and runts. Looking back to the time of Aldrovandi, we find that there existed in Europe, before the year 1600, four breeds which were closely allied to carriers and barbs, but which competent authorities cannot now identify with our present barbs and carriers; nor can Aldrovandi's runts be identified with our present runts. These four breeds certainly did not differ from each other nearly so much as do our existing English carriers, barbs, and runts. All this is exactly what might have been anticipated. If we could collect all the pigeons which have ever lived, from before the time of the Romans to the present day, we should be able to group them in several lines, diverging from the parent rock-pigeon. Each line would consist of almost insensible steps, occasionally broken by some slightly greater variation or sport, and each would culminate in one of our present highly modified forms. Of the many former connecting links, some would be found to have become absolutely extinct without having left any issue, whilst others though extinct would be seen to be the progenitors of the existing races.

I have heard it remarked as a strange circumstance that we occasionally hear of the local or complete extinction of domestic races, whilst we hear nothing of their origin. How, it has been asked, can these losses be compensated, and more than compensated, for we know that with almost all domesticated animals the races have largely increased in number since the time of the Romans? But on the view here given, we can understand this apparent contradiction. The extinction of a race within historical times is an event likely to be noticed; but its gradual and scarcely sensible modification through unconscious selection, and its subsequent divergence, either in the same or more commonly in distant countries, into two or more strains, and their gradual conversion into sub-breeds, and these into well-marked breeds, are events which would rarely be noticed. The death of a tree, that has attained gigantic dimensions, is recorded; the slow growth of smaller trees and their increase in number excite no attention.

In accordance with the belief of the great power of selection, and of the little direct power of changed conditions of life, except in causing general variability or plasticity of organisation, it is not surprising that dovecot-pigeons have remained unaltered from time immemorial; and that some toy-pigeons, which differ in little else besides colour from the dovecot-pigeon, have retained the same character for several centuries. For when one of these toy-pigeons had once become beautifully and symmetrically coloured,—when, for instance, a Spot had been produced with the crown of its head, its tail, and tail-coverts of a uniform colour, the rest of the body being snow-white,—no alteration or improvement would be desired. On the other hand, it is not surprising that during this same interval of time our highly-bred pigeons have undergone an astonishing amount of change; for in regard to them there is no defined limit to the wish of the fancier, and there is no known limit to the variability of their characters. What is there to stop the fancier desiring to give to his carrier a longer and longer beak, or to his tumbler a shorter and shorter beak? nor has the extreme limit of variability in the beak, if there be any such limit, as yet been reached. Notwithstanding the great improvement effected within recent times in the short-faced almond tumbler, Mr. Eaton remarks, "the field is still as open for fresh competitors as it was one hundred years ago;" but this is perhaps an exaggerated assertion, for the young of all highly improved fancy birds are extremely liable to disease and death.

I have heard it objected that the formation of the several domestic races of the pigeon throws no light on the origin of the wild species of the Columbidæ, because their differences are not of the same nature. The domestic races for instance do not differ, or differ hardly at all, in the relative lengths and shapes of the primary wing-feathers, in the relative length of the hind toe, or in habits of life, as in roosting and building in trees. But the above objection shows how completely the principle of selection has been misunderstood. It is not likely that characters selected by the caprice of man should resemble differences preserved under natural conditions, either from being of direct service to each species, or from standing in correlation with other modified and serviceable structures. Until man selects birds differing in the relative length of the wing-feathers or toes, &c., no sensible change in these parts should be expected. Nor could man do anything unless these parts happened to vary under domestication: I do not positively assert that this is the case, although I have seen traces of such variability in the wing-feathers, and certainly in the tail-feathers. It would be a strange fact if the relative length of the hind toe should never vary, seeing how variable the foot is both in size and in the number of the scutellæ. With respect to the domestic races not roosting or building in trees, it is obvious that fanciers would never attend to or select such changes in habits; but we have seen that the pigeons in Egypt, which do not for some reason like settling on the low mud hovels of the natives, are led, apparently by compulsion, to perch in crowds on the trees. We may even affirm that, if our domestic races had become greatly modified in any of the above specified respects, and it could be shown that fanciers had never attended to such points, or that they did not stand in correlation with other selected characters, the fact, on the principles advocated in this chapter, would have offered a serious difficulty.

Let us briefly sum up the last two chapters on the pigeon. We may conclude with confidence that all the domestic races, notwithstanding their great amount of difference, are descended from the *Columba livia*, including under this name certain wild races. But the differences between these latter forms throw no light whatever on the characters which distinguish the domestic races. In each breed or sub-breed the individual birds are more variable than birds in a state of nature; and occasionally they vary in a sudden and strongly-marked manner. This plasticity of organisation apparently results from changed conditions of life. Disuse has reduced certain parts of the body. Correlation of growth so ties the organisation together, that when one part varies other parts vary at the same time. When several breeds have once been formed, their intercrossing aids the progress of modification, and has even produced new sub-breeds. But as, in the construction of a building, mere stones or bricks are of little avail without the builder's art, so, in the production of new races, selection has been the presiding power. Fanciers can act by selection on excessively slight individual differences, as well as on those greater differences which are called sports. Selection is followed methodically when the fancier tries to improve and modify a breed according to a prefixed standard of excellence; or he acts unmethodically and unconsciously, by merely trying to rear as good birds as he can, without any wish or intention to alter the breed. The progress of selection almost inevitably leads to the neglect and ultimate extinction of the earlier and less improved forms, as well as of many intermediate links in each long line of descent. Thus it has come to pass that most of our present races are so marvellously distinct from each other, and from the aboriginal rock-pigeon.

CHAPTER VII.

FOWLS.

BRIEF DESCRIPTIONS OF THE CHIEF BREEDS — ARGUMENTS IN FAVOUR OF THEIR DESCENT FROM SEVERAL SPECIES — ARGUMENTS IN FAVOUR OF ALL THE BREEDS HAVING DESCENDED FROM GALLUS BANKIVA — REVERSION TO THE PARENT-STOCK IN COLOUR — ANALOGOUS VARIATIONS — ANCIENT HISTORY OF THE FOWL — EXTERNAL DIFFERENCES BETWEEN THE SEVERAL BREEDS — EGGS — CHICKENS — SECONDARY SEXUAL CHARACTERS — WING- AND TAIL- FEATHERS, VOICE, DISPOSITION, ETC. — OSTEOLOGICAL DIFFERENCES IN THE SKULL, VERTEBRÆ, ETC. — — EFFECTS OF USE AND DISUSE ON CERTAIN PARTS — CORRELATION OF GROWTH.

As some naturalists may not be familiar with the chief breeds of the fowl, it will be advisable to give a condensed description of them.[1] From what I have read and seen of specimens brought from several quarters of the world, I believe that most of the chief kinds have been imported into England, but many sub breeds are probably still here unknown. The following discussion on the origin of the various breeds and on their characteristic differences does not pretend to completeness, but may be of some interest to the naturalist. The classification of the breeds cannot, as far as I can see, be made natural. They differ from each other in different degrees, and do not afford characters in subordination to each other, by which they can be ranked in group under group. They seem all to have diverged by independent and different roads from a single type. Each chief breed includes differently coloured sub-varieties, most of which can be truly propagated, but it would be superfluous to describe them. I have classed the various crested fowls

[1] I have drawn up this brief synopsis from various sources, but chiefly from information given me by Mr. Tegetmeier. This gentleman has kindly looked through the whole of this chapter; and from his well-known knowledge, the statements here given may be fully trusted. Mr. Tegetmeier has likewise assisted me in every possible way in obtaining for me information and specimens. I must not let this opportunity pass without expressing my cordial thanks to Mr. B. P. Brent, a well-known writer on poultry, for indefatigable assistance and the gift of many specimens.

as sub-breeds under the Polish fowl; but I have great doubts whether this is a natural arrangement, showing true affinity or blood relationship. It is scarcely possible to avoid laying stress on the commonness of a breed; and if certain foreign sub-breeds had been largely kept in this country they would perhaps have been raised to the rank of main-breeds. Several breeds are abnormal in character; that is, they differ in certain points from all wild Gallinaceous birds. At first I made a division of the breeds into normal and abnormal, but the result was wholly unsatisfactory.

Fig. 30.—Spanish Fowl.

1. GAME BREED.—This may be considered as the typical breed, as it deviates only slightly from the wild *Gallus bankiva*, or, as perhaps more correctly named, *ferrugineus*. Beak strong; comb single and upright. Spurs long and sharp. Feathers closely adpressed to the body. Tail with the normal number of 14 feathers. Eggs often pale-buff. Disposition

indomitably courageous, exhibited even in the hens and chickens. An unusual number of differently coloured varieties exist, such as black and brown-breasted reds, duckwings, blacks, whites, piles, &c., with their legs of various colours.

2. MALAY BREED.—Body of great size, with head, neck, and legs elongated; carriage erect; tail small, sloping downwards, generally formed of 16 feathers; comb and wattle small; ear-lobe and face red; skin yellowish; feathers closely adpressed to the body; neck-hackles short, narrow, and hard. Eggs often pale buff. Chickens feather late. Disposition savage. Of Eastern origin.

3. COCHIN, OR SHANGAI BREED.—Size great; wing feathers short, arched, much hidden in the soft downy plumage; barely capable of flight; tail short, generally formed of 16 feathers, developed at a late period in the young males; legs thick, feathered; spurs short, thick; nail of middle toe flat and broad; an additional toe not rarely developed; skin yellowish. Comb and wattle well developed. Skull with deep medial furrow; occipital foramen, sub-triangular, vertically elongated. Voice peculiar. Eggs rough, buff-coloured. Disposition extremely quiet. Of Chinese origin.

4. DORKING BREED.—Size great; body square, compact; feet with an additional toe; comb well developed, but varies much in form; wattles well developed; colour of plumage various. Skull remarkably broad between the orbits. Of English origin.

The white Dorking may be considered as a distinct sub-breed, being a less massive bird.

5. SPANISH BREED.—Tall, with stately carriage; tarsi long; comb single, deeply serrated, of immense size; wattles largely developed; the large ear-lobes and sides of face white. Plumage black glossed with green. Do not incubate. Tender in constitution, the comb being often injured by frost. Eggs white, smooth, of large size. Chickens feather late, but the young cocks show their masculine characters, and crow at an early age. Of Mediterranean origin.

The *Andalusians* may be ranked as a sub-breed: they are of a slaty blue colour, and their chickens are well feathered. A smaller, short-legged Dutch sub-breed has been described by some authors as distinct.

6. HAMBURGH BREED (fig. 31).—Size moderate; comb flat, produced backwards, covered with numerous small points; wattle of moderate dimensions; ear-lobe white; legs blueish, thin. Do not incubate. Skull, with the tips of the ascending branches of the premaxillary and with the nasal bones standing a little separate from each other; anterior margin of the frontal bones less depressed than usual.

There are two sub-breeds; the *spangled* Hamburgh, of English origin, with the tips of the feathers marked with a dark spot; and the *pencilled* Hamburgh, of Dutch origin, with dark transverse lines across each feather, and with the body rather smaller. Both these sub-breeds include gold and silver varieties, as well as some other sub-varieties. Black Hamburghs have been produced by a cross with the Spanish breed.

7. CRESTED OR POLISH BREED (fig. 32).—Head with a large, rounded crest of feathers, supported on a hemispherical protuberance of the frontal bones,

which includes the anterior part of the brain. The ascending branches of the premaxillary bones and the inner nasal processes are much shortened. The orifice of the nostrils raised and crescentic. Beak short. Comb absent, or small and of crescentic shape; wattles either present or replaced by a beard-like tuft of feathers. Legs leaden-blue. Sexual differences appear late in life. Do not incubate. There are several beautiful varieties which differ in colour and slightly in other respects.

The following sub-breeds agree in having a crest, more or less developed,

Fig. 31.—Hamburgh Fowl.

with the comb, when present, of crescentic shape. The skull presents nearly the same remarkable peculiarities of structure as in the true Polish fowl.

Sub-breed (*a*) *Sultans.*—A Turkish breed, resembling white Polish fowls, with a large crest and beard, with short and well-feathered legs. The tail is furnished with additional sickle feathers. Do not incubate.[2]

Sub-breed (*b*) *Ptarmigans.*—An inferior breed closely allied to the last, white, rather small, legs much feathered, with the crest pointed; comb small, cupped; wattles small.

---

[2] The best account of Sultans is by Miss Watts in 'The Poultry Yard,' 1856, p. 79. I owe to Mr. Brent's kindness the examination of some specimens of this breed.

Sub-breed (*c*) *Ghoondooks.*—Another Turkish breed having an extraordinary appearance; black and tailless; crest and beard large; legs feathered. The inner processes of the two nasal bones come into contact with each other, owing to the complete absorption of the ascending branches of the premaxillaries. I have seen an allied, white, tailless breed from Turkey.

Sub-breed (*d*) *Crève-cœur.*—A French breed of large size, barely capable of flight, with short black legs, head crested, comb produced into two

Fig. 32.—Polish Fowl.

points or horns, sometimes a little branched like the horns of a stag; both beard and wattles present. Eggs large. Disposition quiet.[3]

Sub-breed (*e*) *Horned fowl.*—With a small crest; comb produced into two great points, supported on two bony protuberances.

Sub-breed (*f*) *Houdan.*—A French breed; of moderate size, short-legged with five toes, wings well developed; plumage invariably mottled with

---

[3] A good description with figures is given of this sub-breed in the 'Journal of Horticulture,' June 10th, 1862, p. 206.

black, white, and straw-yellow; head furnished with a crest, and a triple comb placed transversely; both wattles and beard present.[4]

Sub-breed (*g*) *Guelderlands.*—No comb, head said to be surmounted by a longitudinal crest of soft velvety feathers; nostrils said to be crescentic; wattles well developed; legs feathered; colour black. From North America. The Breda fowl seems to be closely allied to the Guelderland.

8. BANTAM BREED.—Originally from Japan,[5] characterized by small size alone; carriage bold and erect. There are several sub-breeds, such as the Cochin, Game, and Sebright Bantams, some of which have been recently formed by various crosses. The Black Bantam has a differently shaped skull, with the occipital foramen like that of the Cochin fowl.

9. RUMP-LESS FOWLS.—These are so variable in character[6] that they hardly deserve to be called a breed. Any one who will examine the caudal vertebræ will see how monstrous the breed is.

10. CREEPERS OR JUMPERS.—These are characterized by an almost monstrous shortness of legs, so that they move by jumping rather than by walking; they are said not to scratch up the ground. I have examined a Burmese variety, which had a skull of rather unusual shape.

11. FRIZZLED OR CAFFRE FOWLS.—Not uncommon in India, with the feathers curling backwards, and with the primary feathers of the wing and tail imperfect; periosteum of bones black.

12. SILK FOWLS.—Feathers silky, with the primary wing and tail-feathers imperfect; skin and periosteum of bones black; comb and wattles dark leaden-blue; ear-lappets tinged with blue; legs thin, often furnished with an additional toe. Size rather small.

13. SOOTY FOWLS.—An Indian breed, of a white colour stained with soot, with black skin and periosteum. The hens alone are thus characterized.

From this synopsis we see that the several breeds differ considerably, and they would have been nearly as interesting for us as pigeons, if there had been equally good evidence that all had descended from one parent-species. Most fanciers believe that they are descended from several primitive stocks. The Rev. E. S. Dixon[7] argues strongly on this side of the question; and one fancier even denounces the opposite conclusion by asking, "Do we not perceive pervading this spirit, the spirit of the *Deist?*" Most naturalists, with the exception of a few, such as Temminck, believe that all the breeds have proceeded from a single species; but authority on such a point

[4] A description, with figures, is given of this breed in 'Journal of Horticulture,' June 3rd, 1862, p. 186. Some writers describe the comb as two-horned.

[5] Mr. Crawfurd, 'Descript. Dict. of the Indian Islands,' p. 113. Bantams are mentioned in an ancient native Japanese Encyclopædia, as I am informed by Mr. Birch of the British Museum.

[6] 'Ornamental and Domestic Poultry,' 1848.

[7] 'Ornamental and Domestic Poultry,' 1848.

goes for little. Fanciers look to all parts of the world as the possible sources of their unknown stocks; thus ignoring the laws of geographical distribution. They know well that the several kinds breed truly even in colour. They assert, but, as we shall see, on very weak grounds, that most of the breeds are extremely ancient. They are strongly impressed with the great difference between the chief kinds, and they ask with force, can differences in climate, food, or treatment have produced birds so different as the black stately Spanish, the diminutive elegant Bantam, the heavy Cochin with its many peculiarities, and the Polish fowl with its great top-knot and protuberant skull? But fanciers, whilst admitting and even overrating the effects of crossing the various breeds, do not sufficiently regard the probability of the occasional birth, during the course of centuries, of birds with abnormal and hereditary peculiarities; they overlook the effects of correlation of growth—of the long-continued use and disuse of parts, and of some direct result from changed food and climate, though on this latter head I have found no sufficient evidence; and lastly, they all, as far as I know, entirely overlook the all-important subject of unconscious or unmethodical selection, though they are well aware that their birds differ individually, and that by selecting the best birds for a few generations they can improve their stocks.

An amateur writes[8] as follows. "The fact that poultry have until lately received but little attention at the hands of the fancier, and been entirely confined to the domains of the producer for the market, would alone suggest the improbability of that constant and unremitting attention having been observed in breeding, which is requisite to the consummating, in the offspring of any two birds, transmittable forms not exhibited by the parents." This at first sight appears true. But in a future chapter on Selection, abundant facts will be given showing not only that careful breeding, but that actual selection was practised during ancient periods, and by barely civilised races of man. In the case of the fowl I can adduce no direct facts showing that selection was anciently practised; but the Romans at the commencement of the Christian era kept six or seven breeds, and Columella "particularly recommends as the best, those sorts

[8] Ferguson's 'Illustrated Series of Rare and Prize Poultry,' 1854, p. vi., Preface.

that have five toes and white ears."[9] In the fifteenth century several breeds were known and described in Europe; and in China, at nearly the same period, seven kinds were named. A more striking case is that at present, in one of the Philippine Islands, the semi-barbarous inhabitants have distinct native names for no less than nine sub-breeds of the Game Fowl.[10] Azara,[11] who wrote towards the close of the last century, states that in the interior parts of South America, where I should not have expected that the least care would have been taken of poultry, a black-skinned and black-boned breed is kept, from being considered fertile and its flesh good for sick persons. Now every one who has kept poultry knows how impossible it is to keep several breeds distinct unless the utmost care be taken in separating the sexes. Will it then be pretended that those persons who in ancient times and in semi-civilized countries took pains to keep the breeds distinct, and who therefore valued them, would not occasionally have destroyed inferior birds and occasionally have preserved their best birds? This is all that is required. It is not pretended that any one in ancient times intended to form a new breed, or to modify an old breed according to some ideal standard of excellence. He who cared for poultry would merely wish to obtain, and afterwards to rear, the best birds which he could; but this occasional preservation of the best birds would in the course of time modify the breed, as surely, though by no means as rapidly, as does methodical selection at the present day. If one person out of a hundred or out of a thousand attended to the breeding of his birds, this would be sufficient; for the birds thus tended would soon become superior to others, and would form a new strain; and this strain would, as explained in the last chapter, slowly have its characteristic differences augmented, and at last be converted into a new sub-breed or breed. But breeds would often be for a time neglected and would deteriorate; they would, however, partially retain their character, and afterwards might again come into fashion and be raised to a standard of perfection

[9] Rev. E. S. Dixon, in his 'Ornamental Poultry,' p. 203, gives an account of Columella's work.

[10] Mr. Crawfurd 'On the Relation of the Domesticated Animals to Civilization,' separately printed, p. 6; first read before the Brit. Assoc. at Oxford, 1860.

[11] 'Quadrupèdes du Paraguay,' tom. ii. p. 324.

higher than their former standard; as has actually occurred quite recently with Polish fowls. If, however, a breed were utterly neglected, it would become extinct, as has recently happened with one of the Polish sub-breeds. Whenever in the course of past centuries a bird appeared with some slight abnormal structure, such as with a lark-like crest on its head, it would probably often have been preserved from that love of novelty which leads some persons in England to keep rumpless fowls, and others in India to keep frizzled fowls. And after a time any such abnormal appearance would be carefully preserved, from being esteemed a sign of the purity and excellence of the breed; for on this principle the Romans eighteen centuries ago valued the fifth toe and the white ear-lobe in their fowls.

Thus from the occasional appearance of abnormal characters, though at first only slight in degree; from the effects of the use and the disuse of parts; possibly from the direct effects of changed climate and food; from correlation of growth; from occasional reversions to old and long-lost characters; from the crossing of breeds, when more than one had once been formed; but, above all, from unconscious selection carried on during many generations, there is no insuperable difficulty, to the best of my judgment, in believing that all the breeds have descended from some one parent-source. Can any single species be named from which we may reasonably suppose that all have descended? The *Gallus bankiva* apparently fulfils every requirement. I have already given as fair an account as I could of the arguments in favour of the multiple origin of the several breeds; and now I will give those in favour of their common descent from *G. bankiva.*

But it will be convenient first briefly to describe all the known species of Gallus. The *G. Sonneratii* does not range into the northern parts of India; according to Colonel Sykes,[12] it presents at different heights on the Ghauts, two strongly marked varieties, perhaps deserving to be called species. It was at one time thought to be the primitive stock of all our domestic breeds, and this shows that it closely approaches the common fowl in general structure; but its hackles partially consist of highly peculiar, horny laminæ, transversely banded with three colours; and I have met with no authentic account of any such character having been observed

[12] 'Proc. Zoolog. Soc.' 1832, p. 151.

in any domestic breed.[13] This species also differs greatly from the common fowl, in the comb being finely serrated, and in the loins being destitute of true hackles. Its voice is utterly different. It crosses readily in India with domestic hens; and Mr. Blyth[14] raised nearly 100 hybrid chickens; but they were tender and mostly died whilst young. Those which were reared were absolutely sterile when crossed *inter se* or with either parent. At the Zoological Gardens, however, some hybrids of the same parentage were not quite so sterile: Mr. Dixon, as he informed me, made, with Mr. Yarrell's aid, particular inquiries on this subject, and was assured that out of 50 eggs only five or six chickens were reared. Some, however, of these half-bred birds were crossed with one of their parents, namely, a Bantam, and produced a few extremely feeble chickens. Mr. Dixon also procured some of these same birds and crossed them in several ways, but all were more or less infertile. Nearly similar experiments have recently been tried on a great scale in the Zoological Gardens with almost the same result.[15] Out of 500 eggs, raised from various first crosses and hybrids, between *G. Sonneratii*, *bankiva*, and *varius*, only 12 chickens were reared, and of these only three were the product of hybrids *inter se*. From these facts, and from the above-mentioned strongly-marked differences in structure between the domestic fowl and *G. Sonneratii*, we may reject this latter species as the parent of any domestic breed.

Ceylon possesses a fowl peculiar to the island, viz. *G. Stanleyii*; this species approaches so closely (except in the colouring of the comb) to the domestic fowl, that Messrs. E. Layard and Kellaert[16] would have considered it, as they inform me, as one of the parent-stocks, had it not been for its singularly different voice. This bird, like the last, crosses readily with tame hens, and even visits solitary farms and ravishes them. Two hybrids, a male and female, thus produced, were found by Mr. Mitford to be quite sterile: both inherited the peculiar voice of *G. Stanleyii*. This species, then, may in all probability be rejected as one of the primitive stocks of the domestic fowl.

Java and the islands eastward as far as Flores are inhabited by *G. varius* (or *furcatus*), which differs in so many characters—green plumage, unserrated comb, and single median wattle—that no one supposes it to have been the parent of any one of our breeds; yet, as I am informed by Mr. Crawfurd,[17] hybrids are commonly raised between the male *G. varius* and the common hen, and are kept for their great beauty, but are invariably sterile; this, however, was not the case with some bred in the Zoological Gardens. These hybrids were at one time thought to

---

[13] I have examined the feathers of some hybrids raised in the Zoological Gardens between the male *G. Sonneratii* and a red game-hen, and these feathers exhibited the true character of those of *G. Sonneratii*, except that the horny laminæ were much smaller.

[14] *See* also an excellent letter on the Poultry of India, by Mr. Blyth, in 'Gardener's Chronicle,' 1851, p. 619.

[15] Mr. S. J. Salter, in 'Natural History Review,' April, 1863, p. 276.

[16] *See* also Mr. Layard's paper in 'Annals and Mag. of Nat. History,' 2nd series, vol. xiv. p. 62.

[17] *See* also Mr. Crawfurd's 'Descriptive Dict. of the Indian Islands,' 1856, p. 113.

be specifically distinct, and were named *G. æneus*. Mr. Blyth and others believe that the *G. Temminckii*[18] (of which the history is not known) is a similar hybrid. Sir J. Brooke sent me some skins of domestic fowls from Borneo, and across the tail of one of these, as Mr. Tegetmeier observed, there were transverse blue bands like those which he had seen on the tail-feathers of hybrids from *G. varius*, reared in the Zoological Gardens. This fact apparently indicates that some of the fowls of Borneo have been slightly affected by crosses with *G. varius*, but the case may possibly be one of analogous variation. I may just allude to the *G. giganteus*, so often referred to in works on poultry as a wild species; but Marsden,[19] the first describer, speaks of it as a tame breed; and the specimen in the British Museum evidently has the aspect of a domestic variety.

The last species to be mentioned, namely, *Gallus bankiva*, has a much wider geographical range than the three previous species; it inhabits Northern India as far west as Sinde, and ascends the Himalaya to a height of 4000 ft.; it inhabits Burmah, the Malay peninsula, the Indo-Chinese countries, the Philippine Islands, and the Malayan archipelago as far eastward as Timor. This species varies considerably in the wild state. Mr. Blyth informs me that the specimens, both male and female, brought from near the Himalaya, are rather paler coloured than those from other parts of India; whilst those from the Malay peninsula and Java are brighter coloured than the Indian birds. I have seen specimens from these countries, and the difference of tint in the hackles was conspicuous. The Malayan hens were a shade redder on the breast and neck than the Indian hens. The Malayan males generally had a red ear-lappet, instead of a white one as in India; but Mr. Blyth has seen one Indian specimen without the white ear-lappet. The legs are leaden blue in the Indian, whereas they show some tendency to be yellowish in the Malayan and Javan specimens. In the former Mr. Blyth finds the tarsus remarkably variable in length. According to Temminck[20] the Timor specimens differ as a local race from that of Java. These several wild varieties have not as yet been ranked as distinct species; if they should, as is not unlikely, be hereafter thus ranked, the circumstance would be quite immaterial as far as the parentage and differences of our domestic breeds are concerned. The wild *G. bankiva* agrees most closely with the black-breasted red Game-breed; in colouring and in all other respects, except in being smaller, and in the tail being carried more horizontally. But the manner in which the tail is carried is highly variable in many of our breeds, for, as Mr. Brent informs me, the tail slopes much in the Malays, is erect in the Games and some other breeds, and is more than erect in Dorkings, Bantams, &c. There is one other difference, namely, that in *G. bankiva*, according to Mr. Blyth, the neck-hackles when first moulted are replaced during two or three months, not by other

---

[18] Described by Mr. G. R. Gray, 'Proc. Zoolog. Soc.,' 1849, p. 62.

[19] The passage from Marsden is given by Mr. Dixon in his 'Poultry Book,' p. 176. No ornithologist now ranks this bird as a distinct species.

[20] Coup-d'œil général sur l'Inde Archipélagique,' tom. iii. (1849), p. 177; *see* also Mr. Blyth in 'Indian Sporting Review,' vol. ii. p. 5, 1856.

hackles, as with our domestic poultry, but by short blackish feathers.[21] Mr. Brent, however, has remarked that these black feathers remain in the wild bird after the development of the lower hackles, and appear in the domestic bird at the same time with them; so that the only difference is that the lower hackles are replaced more slowly in the wild than in the tame bird; but as confinement is known sometimes to affect the masculine plumage, this slight difference cannot be considered of any importance. It is a significant fact that the voice of both the male and female *G. bankiva* closely resembles, as Mr. Blyth and others have noted, the voice of both sexes of the common domestic fowl; but the last note of the crow of the wild bird is rather less prolonged. Captain Hutton, well known for his researches into the natural history of India, informs me that he has seen several crossed fowls from the wild species and the Chinese bantam; these crossed fowls *bred freely* with bantams, but unfortunately were not crossed *inter se*. Captain Hutton reared chickens from the eggs of the *Gallus bankiva*; and these, though at first very wild, afterwards became so tame that they would crowd round his feet. He did not succeed in rearing them to maturity; but, as he remarks, "no wild gallinaceous bird thrives well at first on hard grain." Mr. Blyth also found much difficulty in keeping *G. bankiva* in confinement. In the Philippine Islands, however, the natives must succeed better, as they keep wild cocks to fight with their domestic game-birds.[22] Sir Walter Elliot informs me that the hen of a native domestic breed of Pegu is undistinguishable from the hen of the wild *G. bankiva*; and the natives constantly catch wild cocks by taking tame cocks to fight with them in the woods.[23] Mr. Crawfurd remarks that from etymology it might be argued that the fowl was first domesticated by the Malays and Javanese.[24] It is also a curious fact, of which I have been assured by Mr. Blyth, that wild specimens of the *Gallus bankiva*, brought from the countries east of the Bay of Bengal, are far more easily tamed than those of India; nor is this an unparalleled fact, for, as Humboldt long ago remarked, the same species sometimes evinces a more tameable disposition in one country than in another. If we suppose that the *G. bankiva* was first tamed in Malaya and afterwards imported into India, we can understand an observation made to me by Mr. Blyth, that the domestic fowls of India do not resemble the wild *G. bankiva* more closely than do those of Europe.

From the extremely close resemblance in colour, general structure, and especially in voice, between *Gallus bankiva* and the Game fowl; from their fertility, as far as this has been ascertained, when crossed; from the possibility of the wild species being tamed, and from its varying in the wild state, we may confidently look at it as the parent of the most typical of all the

[21] Mr. Blyth, in 'Annals and Mag. of Nat. Hist.,' 2nd ser., vol. i. (1848), p. 455.

[22] Crawfurd. 'Desc. Dict. of Indian Islands,' 1856, p. 112.

[23] In Burmah, as I hear from Mr. Blyth, the wild and tame poultry constantly cross together, and irregular transitional forms may be seen.

[24] Idem, p. 113.

domestic breeds, namely, the Game-fowl. It is a significant fact, that almost all the naturalists in India, namely, Sir W. Elliot, Mr. S. N. Ward, Mr. Layard, Mr. J. C. Jerdon, and Mr. Blyth,[25] who are familiar with *G. bankiva*, believe that it is the parent of most or all our domestic breeds. But even if it be admitted that *G. bankiva* is the parent of the Game breed, yet it may be urged that other wild species have been the parents of the other domestic breeds; and that these species still exist, though unknown, in some country, or have become extinct. The extinction, however, of several species of fowls, is an improbable hypothesis, seeing that the four known species have not become extinct in the most anciently and thickly peopled regions of the East. There is, in fact, only one kind of domesticated bird, namely, the Chinese goose or *Anser cygnoides*, of which the wild parent-form is said to be still unknown, or extinct. For the discovery of new, or the rediscovery of old species of Gallus, we must not look, as fanciers often look, to the whole world. The larger gallinaceous birds, as Mr. Blyth has remarked,[26] generally have a restricted range: we see this well illustrated in India, where the genus Gallus inhabits the base of the Himalaya, and is succeeded higher up by Gallophasis, and still higher up by Phasianus. Australia, with its islands, is out of the question as the home for unknown species of the genus. It is, also, as improbable that Gallus should inhabit South America[27] as that a humming-bird should be found in the Old World. From the character of the other gallinaceous

[25] Mr. Jerdon, in the 'Madras Journ. of Lit. and Science,' vol. xxii. p. 2, speaking of *G. bankiva*, says, "unquestionably the origin of most of the varieties of our common fowls." For Mr. Blyth, *see* his excellent article in 'Gardener's Chron.' 1851, p. 619; and in 'Annals and Mag. of Nat. Hist.,' vol. xx., 1847, p. 388.

[26] 'Gardener's Chronicle,' 1851, p. 619.

[27] I have consulted an eminent authority, Mr. Sclater, on this subject, and he thinks that I have not expressed myself too strongly. I am aware that one ancient author, Acosta, speaks of fowls as having inhabited S. America at the period of its discovery; and more recently, about 1795, Olivier de Serres speaks of wild fowls in the forests of Guiana; these were probably feral birds. Dr. Daniell tells me, he believes that fowls have become wild on the west coast of Equatorial Africa; they may, however, not be true fowls, but gallinaceous birds belonging to the genus Phasidus. The old voyager Barbut says that poultry are not natural to Guinea. Capt. W. Allen ('Narrative of Niger Expedition,' 1848, vol. ii. p. 42) describes wild fowls on Ilha dos Rollas, an island near St. Thomas's, on the west coast of Africa: the natives informed him that they had escaped from

birds of Africa, it is not probable that Gallus is an African genus. We need not look to the western parts of Asia, for Messrs. Blyth and Crawfurd, who have attended to this subject, doubt whether Gallus ever existed in a wild state even as far west as Persia. Although the earliest Greek writers speak of the fowl as a Persian bird, this probably merely indicates its line of importation. For the discovery of unknown species we must look to India, to the Indo-Chinese countries, and to the northern parts of the Malay Archipelago. The southern portion of China is the most likely country; but as Mr. Blyth informs me, skins have been exported from China during a long period, and living birds are largely kept there in aviaries, so that any native species of Gallus would probably have become known. Mr. Birch, of the British Museum, has translated for me passages from a Chinese Encyclopædia published in 1609, but compiled from more ancient documents, in which it is said that fowls are creatures of the West, and were introduced into the East (*i.e.* China) in a dynasty 1400 B.C. Whatever may be thought of so ancient a date, we see that the Indo-Chinese and Indian regions were formerly considered by the Chinese as the source of the domestic fowl. From these several considerations we must look to the present metropolis of the genus, namely, to the south-eastern parts of Asia, for the discovery of species which were formerly domesticated, but are now unknown in the wild state; and the most experienced ornithologists do not consider it probable that such species will be discovered.

In considering whether the domestic breeds are descended from one species, namely, *G. bankiva*, or from several, we must

a vessel wrecked there many years ago; they were extremely wild, and had "a cry quite different to that of the domestic fowl,' and their appearance was somewhat changed. Hence it is not a little doubtful, notwithstanding the statement of the natives, whether these birds really were fowls. That the fowl has become feral on several islands is certain. Mr. Fry, a very capable judge, informed Mr. Layard, in a letter, that the fowls which have run wild on Ascension "had nearly all got back to their primitive colours, red and black cocks, and smoky-grey hens." But unfortunately we do not know the colour of the poultry which were turned out. Fowls have become feral on the Nicobar Islands (Blyth in the 'Indian Field,' 1858, p. 62), and in the Ladrones (Anson's Voyage). Those found in the Pellew Islands (Crawfurd) are believed to be feral; and lastly, it is asserted that they have become feral in New Zealand, but whether this is correct I know not.

not quite overlook, though we must not exaggerate, the importance of the test of fertility. Most of our domestic breeds have been so often crossed, and their mongrels so largely kept, that it is almost certain, if any degree of infertility had existed between them, it would have been detected. On the other hand, the four known species of Gallus when crossed with each other, or when crossed, with the exception of *G. bankiva*, with the domestic fowl, produce infertile hybrids.

Finally, we have not such good evidence with fowls as with pigeons, of all the breeds having descended from a single primitive stock. In both cases the argument of fertility must go for something; in both we have the improbability of man having succeeded in ancient times in thoroughly domesticating several supposed species,—most of these supposed species being extremely abnormal as compared with their natural allies,—all being now either unknown or extinct, though the parent-form of scarcely any other domesticated bird has been lost. But in searching for the supposed parent-stocks of the various breeds of the pigeon, we were enabled to confine our search to species having peculiar habits of life; whilst with fowls there is nothing in their habits in any marked manner distinct from those of other gallinaceous birds. In the case of pigeons, I have shown that purely-bred birds of every race and the crossed offspring of distinct races frequently resemble, or revert to, the wild rock-pigeon in general colour and in each characteristic mark. With fowls we have facts of a similar nature, but less strongly pronounced, which we will now discuss.

Reversion and Analogous Variation.—Purely-bred Game, Malay, Cochin, Dorking, Bantam, and, as I hear from Mr. Tegetmeier, Silk fowls, may frequently or occasionally be met with, which are almost identical in plumage with the wild *G. bankiva*. This is a fact well deserving attention, when we reflect that these breeds rank amongst the most distinct. Fowls thus coloured are called by amateurs black-breasted reds. Hamburghs properly have a very different plumage; nevertheless, as Mr. Tegetmeier informs me, "the great difficulty in breeding cocks of the golden-spangled variety is their tendency to have black breasts and red backs." The males of white Bantams and white Cochins, as they come to maturity, often assume a yellowish or saffron tinge; and the longer neck hackles of black bantam cocks,[28] when two or three years old, not uncommonly become ruddy; these latter bantams occasionally "even moult brassy winged, or actually red shouldered." So that in these several cases we see a plain tendency to reversion to the hues of *G. bankiva*, even during the lifetime of the individual bird. With Spanish, Polish, pencilled Hamburgh, silver-spangled Hamburgh fowls, and with some other less common breeds, I have never heard of a black-breasted red bird having appeared.

From my experience with pigeons, I made the following crosses. I first killed all my own poultry, no others living near my house, and then procured, by Mr. Tegetmeier's assistance, a first-rate black Spanish cock, and hens of the following pure breeds,—white Game, white Cochin, silver-spangled Polish, silver-spangled Hamburgh, silver-pencilled Hamburgh, and white Silk. In none of these breeds is there a trace of red, nor when kept pure have I ever heard of the appearance of a red feather; though such an occurrence would perhaps not be very improbable with white Games and white Cochins. Of the many chickens reared from the above six crosses the majority were black, both in the down and in the first plumage; some were white, and a very few were mottled black and white. In one lot of eleven mixed eggs from the white Game and white Cochin by the black Spanish cock, seven of the chickens were white, and only four black: I mention this fact to show that whiteness of plumage is strongly inherited, and that the belief in the prepotent power in the male to transmit his colour is not always correct. The chickens were hatched in the spring, and in the latter part of August several of the young cocks began to exhibit a change, which with some of them increased during the following years. Thus a young male bird from the silver-spangled Polish hen was in its first plumage coal-black, and combined in its comb, crest, wattle, and beard, the characters of both parents; but when two years old the secondary wing-feathers became largely and symmetrically marked with white, and, wherever in *G. bankiva* the hackles are red, they were in this bird greenish-black along the shaft, narrowly bordered

[28] Mr. Hewitt, in 'The Poultry Book,' by W. B. Tegetmeier, 1866, p. 248.

with brownish-black, and this again broadly bordered with very pale yellowish-brown; so that in general appearance the plumage had become pale-coloured instead of black. In this case, with advancing age there was a great change, but no reversion to the red colour of *G. bankiva.*

A cock with a regular rose comb derived either from the spangled or pencilled silver Hamburgh was likewise at first quite black; but in less than a year the neck-hackles, as in the last case, became whitish, whilst those on the loins assumed a decided reddish-yellow tint; and here we see the first symptom of reversion; this likewise occurred with some other young cocks, which need not here be described. It has also been recorded[29] by a breeder, that he crossed two silver-pencilled Hamburgh hens with a Spanish cock, and reared a number of chickens, all of which were black, the cocks having *golden* and the hens brownish hackles; so that in this instance likewise there was a clear tendency to reversion.

Two young cocks from my white Game hen were at first snow white; of these, one subsequently assumed pale orange-coloured hackles, chiefly on the loins, and the other an abundance of fine orange-red hackles on the neck, loins, and upper wing-coverts. Here again we have a more decided, though partial, reversion to the colours of *G. bankiva.* This second cock was in fact coloured like an inferior "pile Game cock;"—now this sub-breed can be produced, as I am informed by Mr. Tegetmeier, by crossing a black-breasted red Game cock with a white Game hen, and the "pile" sub-breed thus produced can afterwards be truly propagated. So that we have the curious fact of the glossy-black Spanish cock and the black-breasted red Game cock when crossed with white Game-hens producing offspring of nearly the same colours.

I reared several birds from the white Silk-hen by the Spanish cock: all were coal-black, and all plainly showed their parentage in having blackish combs and bones; none inherited the so-called silky feathers, and the non-inheritance of this character has been observed by others. The hens never varied in their plumage. As the young cocks grew old, one of them assumed yellowish-white hackles, and thus resembled in a considerable

[29] 'Journal of Horticulture,' Jan. 14th, 1862, p. 325.

degree the cross from the Hamburgh hen; the other became a gorgeous bird, so much so that an acquaintance had it preserved and stuffed simply from its beauty. When stalking about it closely resembled the wild *Gallus bankiva*, but with the red feathers rather darker. On close comparison one considerable difference presented itself, namely, that the primary and secondary wing-feathers were edged with greenish-black, instead of being edged, as in *G. bankiva*, with fulvous and red tints. The space, also, across the back, which bears dark-green feathers, was broader, and the comb was blackish. In all other respects, even in trifling details of plumage, there was the closest accordance. Altogether it was a marvellous sight to compare this bird first with *G. bankiva*, and then with its father, the glossy green-black Spanish cock, and with its diminutive mother, the white Silk hen. This case of reversion is the more extraordinary as the Spanish breed has long been known to breed true, and no instance is on record of its throwing a single red feather. The Silk hen likewise breeds true, and is believed to be ancient, for Aldrovandi, before 1600, alludes probably to this breed, and describes it as covered with wool. It is so peculiar in many characters that some writers have considered it as specifically distinct; yet, as we now see, when crossed with the Spanish fowl, it yields offspring closely resembling the wild *G. bankiva*.

Mr. Tegetmeier has been so kind as to repeat, at my request, the cross between a Spanish cock and Silk hen, and he obtained similar results; for he thus raised, besides a black hen, seven cocks, all of which were dark-bodied with more or less orange-red hackles. In the ensuing year he paired the black hen with one of her brothers, and raised three young cocks, all coloured like their father, and a black hen mottled with white.

The hens from the six above-described crosses showed hardly any tendency to revert to the mottled-brown plumage of the female *G. bankiva*: one hen, however, from the white Cochin, which was at first coal-black, became slightly brown or sooty. Several hens, which were for a long time snow-white, acquired as they grew old a few black feathers. A hen from the white Game, which was for a long time entirely black glossed with green, when two years old had some of the primary wing-feathers greyish-white, and a multitude of feathers over her body narrowly and symmetrically tipped or laced with white. I had expected that some of the chickens whilst covered with down would have assumed the longitudinal stripes so general with gallinaceous birds; but this did not occur in a single instance. Two or three alone were reddish-brown about their heads. I was unfortunate in losing nearly all the white chickens from the first crosses; so that black prevailed with the grandchildren; but they were much diversified in colour, some being sooty, others mottled, and one blackish chicken had its feathers oddly tipped and barred with brown.

I will here add a few miscellaneous facts connected with reversion, and with the law of analogous variation. This law implies, as stated in a previous chapter, that the varieties of one species frequently mock distinct but allied species; and this fact is explained, according to the views which I maintain, on the principle of allied species having descended from one primitive form. The white Silk fowl with black skin and bones degenerates, as has been observed by Mr. Hewitt and Mr. R. Orton, in our climate; that is, it reverts to the ordinary colour of the common fowl in its skin and bones, due care having been taken to prevent any cross. In Germany [30] a distinct breed with black bones, and with black, not silky plumage, has likewise been observed to degenerate.

Mr. Tegetmeier informs me that, when distinct breeds are crossed, fowls are frequently produced with their feathers marked or pencilled by narrow transverse lines of a darker colour. This may be in part explained by direct reversion to the parent-form, the Bankiva hen; for this bird has all its upper plumage finely mottled with dark and rufous brown, with the mottling partially and obscurely arranged in transverse lines. But the tendency to pencilling is probably much strengthened by the law of analogous variation, for the hens of some other species of Gallus are more plainly pencilled, and the hens of many gallinaceous birds belonging to other genera, as the partridge, have pencilled feathers. Mr. Tegetmeier has

[30] 'Die Hühner und Pfauenzucht.' Ulm, 1827, s. 17. For Mr. Hewitt's statement with respect to the white Silk fowl, *see* the 'Poultry Book,' by W. B. Tegetmeier, 1866, p. 222. I am indebted to Mr. Orton for a letter on the same subject.

also remarked to me, that, although with domestic pigeons we have so great a diversity of colouring, we never see either pencilled or spangled feathers; and this fact is intelligible on the law of analogous variation, as neither the wild rock-pigeon nor any closely-allied species has such feathers. The frequent appearance of pencilling in crossed birds probably accounts for the existence of "cuckoo" sub-breeds in the Game, Polish, Dorking, Cochin, Andalusian, and Bantam breeds. The plumage of these birds is slaty-blue or grey, with each feather transversely barred with darker lines, so as to resemble in some degree the plumage of the cuckoo. It is a singular fact, considering that the male of no species of Gallus is in the least barred, that the cuckoo-like plumage has often been transferred to the male, more especially in the cuckoo Dorking; and the fact is all the more singular, as in gold and silver pencilled Hamburghs, in which pencilling is characteristic of the breed, the male is hardly at all pencilled, this kind of plumage being confined to the female.

Another case of analogous variation is the occurrence of spangled sub-breeds of Hamburgh, Polish, Malay, and Bantam fowls. Spangled feathers have a dark mark, properly crescent-shaped, on their tips; whilst pencilled feathers have several transverse bars. The spangling cannot be due to reversion to *G. bankiva;* nor does it often follow, as I hear from Mr. Tegetmeier, from crossing distinct breeds; but it is a case of analogous variation, for many gallinaceous birds have spangled feathers,—for instance, the common pheasant. Hence spangled breeds are often called "pheasant"-fowls. Another case of analogous variation in several domestic breeds is inexplicable; it is, that the chickens, whilst covered with down, of the black Spanish, black Game, black Polish, and black Bantam, all have white throats and breasts, and often have some white on their wings.[31] The editor of the 'Poultry Chronicle'[32] remarks that all the breeds which properly have red ear-lappets occasionally produce birds with white ear-lappets. This remark more especially applies to the Game breed, which of all comes nearest to the

[31] Dixon, 'Ornamental and Domestic Poultry,' pp. 253, 324, 335. For game fowls, *see* Ferguson on 'Prize Poultry,' p. 260.

[32] 'Poultry Chronicle,' vol. ii. p. 71.

G. bankiva; and we have seen that with this species living in a state of nature, the ear-lappets vary in colour, being red in the Malayan countries, and generally, but not invariably, white in India.

In concluding this part of my subject I may repeat that there exists one widely-ranging, varying, and common species of Gallus, namely *G. bankiva*, which can be tamed, produces fertile offspring when crossed with common fowls, and closely resembles in its whole structure, plumage, and voice the Game breed; hence it may be safely ranked as the parent of this, the most typical domesticated breed. We have seen that there is much difficulty in believing that other, now unknown, species have been the parents of the other domestic breeds. We know that all the breeds are most closely allied, as shown by their similarity in most points of structure and in habits, and by the analogous manner in which they vary. We have also seen that several of the most distinct breeds occasionally or habitually closely resemble in plumage *G. bankiva*, and that the crossed offspring of other breeds, which are not thus coloured, show a stronger or weaker tendency to revert to this same plumage. Some of the breeds, which appear the most distinct and the least likely to have proceeded from *G. bankiva*, such as Polish fowls, with their protuberant and little ossified skulls, and Cochins, with their imperfect tail and small wings, bear in these characters the plain marks of their artificial origin. We know well that of late years methodical selection has greatly improved and fixed many characters; and we have every reason to believe that unconscious selection, carried on for many generations, will have steadily augmented each new peculiarity and thus have given rise to new breeds. As soon as two or three breeds had once been formed, crossing would come into play in changing their character and in increasing their number. Brahma Pootras, according to an account lately published in America, offer a good instance of a breed, lately formed by a cross, which can be truly propagated. The well-known Sebright Bantams offer another and similar instance. Hence it may be concluded that not only the Game-breed but that all our breeds are probably the descendants of the

Malayan or Indian variety of *G. bankiva*. If so, this species has varied greatly since it was first domesticated; but there has been ample time, as we shall now show.

History of the Fowl.—Rütimeyer found no remains of the fowl in the ancient Swiss lake-dwellings. It is not mentioned in the Old Testament; nor is it figured on the ancient Egyptian monuments.[33] It is not referred to by Homer or Hesiod (about 900 B.C.); but is mentioned by Theognis and Aristophanes between 400 and 500 B.C. It is figured on some of the Babylonian cylinders, of which Mr. Layard sent me an impression, between the sixth and seventh centuries B.C.; and on the Harpy Tomb in Lycia, about 600 B.C.: so that we may feel pretty confident that the fowl reached Europe somewhere near the sixth century B.C. It had travelled still farther westward by the time of the Christian era, for it was found in Britain by Julius Cæsar. In India it must have been domesticated when the Institutes of Manu were written, that is, according to Sir W. Jones, 1200 B.C., but, according to the later authority of Mr. H. Wilson, only 800 B.C., for the domestic fowl is forbidden, whilst the wild is permitted to be eaten. If, as before remarked, we may trust the old Chinese Encyclopædia, the fowl must have been domesticated several centuries earlier, as it is said to have been introduced from the West into China 1400 B.C.

Sufficient materials do not exist for tracing the history of the separate breeds. About the commencement of the Christian era,

[33] Dr. Pickering, in his 'Races of Man,' 1850, p. 374, says that the head and neck of a fowl is carried in a Tribute-procession to Thoutmousis III. (1445 B.C.); but Mr. Birch of the British Museum doubts whether the figure can be identified as the head of a fowl. Some caution is necessary with reference to the absence of figures of the fowl on the ancient Egyptian monuments, on account of the strong and widely prevalent prejudice against this bird. I am informed by the Rev. S. Erhardt that on the east coast of Africa, from 4° to 6° south of the equator, most of the pagan tribes at the present day hold the fowl in aversion. The natives of the Pellew Islands would not eat the fowl, nor will the Indians in some parts of S. America. For the ancient history of the fowl, *see* also Volz, 'Beitrage zur Culturgeschichte,' 1852, s. 77; and Isid. Geoffroy St. Hilaire, 'Hist. Nat. Gén.,' tom. iii. p. 61. Mr. Crawfurd has given an admirable history of the fowl in his paper 'On the Relation of Domesticated Animals to Civilisation,' read before the Brit. Assoc. at Oxford in 1860, and since printed separately. I quote from him on the Greek poet Theognis, and on the Harpy Tomb described by Sir C. Fellowes. I quote from a letter of Mr. Blyth's with respect to the Institutes of Manu.

Columella mentions a five-toed fighting breed, and some provincial breeds; but we know nothing more about them. He also alludes to dwarf fowls; but these cannot have been the same with our Bantams, which, as Mr. Crawfurd has shown, were imported from Japan into Bantam in Java. A dwarf fowl, probably the true Bantam, is referred to in an old Japanese Encyclopædia, as I am informed by Mr. Birch. In the Chinese Encyclopædia published in 1596, but compiled from various sources, some of high antiquity, seven breeds are mentioned, including what we should now call jumpers or creepers, and likewise fowls with black feathers, bones, and flesh. In 1600 Aldrovandi describes seven or eight breeds of fowls, and this is the most ancient record from which the age of our European breeds can be inferred. The *Gallus Turcicus* certainly seems to be a pencilled Hamburgh; but Mr. Brent, a most capable judge, thinks that Aldrovandi "evidently figured what he happened to see, and not the best of the breed." Mr. Brent, indeed, considers all Aldrovandi's fowls as of impure breed; but it is a far more probable view that all our breeds since his time have been much improved and modified; for, as he went to the expense of so many figures, he probably would have secured characteristic specimens. The Silk fowl, however, probably then existed in its present state, as did almost certainly the fowl with frizzled or reversed feathers. Mr. Dixon[34] considers Aldrovandi's Paduan fowl as "a variety of the Polish," whereas Mr. Brent believes it to have been more nearly allied to the Malay. The anatomical peculiarities of the skull of the Polish breed were noticed by P. Borelli in 1656. I may add that in 1737 one Polish sub-breed, viz. the golden spangled, was known; but judging from Albin's description, the comb was then larger, the crest of feathers much smaller, the breast more coarsely spotted, and the stomach and thighs much blacker: a golden-spangled Polish fowl in this condition would now be of no value.

Differences in External and Internal Structure between the

[34] 'Ornamental and Domestic Poultry,' 1847, p. 185; for passages translated from Columella, *see* p. 312. For Golden Hamburghs, *see* Albin's 'Natural History of Birds,' 3 vols., with plates, 1731-38.

Breeds: Individual Variability.—Fowls have been exposed to diversified conditions of life, and as we have just seen there has been ample time for much variability and for the slow action of unconscious selection. As there are good grounds for believing that all the breeds are descended from *Gallus bankiva*, it will be worth while to describe in some detail the chief points of difference. Beginning with the eggs and chickens, I will pass on to the secondary sexual characters, and then to the differences in external structure and in the skeleton. I enter on the following details chiefly to show how variable almost every character has become under domestication.

Eggs.—Mr. Dixon remarks[35] that "to every hen belongs an individual peculiarity in the form, colour, and size of her egg, which never changes during her life time, so long as she remains in health, and which is as well known to those who are in the habit of taking her produce, as the handwriting of their nearest acquaintance." I believe that this is generally true, and that, if no great number of hens be kept, the eggs of each can almost always be recognised. The eggs of differently sized breeds naturally differ much in size; but, apparently, not always in strict relation to the size of the hen: thus the Malay is a larger bird than the Spanish, but *generally* she produces not such large eggs; white Bantams are said to lay smaller eggs than other Bantams;[36] white Cochins, on the other hand, as I hear from Mr. Tegetmeier, certainly lay larger eggs than buff Cochins. The eggs, however, of the different breeds vary considerably in character; for instance, Mr. Ballance states[37] that his Malay "pullets of last year laid eggs equal in size to those of any duck, and other Malay hens, two or three years old, laid eggs very little larger than a good-sized Bantam's egg. Some were as white as a Spanish hen's egg, and others varied from a light cream-colour to a deep rich buff, or even to a brown." The shape also varies, the two ends being much more equally rounded in Cochins than in Games or Polish. Spanish fowls lay smoother eggs than Cochins, of which the eggs are generally granulated. The shell in this latter breed, and more especially in Malays, is apt to be thicker than in Games or Spanish; but the Minorcas, a sub-breed of Spanish, are said to lay harder eggs than true Spanish.[38] The colour differs considerably,—the Cochins laying buff-coloured eggs; the Malays

[35] 'Ornamental and Domestic Poultry,' p. 152.

[36] Ferguson on 'Rare Prize Poultry,' p. 297. This writer, I am informed, cannot generally be trusted. He gives, however, figures and much information on eggs. *See* pp. 34 and 235 on the eggs of the Game fowl.

[37] *See* 'Poultry Book,' by Mr. Tegetmeier, 1866, pp. 81 and 78.

[38] 'The Cottage Gardener,' Oct. 1855, p. 13. On the thinness of the eggs of Game-fowls, *see* Mowbray on Poultry, 7th edit., p. 13.

a paler variable buff; and Games a still paler buff. It would appear that darker-coloured eggs characterise the breeds which have lately come from the East, or are still closely allied to those now living there. The colour of the yolk, according to Ferguson, as well as of the shell, differs slightly in the sub-breeds of the Game, and stands in some degree of correlation with the colour of the plumage. I am also informed by Mr. Brent that dark partridge-coloured Cochin hens lay darker coloured eggs than the other Cochin sub-breeds. The flavour and richness of the egg certainly differ in different breeds. The productiveness of the several breeds is very different. Spanish, Polish, and Hamburgh hens have lost the incubating instinct.

Chickens.—As the young of almost all gallinaceous birds, even of the black curassow and black grouse, whilst covered with down, are longitudinally striped on the back,—of which character, when adult, neither sex retains a trace,—it might have been expected that the chickens of all our domestic fowls would have been similarly striped.[39] This could, however, hardly have been expected, when the adult plumage in both sexes has undergone so great a change as to be wholly white or black. In white fowls of various breeds the chickens are uniformly yellowish white, passing in the black-boned Silk fowl into bright canary-yellow. This is also generally the case with the chickens of white Cochins, but I hear from Mr. Zurhost that they are sometimes of a buff or oak colour, and that all those of this latter colour, which were watched, turned out males. The chickens of buff Cochins are of a golden-yellow, easily distinguishable from the paler tint of the white Cochins, and are often longitudinally streaked with dark shades: the chickens of silver-cinnamon Cochins are almost always of a buff colour. The chickens of the white Game and white Dorking breeds, when held in particular lights, sometimes exhibit (on the authority of Mr. Brent) faint traces of longitudinal stripes. Fowls which are entirely black, namely Spanish, black Game, black Polish, and black Bantams, display a new character, for their chickens have their breasts and throats more or less white, with sometimes a little white elsewhere. Spanish chickens also, occasionally (Brent), have, where the down was white, their first true feathers tipped for a time with white. The primordially striped character is retained by the chickens of most of the Game sub-breeds (Brent, Dixon); by Dorkings; by the partridge and grouse-coloured sub-breeds of Cochins (Brent), but not, as we have seen, by all the other sub-breeds; by the pheasant-Malay (Dixon), but apparently not (at which I am much surprised) by other Malays. The following breeds and sub-breeds are barely, or not at all, longitudinally striped; viz. gold and silver pencilled Hamburghs, which can hardly be distinguished from each other (Brent) in the down, both having a few

[39] My information, which is very far from perfect, on chickens in the down, is derived chiefly from Mr. Dixon's 'Ornamental and Domestic Poultry.' Mr. B. P. Brent has also communicated to me many facts by letter, as has Mr. Tegetmeier. I will in each case mark my authority by the name within brackets. For the chickens of white Silk-fowls, *see* Tegetmeier's 'Poultry Book,' 1866, p. 221.

dark spots on the head and rump, with occasionally a longitudinal stripe (Dixon) on the back of the neck. I have seen only one chicken of the silver-spangled Hamburgh, and this was obscurely striped along the back. Gold-spangled Polish chickens (Tegetmeier) are of a warm russet brown; and silver-spangled Polish chickens are grey, sometimes (Dixon) with dashes of ochre on the head, wings, and breast. Cuckoo and blue-dun fowls (Dixon) are grey in the down. The chickens of Sebright Bantams (Dixon) are uniformly dark brown, whilst those of the brown-breasted red Game Bantam are black, with some white on the throat and breast. From these facts we see that the chickens of the different breeds, and even of the same main breed, differ much in their downy plumage; and, although longitudinal stripes characterise the young of all wild gallinaceous birds, they disappear in several domestic breeds. Perhaps it may be accepted as a general rule that the more the adult plumage differs from that of the adult *G. bankiva*, the more completely the chickens have lost their proper stripes.

With respect to the period of life at which the characters proper to each breed first appear, it is obvious that such structures as additional toes must be formed long before birth. In Polish fowls, the extraordinary protuberance of the anterior part of the skull is well developed before the chickens come out of the egg;[40] but the crest, which is supported on the protuberance, is at first feebly developed, nor does it attain its full size until the second year. The Spanish cock is pre-eminent for his magnificent comb, and this is developed at an unusually early age; so that the young males can be distinguished from the females when only a few weeks old, and therefore earlier than in other breeds; they likewise crow very early, namely, when about six weeks old. In the Dutch sub-breed of the Spanish fowl the white ear-lappets are developed earlier than in the common Spanish breed.[41] Cochins are characterised by a small tail, and in the young cocks the tail is developed at an unusually late period.[42] Game fowls are notorious for their pugnacity; and the young cocks crow, clap their little wings, and obstinately fight with each other, even whilst under their mother's care.[43] "I have often had," says one

[40] As I hear from Mr. Tegetmeier; *see* also 'Proc. Zoolog. Soc.' 1856, p. 366. On the late development of the crest, *see* 'Poultry Chronicle,' vol. ii. p. 132.

[41] On these points, *see* 'Poultry Chronicle,' vol. iii. p. 166; and Tegetmeier's 'Poultry Book,' 1866, pp. 105 and 121.

[42] Dixon, 'Ornamental and Domestic Poultry,' p. 273.

[43] Ferguson on Rare and Prize Poultry, p. 261.

author,[44] "whole broods, scarcely feathered, stone-blind from fighting; the rival couples moping in corners, and renewing their battles on obtaining the first ray of light." With the males of all gallinaceous birds the use of their weapons and pugnacity is to fight for the possession of the females; so that the tendency in our Game chickens to fight at an extremely early age is not only useless, but is injurious, as they suffer so much from their wounds. The training for battle during an early period may be natural to the wild *Gallus bankiva;* but as man during many generations has gone on selecting the most obstinately pugnacious cocks, it is more probable that their pugnacity has been unnaturally increased, and unnaturally transferred to the young male chickens. In the same manner, it is probable that the extraordinary development of the comb in the Spanish cock has been unintentionally transferred to the young cocks; for fanciers would not care whether their young birds had large combs, but would select for breeding the adults which had the finest combs, whether or not developed at an early period. The last point which need here be noticed is that, though the chickens of Spanish and Malay fowls are well covered with down, the true feathers are acquired at an unusually late age; so that for a time the young birds are partially naked, and are liable to suffer from cold.

Secondary Sexual Characters.—The two sexes in the parent-form, the *Gallus bankiva,* differ much in colour. In our domestic breeds the difference is never greater, but is often less, and varies much in degree even in the sub-breeds of the same main breed. Thus in certain Game fowls the difference is as great as in the parent-form, whilst in the black and white sub-breeds there is no difference in plumage. Mr. Brent informs me that he has seen two strains of black-breasted red Games, in which the cocks could not be distinguished, whilst the hens in one were partridge-brown and in the other fawn-brown. A similar case has been observed in the strains of the brown-breasted red Game. The hen of the "duck-winged Game" is "extremely beautiful," and differs much from the hens of all the other Game sub-breeds; but generally, as with the blue and grey Game and

[44] Mowbray on Poultry, 7th edit. 1834, p. 13.

with some sub-varieties of the pile-game, a moderately close relation may be observed between the males and females in the variation of their plumage.[45] A similar relation is also evident when we compare the several varieties of Cochins. In the two sexes of gold and silver-spangled and of buff Polish fowls, there is much general similarity in the colouring and marks of the whole plumage, excepting of course in the hackles, crest, and beard. In spangled Hamburghs, there is likewise a considerable degree of similarity between the two sexes. In pencilled Hamburghs, on the other hand, there is much dissimilarity; the pencilling which is characteristic of the hens being almost absent in the males of both the golden and silver varieties. But, as we have already seen, it cannot be given as a general rule that male fowls never have pencilled feathers, for Cuckoo Dorkings are "remarkable from having nearly similar markings in both sexes."

It is a singular fact that the males in certain sub-breeds have lost some of their secondary masculine characters, and, from their close resemblance in plumage to the females, are often called hennies. There is much diversity of opinion whether these males are in any degree sterile; that they sometimes are partially sterile seems clear,[46] but this may have been caused by too close interbreeding. That they are not quite sterile, and that the whole case is widely different from that of old females assuming masculine characters, is evident from several of these hen-like sub-breeds having been long propagated. The males and females of gold and silver-laced Sebright Bantams can be barely distinguished from each other, except by their combs, wattles, and spurs, for they are coloured alike, and the males have not hackles, nor the flowing sickle-like tail-feathers. A hen-tailed sub-breed of Hamburghs was recently much esteemed. There is also a breed of Game-fowls, in which the males and females resemble each other so closely that the cocks have often mistaken their hen-feathered opponents in the cock-pit for real hens, and by the mistake have lost their lives.[47] The cocks,

[45] *See* the full description of the varieties of the Game-breed, in Tegetmeier's 'Poultry Book,' 1866, p. 131. For Cuckoo Dorkings, p. 97.

[46] Mr. Hewitt in Tegetmeier's 'Poultry Book,' 1866, pp. 246 and 156. For hen-tailed game-cocks, *see* p. 131.

[47] 'The Field,' April 20th, 1861. The writer says he has seen half-a-dozen cocks thus sacrificed.

though dressed in the feathers of the hen, "are high-spirited birds, and their courage has been often proved:" an engraving even has been published of one celebrated hen-tailed victor. Mr. Tegetmeier[48] has recorded the remarkable case of a brown-breasted red Game-cock which, after assuming its perfect masculine plumage, became hen-feathered in the autumn of the following year; but he did not lose voice, spurs, strength, nor productiveness. This bird has now retained the same character during five seasons, and has begot both hen-feathered and male-feathered offspring. Mr. Grantley F. Berkeley relates the still more singular case of a celebrated strain of "polecat Game-fowls," which produced in nearly every brood a single hen-cock. "The great peculiarity in one of these birds was that he, as the seasons succeeded each other, was not always a hen-cock, and not always of the colour called the polecat, which is black. From the polecat and hen-cock feather in one season he moulted to a full male-plumaged black-breasted red, and in the following year he returned to the former feather."[49]

I have remarked in my 'Origin of Species' that secondary sexual characters are apt to differ much in the species of the same genus, and to be unusually variable in the individuals of the same species. So it is with the breeds of the fowl, as we have already seen, as far as the colour of plumage is concerned, and so it is with the other secondary sexual characters. Firstly, the comb differs much in the various breeds,[50] and its form is eminently characteristic of each kind, with the exception of the Dorkings, in which the form has not been as yet determined on by fanciers, and fixed by selection. A single, deeply-serrated comb is the typical and most common form. It differs much in size, being immensely developed in Spanish fowls; and in a local breed called Red-caps, it is sometimes "upwards of three inches in breadth at the front, and more than four inches in length, measured to the end of the peak behind."[51] In some breeds the comb is double, and when the two ends are cemented

[48] 'Proceedings of Zoolog. Soc.' March, 1861, p. 102. The engraving of the hen-tailed cock just alluded to was exhibited at the Society.

[49] 'The Field,' April 20th, 1861.

[50] I am much indebted to Mr. Brent for an account, with sketches, of all the variations of the comb known to him, and likewise with respect to the tail, as presently to be given.

[51] The 'Poultry Book,' by Tegetmeier, 1866, p. 234.

together it forms a "cup-comb;" in the "rose-comb" it is depressed, covered with small projections, and produced backwards; in the horned and crève-cœur fowl it is produced into two horns; it is triple in the pea-combed Brahmas, short and truncated in the Malays, and absent in the Guelderlands. In the tasselled Game a few long feathers arise from the back of the comb; in many breeds a crest of feathers replaces the comb. The crest, when little developed, arises from a fleshy mass, but, when much developed, from a hemispherical protuberance of the skull. In the best Polish fowls it is so largely developed, that I have seen birds which could hardly pick up their food; and a German writer asserts [52] that they are in consequence liable to be struck by hawks. Monstrous structures of this kind would thus be suppressed in a state of nature. The wattles, also, vary much in size, being small in Malays and some other breeds; they are replaced in certain Polish subbreeds by a great tuft of feathers called a beard.

The hackles do not differ much in the various breeds, but are short and stiff in Malays, and absent in Hennies. As in some orders of birds the males display extraordinarily-shaped feathers, such as naked shafts with discs at the end, &c., the following case may be worth giving. In the wild *Gallus bankiva* and in our domestic fowls, the barbs which arise from each side of the extremities of the hackles are naked or not clothed with barbules, so that they resemble bristles; but Mr. Brent sent me some scapular hackles from a young Birchen Duckwing Game cock, in which the naked barbs became densely reclothed with barbules towards their tips; so that these tips, which were dark coloured with a metallic lustre, were separated from the lower parts by a symmetrically-shaped transparent zone formed of the naked portions of the barbs. Hence the coloured tips appeared like little separate metallic discs.

The sickle-feathers in the tail, of which there are three pair, and which are eminently characteristic of the male sex, differ much in the various breeds. They are scimitar-shaped in some Hamburghs, instead of being long and flowing as in the typical breeds. They are extremely short in Cochins, and are not at

[52] 'Die Hühner und Pfauenzucht,' 1827, s. 11.

all developed in Hennies. They are carried, together with the whole tail, erect in Dorkings and Games; but droop much in Malays and in some Cochins. Sultans are characterized by an additional number of lateral sickle-feathers. The spurs vary much, being placed higher or lower on the shank; being extremely long and sharp in Games, and blunt and short in Cochins. These latter birds seem aware that their spurs are not efficient weapons; for though they occasionally use them, they more frequently fight, as I am informed by Mr. Tegetmeier, by seizing and shaking each other with their beaks. In some Indian Game-cocks, received by Mr. Brent from Germany, there are, as he informs me, three, four, or even five spurs on each leg. Some Dorkings also have two spurs on each leg;[53] and in birds of this breed the spur is often placed almost on the outside of the leg. Double spurs are mentioned in the ancient Chinese Encyclopædia. Their occurrence may be considered as a case of analogous variation, for some wild gallinaceous birds, for instance, the Polyplectron, have double spurs.

Judging from the differences which generally distinguish the sexes in the Gallinaceæ, certain characters in our domestic fowls appear to have been transferred from the one sex to the other. In all the species (except in Turnix), when there is any conspicuous difference in plumage between the male and female, the male is always the most beautiful; but in golden-spangled Hamburghs the hen is equally beautiful with the cock, and incomparably more beautiful than the hen in any natural species of Gallus; so that here a masculine character has been transferred to the female. On the other hand, in cuckoo Dorkings and in other cuckoo breeds the pencilling, which in Gallus is a female attribute, has been transferred to the male: nor, on the principle of analogous variation, is this transference surprising, as the males in many gallinaceous genera are barred or pencilled. With most of these birds head ornaments of all kinds are more fully developed in the male than in the female; but in Polish fowls the crest or top-knot, which in the male replaces the comb, is equally developed in both sexes. In certain sub-

[53] 'Poultry Chronicle,' vol. i. p. 595. Mr. Brent has informed me of the same fact. With respect to the position of the spurs in Dorkings, *see* 'Cottage Gardener,' Sept. 18th, 1860, p. 380.

breeds, which, from the hen having a small crest, are called lark-crested, "a single upright comb sometimes almost entirely takes the place of the crest in the male."[54] From this latter case, and from some facts presently to be given with respect to the protuberance of the skull in Polish fowls, the crest in this breed ought perhaps to be viewed as a feminine character which has been transferred to the male. In the Spanish breed the male, as we know, has an immense comb, and this has been partially transferred to the female, for her comb is unusually large, though not upright. In Game-fowls the bold and savage disposition of the male has likewise been largely transferred to the female;[55] and she sometimes even possesses the eminently masculine character of spurs. Many cases are on record of hens being furnished with spurs; and in Germany, according to Bechstein,[56] the spurs in the Silk-hen are sometimes very long. He mentions also another breed similarly characterized, in which the hens are excellent layers, but are apt to disturb and break their eggs owing to their spurs.

Mr. Layard[57] has given an account of a breed of fowls in Ceylon with black skin, bones, and wattle, but with ordinary feathers, and which cannot "be more aptly described than by comparing them to a white fowl drawn down a sooty chimney; it is, however," adds Mr. Layard, "a remarkable fact that a male bird of the pure sooty variety is almost as rare as a tortoise-shell tom-cat." Mr. Blyth finds that the same rule holds good with this breed near Calcutta. The males and females, on the other hand, of the black-boned European breed, with silky feathers, do not differ from each other; so that in the one breed black skin and bones, and the same kind of plumage, are common to both sexes, whilst in the other breed these characters are confined to the female sex.

At the present day all the breeds of Polish fowls have the great bony protuberance on their skulls, which includes part of the brain and supports the crest, equally developed in both sexes.

[54] Dixon, 'Ornamental and Domestic Poultry,' p. 320.

[55] Mr. Tegetmeier informs me that Game hens have been found so combative, that it is now generally the practice to exhibit each hen in a separate pen.

[56] 'Naturgeschichte Deutschlands,' Band iii. (1793), s. 339, 407.

[57] On the Ornithology of Ceylon in 'Annals and Mag. of Nat. History,' 2nd series, vol. xiv. (1854), p. 63.

But formerly in Germany the skull of the hen alone was protuberant: Blumenbach,[58] who particularly attended to abnormal peculiarities in domestic animals, states, in 1813, that this was the case; and Bechstein had previously, in 1793, observed the same fact. This latter author has carefully described the effects of a crest on the skull not only in fowls, but in ducks, geese, and canaries. He states that with fowls, when the crest is not much developed, it is supported on a fatty mass; but when much developed, it is always supported on a bony protuberance of variable size. He well describes the peculiarities of this protuberance, and he attended to the effects of the modified shape of the brain on the intellect of these birds, and disputes Pallas' statement that they are stupid. He then expressly states that he never observed this protuberance in male fowls. Hence there can be no doubt that this remarkable character in the skulls of Polish fowls was formerly in Germany confined to the female sex, but has now been transferred to the males, and has thus become common to both sexes.

External Differences, not connected with the sexes, between the breeds and between individual birds.

The size of the body differs greatly. Mr. Tegetmeier has known a Brahma to weigh 17 pounds; a fine Malay cock 10 pounds; whilst a first-rate Sebright Bantam weighs hardly more than 1 pound. During the last 20 years the size of some of our breeds has been largely increased by methodical selection, whilst that of other breeds has been much diminished. We have already seen how greatly colour varies even within the same breed; we know that the wild *G. bankiva* varies slightly in colour; we know that colour is variable in all our domestic animals; nevertheless some eminent fanciers have so little faith in variability, that they have actually argued that the chief Game sub-breeds, which differ from each other in nothing but colour, are descended from distinct wild species! Crossing often causes strange modifications of colour. Mr. Tegetmeier informs me that when buff and white Cochins are crossed, some of the

[58] I quote Blumenbach on the authority of Mr. Tegetmeier, who gives in 'Proc. Zoolog. Soc.,' Nov. 25th, 1856, a very interesting account of the skulls of Polish fowls. Mr. Tegetmeier, not knowing of Bechstein's account, disputed the accuracy of Blumenbach's statement. For Bechstein, *see* 'Naturgeschichte Deutschlands,' Band iii. (1793), s. 399, note. I may add that at the first exhibition of poultry at the Zoological Gardens, in May, 1845, I saw some fowls, called Friezland fowls, of which the hens were crested, and the cocks were furnished with a comb.

chickens are almost invariably black. According to Mr. Brent, black and white Cochins occasionally produce chickens of a slaty-blue tint; and this same tint appears, as Mr. Tegetmeier tells me, from crossing white Cochins with black Spanish fowls, or white Dorkings with black Minorcas.[59] A good observer[60] states that a first-rate silver-spangled Hamburgh hen gradually lost the most characteristic qualities of the breed, for the black lacing to her feathers disappeared, and her legs changed from leaden-blue to white; but what makes the case remarkable is, that this tendency ran in the blood, for her sister changed in a similar but less strongly marked manner; and chickens produced from this latter hen were at first almost pure white, "but on moulting acquired black collars and some spangled feathers with almost obliterated markings;" so that a new variety arose in this singular manner. The skin in the different breeds differs much in colour, being white in common kinds, yellow in Malays and Cochins, and black in Silk fowls; thus mocking, as M. Godron[61] remarks, the three principal types of skin in mankind. The same author adds, that, as different kinds of fowls living in distant and isolated parts of the world have black skin and bones, this colour must have appeared at various times and places.

The shape and carriage of the body and the shape of the head differ much. The beak varies slightly in length and curvature, but incomparably less than with pigeons. In most crested fowls the nostrils offer a remarkable peculiarity in being raised with a crescentic outline. The primary wing-feathers are short in Cochins; in a male, which must have been more than twice as heavy as *G. bankiva*, these feathers were in both birds of the same length. I have counted, with Mr. Tegetmeier's aid, the primary wing-feathers in thirteen cocks and hens of various breeds; in four of them, namely in two Hamburghs, a Cochin, and Game Bantam, there were 10, instead of the normal number 9; but in counting these feathers I have followed the practice of fanciers, and have *not* included the first minute primary feather, barely three-quarters of an inch in length. These feathers differ considerably in relative length, the fourth, or the fifth, or the sixth, being the longest; with the third either equal to, or considerably shorter than the fifth. In wild gallinaceous species the relative length and number of the main wing and tail-feathers are extremely constant.

The tail differs much in erectness and size, being small in Malays and very small in Cochins. In thirteen fowls of various breeds which I have examined, five had the normal number of 14 feathers, including in this number the two middle sickle-feathers; six others (viz. a Caffre cock, Gold-spangled Polish cock, Cochin hen, Sultan hen, Game hen, and Malay hen) had 16;

---

[59] 'Cottage Gardener,' Jan. 3rd, 1860, p. 218.

[60] Mr. Williams, in a paper read before the Dublin Nat. Hist. Soc., quoted in 'Cottage Gardener,' 1856, p. 161.

[61] 'De l'Espèce,' 1859, p. 442. For the occurrence of black-boned fowls in South America, *see* Roulin, in 'Mém. de l'Acad. des Sciences,' tom. vi. p. 351; and Azara, 'Quadrupèdes du Paraguay,' tom. ii. p. 324. A frizzled fowl sent to me from Madras had black bones.

and two (an old Cochin cock and Malay hen) had 17 feathers. The rumpless fowl has no tail, and in a bird which I kept alive the oil-gland had aborted; but this bird, though the os coccygis was extremely imperfect, had a vestige of a tail with two rather long feathers in the position of the outer caudals. This bird came from a family where, as I was told, the breed had kept true for twenty years; but rumpless fowls often produce chickens with tails.[62] An eminent physiologist[63] has recently spoken of this breed as a distinct species; had he examined the deformed state of the os coccyx he would never have come to this conclusion; he was probably misled by the statement, which may be found in some works, that tailless fowls are wild in Ceylon; but this statement, as I have been assured by Mr. Layard and Dr. Kellaert, who have so closely studied the birds of Ceylon, is utterly false.

The tarsi vary considerably in length, being relatively to the femur considerably longer in the Spanish and Frizzled, and shorter in the Silk and Bantam breeds, than in the wild *G. bankiva;* but in the latter, as we have seen, the tarsi vary in length. The tarsi are often feathered. The feet in many breeds are furnished with additional toes. Golden-spangled Polish fowls are said[64] to have the skin between their toes much developed; Mr. Tegetmeier observed this in one bird, but it was not so in one which I examined. In Cochins the middle toe is said[65] to be nearly double the length of the lateral toes, and therefore much longer than in *G. bankiva* or in other fowls; but this was not the case in two which I examined. The nail of the middle toe in this same breed is surprisingly broad and flat, but in a variable degree in two birds which I examined; of this structure in the nail there is only a trace in *G. bankiva.*

The voice differs slightly, as I am informed by Mr. Dixon, in almost every breed. The Malays[66] have a loud, deep, somewhat prolonged crow, but with considerable individual differences. Colonel Sykes remarks that the domestic Kulm cock in India has not the shrill clear pipe of the English bird, and "his scale of notes appears more limited." Dr. Hooker was struck with the "prolonged howling screech" of the cocks in Sikhim.[67] The crow of the Cochin is notoriously and ludicrously different from that of the common cock. The disposition of the different breeds is widely different, varying from the savage and defiant temper of the Game-cock to the extremely peaceable temper of the Cochin. The latter, it has been asserted, "graze to a much greater extent than any other varieties." The Spanish fowls suffer more from frost than other breeds.

Before we pass on to the skeleton, the degree of distinctness of the several breeds from *G. bankiva* ought to be noticed. Some

[62] Mr. Hewitt, in Tegetmeier's 'Poultry Book,' 1866, p. 231.

[63] Dr. Broca, in Brown-Sequard's 'Journal de Phys.,' tom. ii. p. 361.

[64] Dixon's 'Ornamental Poultry,' p. 325.

[65] 'Poultry Chronicle,' vol. i. p. 485. Tegetmeier's 'Poultry Book,' 1866, p. 41. On Cochins grazing, idem, p. 46.

[66] Ferguson on 'Prize Poultry,' p. 187.

[67] Col. Sykes in 'Proc. Zoolog. Soc.,' 1832, p. 151. Dr. Hooker's 'Himalayan Journals,' vol. i. p. 314.

writers speak of the Spanish as one of the most distinct breeds, and so it is in general aspect; but its characteristic differences are not important. The Malay appears to me more distinct, from its tall stature, small drooping tail with more than fourteen tail-feathers, and from its small comb and wattles; nevertheless one Malay sub-breed is coloured almost exactly like *G. bankiva.* Some authors consider the Polish fowl as very distinct; but this is a semi-monstrous breed, as shown by the protuberant and irregularly perforated skull. The Cochin, with its deeply furrowed frontal bones, peculiarly shaped occipital foramen, short wing-feathers, short tail containing more than fourteen feathers, broad nail to the middle toe, fluffy plumage, rough and dark-coloured eggs, and especially from its peculiar voice, is probably the most distinct of all the breeds. If any one of our breeds has descended from some unknown species, distinct from *G. bankiva,* it is probably the Cochin; but the balance of evidence does not favour this view. All the characteristic differences of the Cochin breed are more or less variable, and may be detected in a greater or lesser degree in other breeds. One sub-breed is coloured closely like *G. bankiva.* The feathered legs, often furnished with an additional toe, the wings incapable of flight, the extremely quiet disposition, indicate a long course of domestication; and these fowls come from China, where we know that plants and animals have been tended from a remote period with extraordinary care, and where consequently we might expect to find profoundly modified domestic races.

Osteological Differences.—I have examined twenty-seven skeletons and fifty-three skulls of various breeds, including three of *G. bankiva:* nearly half of these skulls I owe to the kindness of Mr. Tegetmeier, and three of the skeletons to Mr. Eyton.

The *Skull* differs greatly in size in different breeds, being nearly twice as long in the largest Cochins, but not nearly twice as broad, as in Bantams. The bones at the base, from the occipital foramen to the anterior end (including the quadrates and pterygoids), are absolutely identical in *shape* in all the skulls. So is the lower jaw. In the forehead slight differences are often perceptible between the males and females, evidently caused by the presence of the comb. In every case I take the skull of *G. bankiva* as the standard of comparison. In four Games, in one Malay hen, in an

African cock, in a Frizzled cock from Madras, in two black-boned Silk hens, no differences occur worth notice. In three *Spanish* cocks, the form of the forehead between the orbits differs considerably; in one it is considerably depressed, whilst in the two others it is rather prominent, with a deep medial furrow; the skull of the hen is smooth. In three skulls of *Sebright Bantams* the crown is more globular, and slopes more abruptly to the occiput, than in *G. bankiva.* In a Bantam or Jumper from Burmah these same characters are more strongly pronounced, and the supra-occiput is more pointed. In a black Bantam the skull is not so globular, and the occipital foramen is very large, and has nearly the same sub-triangular outline presently to be described in Cochins; and in this skull the two ascending branches of the premaxillary are overlapped in a singular manner by the processes of the nasal bone, but, as I have seen only one specimen, some of these differences may be individual. Of Cochins and Brahmas (the latter a crossed race approaching closely to Cochins) I have examined seven skulls; at the point where the ascending branches of the premaxillary rest on the frontal bone the surface is much depressed, and from this depression a deep medial furrow extends backwards to a variable distance; the edges of this fissure are rather prominent, as is the top of the skull behind and over the orbits. These characters are less developed in the hens. The pterygoids, and the processes of the lower jaw, relatively to the size of the head, are broader than in *G. bankiva*; and this is likewise the case with Dorkings when of large size. The terminal fork of the hyoid bone in Cochins is twice as wide as in *G. bankiva*, whereas the length of the other hyoid bones is only as three to two. But the most remarkable character is the shape of the occipital foramen: in *G. bankiva* (A) the breadth in a horizontal line exceeds the height in a vertical line, and the outline is nearly circular; whereas in Cochins (B) the outline is sub-triangular, and the vertical line exceeds the horizontal line in length. This same form likewise occurs in the black Bantam above referred to, and an approach to it may be seen in some Dorkings, and in a slight degree in certain other breeds.

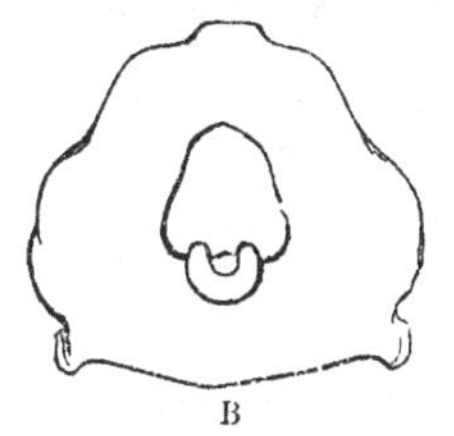

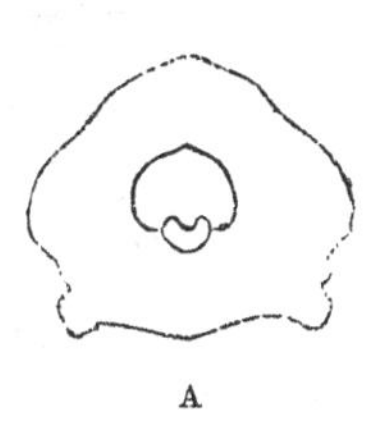

Fig. 33.—Occipital Foramen, of natural size. A. Wild *Gallus bankiva*. B. Cochin Cock.

Of *Dorkings* I have examined three skulls, one belonging to the white sub-breed; the one character deserving notice is the breadth of the frontal bones, which are moderately furrowed in the middle; thus in a skull which was less than once and a half the length of that of *G. bankiva*, the breadth between the orbits was exactly double. Of *Hamburghs* I have examined four skulls (male and female) of the pencilled sub-breed, and one (male) of the spangled sub-breed; the nasal bones stand remarkably wide apart, but in a variable degree; consequently narrow membrane-covered spaces are left between the tips of the two ascending branches of the premaxillary bones, which are rather short, and between these branches and the nasal bones. The surface of the frontal bone, on which the branches of the premaxillary rest, is very little depressed. These peculiarities no doubt stand in close relation with the broad flattened rose-comb characteristic of the Hamburgh breed.

I have examined fourteen skulls of *Polish and other crested breeds.* Their differences are extraordinary. First for nine skulls of different sub-breeds of English Polish fowls. The hemispherical protuberance of the frontal bones[68] may be seen in the accompanying drawings, in which (B) the skull of a white-crested Polish fowl is shown obliquely from above, with the skull (A) of *G. bankiva* in the same position. In fig. 35 longitudinal sections are given of the skulls of a Polish fowl, and, for comparison, of a Cochin of the same size. The protuberance in all Polish fowls occupies the same position, but differs much in size. In one of my nine specimens it was extremely slight. The degree to which the protuberance is ossified varies greatly, larger or smaller portions of bone being replaced by membrane. In one specimen there was only a single open pore; generally, there are many variously-shaped open spaces, the bone forming an irregular reticulation. A medial, longitudinal, arched ribbon of bone is generally retained, but in one specimen there was no bone whatever over the whole protuberance, and the skull when cleaned and viewed from above presented the appearance of an open basin. The change in the whole internal form of the skull is surprisingly great. The brain is modified in a corresponding manner, as is shown in the two longitudinal sections,

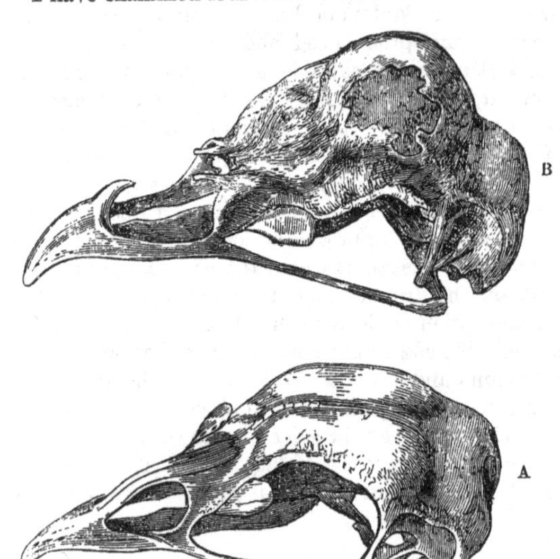

Fig. 34.—Skulls of natural size, viewed from above, a little obliquely. A. Wild *Gallus bankiva*. B. White-crested Polish Cock.

[68] *See* Mr. Tegetmeier's account, with woodcuts, of the skull of Polish fowls, in 'Proc. Zoolog. Soc.,' Nov. 25th, 1856. For other references, *see* Isid. Geoffroy Saint Hilaire, 'Hist. Gén. des Anomalies,' tom. i. p. 287. M. C. Dareste suspects ('Recherches sur les Conditions de la Vie,' &c., Lille, 1863, p. 36) that the protuberance is not formed by the frontal bones, but by the ossification of the dura mater.

which deserve attentive consideration. The upper and anterior cavity of the three into which the skull may be divided, is the one which is so greatly modified; it is evidently much larger than in the Cochin skull of the same size, and extends much further beyond the interorbital septum, but laterally is less deep. Whether this cavity is entirely filled by the brain, may be doubted. In the skull of the Cochin and of all

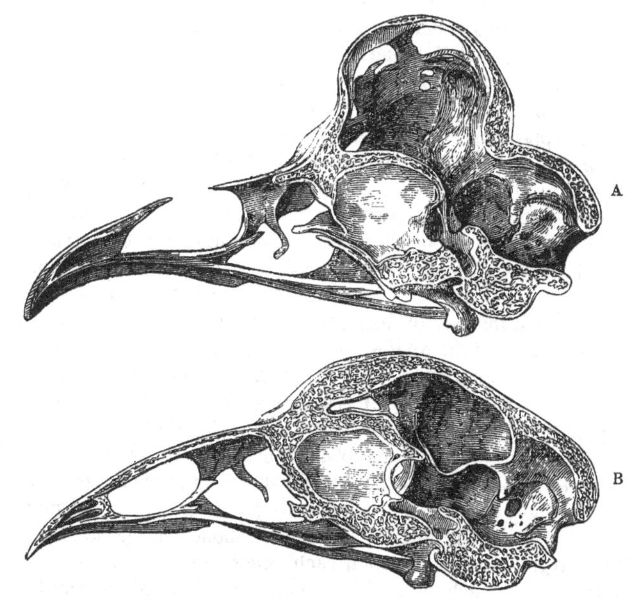

Fig. 35.—Longitudinal sections of Skull, of natural size, viewed laterally. A. Polish Cock. B. Cochin Cock, selected for comparison with the above from being of nearly the same size.

ordinary fowls a strong internal ridge of bone separates the anterior from the central cavity; but this ridge is entirely absent in the Polish skull here figured. The shape of the central cavity is circular in the Polish, and lengthened in the Cochin skull. The shape of the posterior cavity, together with the position, size, and number of the pores for the nerves, differ much in these two skulls. A pit deeply penetrating the occipital bone of the Cochin is entirely absent in this Polish skull, whilst in another specimen it was well developed. In this second specimen the whole internal surface of the posterior cavity likewise differs to a certain extent in shape. I made sections of two other skulls,—namely, of a Polish fowl with the protuberance singularly little developed, and of a Sultan in which it was a little more developed; and when these two skulls were placed between the two above figured (fig. 35), a perfect gradation in the configuration of each part of the internal surface could be traccd. In the Polish skull, with a small protuberance, the ridge between the anterior and middle cavities was present, but low; and in the Sultan this ridge was replaced by a narrow furrow standing on a broad raised eminence.

It may naturally be asked whether these remarkable modifications in the form of the brain affect the intellect of Polish fowls; some writers have stated that they are extremely stupid, but Bechstein and Mr. Tegetmeier have shown that this is by no means generally the case. Nevertheless Bechstein[69] states that he had a Polish hen which "was crazy, and anxiously wandered about all day long." A hen in my possession was solitary in her habits, and was often so absorbed in reverie that she could be touched; she was also deficient in the most singular manner in the faculty of finding her way, so that, if she strayed a hundred yards from her feeding-place, she was completely lost, and would then obstinately try to proceed in a wrong direction. I have received other and similar accounts of Polish fowls appearing stupid or half-idiotic.[70]

To return to the skull. The posterior part, viewed externally, differs little from that of *G. bankiva*. In most fowls the posterior-lateral process of the frontal bone and the process of the squamosal bone run together and are ossified near their extremities: this union of the two bones, however, is not constant in any breed; and in eleven out of fourteen skulls of crested breeds, these processes were quite distinct. These processes, when not united, instead of being inclined anteriorly as in all common breeds, descend at right angles to the lower jaw; and in this case the longer axis of the bony cavity of the ear is likewise more perpendicular than in other breeds. When the squamosal process is free, instead of expanding at the tip, it is reduced to an extremely fine and pointed style, of variable length. The pterygoid and quadrate bones present no difference. The palatine bones are a little more curved upwards at their posterior ends. The frontal bones, anteriorly to the protuberance, are, as in Dorkings, very broad, but in a variable degree. The nasal bones either stand far apart, as in Hamburghs, or almost touch each other, and in one instance were ossified together. Each nasal bone properly sends out in front two long processes of equal lengths, forming a fork; but in all the Polish skulls, except one, the inner process was considerably, but in a variable degree, shortened and somewhat upturned. In all the skulls, except one, the two ascending branches of the premaxillary, instead of running up between the processes of the nasal bones and resting on the ethmoid bone, are much shortened and terminate in a blunt, somewhat upturned point. In those skulls in which the nasal bones approach quite close to each other or are ossified together, it would be impossible for the ascending branches of the premaxillary to reach the ethmoid and frontal bones; hence we see that even the relative connection of the bones has been changed. Apparently in consequence of the branches of the premaxillary and of the inner processes of the nasal bones being somewhat upturned, the external orifices of the nostrils are upraised and assume a crescentic outline.

I must still say a few words on some of the foreign Crested breeds. The skull of a crested, rumpless, white Turkish fowl is very slightly protuberant, and but little perforated; the ascending branches of the premaxillary

---

69 'Naturgeschichte Deutschlands,' Band iii. (1793), s. 400.

70 The 'Field,' May 11th, 1861. I have received communications to a similar effect from Messrs. Brent and Tegetmeier.

are well developed. In another Turkish breed, called Ghoondooks, the skull is considerably protuberant and perforated; the ascending branches of the premaxillary are so much aborted that they project only $\frac{1}{13}$th of an inch; and the inner processes of the nasal bone are so completely aborted, that the surface where they should have projected is quite smooth. Here then we see these two bones modified to an extreme degree. Of Sultans (another Turkish breed) I examined two skulls; in that of the female the protuberance was much larger than in the male. In both skulls the ascending branches of the premaxillary were very short, and in both the basal portion of the inner processes of the nasal bones were ossified together. These Sultan skulls differed from those of English Polish fowls in the frontal bones, anteriorly to the protuberance, not being broad.

The last skull which I need describe is a unique one, lent to me by Mr. Tegetmeier: it resembles a Polish skull in most of its characters, but has not the great frontal protuberance; it has, however, two rounded knobs of a different nature, which stand more in front, above the lachrymal bones.

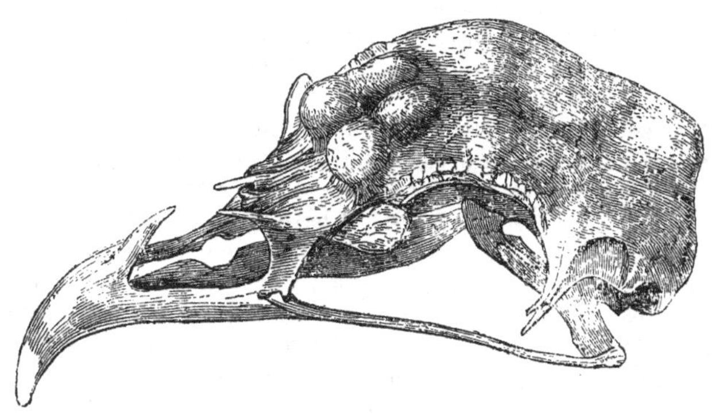

Fig. 36.—Skull of Horned Fowl, of natural size, viewed from above, a little obliquely. (In the possession of Mr. Tegetmeier.)

These curious knobs, into which the brain does not enter, are separated from each other by a deep medial furrow; and this is perforated by a few minute pores. The nasal bones stand rather wide apart, with their inner processes, and the ascending branches of the premaxillary, upturned and shortened. The two knobs no doubt supported the two great horn-like projections of the comb.

From the foregoing facts we see in how astonishing a manner some of the bones of the skull vary in Crested fowls. The protuberance may certainly be called in one sense a monstrosity, as being wholly unlike anything observed in nature: but as in ordinary cases it is not injurious to the bird, and as it is strictly inherited, it can hardly in another sense be called a monstrosity. A series may be formed commencing with the black-boned Silk fowl, which has a very small crest with the skull beneath penetrated only by a few minute orifices, but with no other change in its structure; and from this first stage we may proceed to fowls with a moderately large crest, which rests, according to Bechstein, on a fleshy mass, but without any protuberance in the skull. I may add that I have seen a similar fleshy or fibrous mass beneath the tuft of feathers on the head of the Tufted duck; and in this case there was no actual protuberance in the skull, but it had become a little more globular. Lastly, when we come to fowls with a largely developed crest, the skull becomes largely protuberant and is perforated by a multitude of irregular open spaces. The close relation between the crest and the size of the bony protuberance is shown in another way; for Mr. Tegetmeier informs me that if chickens lately hatched be selected with a large bony protuberance, when adult they will have a large crest. There can be no doubt that in former times the breeder of Polish fowls attended solely to the crest, and not to the skull; nevertheless, by increasing the crest, in which he has wonderfully succeeded, he has unintentionally made the skull protuberant to an astonishing degree; and through correlation of growth, he has at the same time affected the form and relative connexion of the premaxillary and nasal bones, the shape of the orifice of the nose, the breadth of the frontal bones, the shape of the post-lateral processes of the frontal and squamosal bones, the direction of the axis of the bony cavity of the ear, and lastly the internal configuration of the whole skull together with the shape of the brain.

Vertebræ.—In *G. bankiva* there are fourteen cervical, seven dorsal with ribs, apparently fifteen lumbar and sacral, and six caudal vertebræ;[71] but the lumbar and sacral are so much anchylosed that I am not sure of their number, and this makes the comparison of the total number of vertebræ in the several breeds difficult. I have spoken of six caudal vertebræ, because the basal one is almost completely anchylosed with the pelvis; but if we consider the number as seven, the caudal vertebræ agree in all the skeletons. The cervical vertebræ are, as just stated, in appearance fourteen; but out of twenty-three skeletons in a fit state for examination, in five of them, namely, in two Games, in two pencilled Hamburghs, and in a Polish, the fourteenth vertebra bore ribs, which, though small, were perfectly developed with a double articulation. The presence of these little ribs cannot be considered as a fact of much importance, for all the cervical vertebræ bear representatives of ribs; but their development in the fourteenth vertebra reduces the size of the passages in the transverse processes, and makes this vertebra exactly like the first dorsal vertebra. The addition of these little ribs does not affect the fourteenth cervical alone, for properly the ribs of the first true dorsal vertebra are destitute of processes; but in some of the skeletons in which the fourteenth cervical bore little ribs, the first pair of true ribs had well-developed processes. When we know that the sparrow has only nine, and the swan twenty-three cervical vertebræ,[72] we need feel no surprise at the number of the cervical vertebræ in the fowl being, as it appears, variable.

There are seven dorsal vertebræ bearing ribs; the first dorsal is never

---

[71] It appears that I have not correctly designated the several groups of vertebræ, for a great authority, Mr. W. K. Parker ('Transact. Zoolog. Soc.,' vol. v. 198), specifies 16 cervical, 4 dorsal, 15 lumbar, and 6 caudal vertebræ in this genus. But I have used the same terms in all the following descriptions.

[72] Macgillivray, 'British Birds,' vol. i. p. 25.

anchylosed with the succeeding four, which are generally anchylosed together. In one Sultan fowl, however, the two first dorsal vertebræ were free. In two skeletons, the fifth dorsal was free; generally the sixth is free (as in *G. bankiva*), but sometimes only at its posterior end, where in contact with the seventh. The seventh dorsal vertebra, in every case excepting in one Spanish cock, was anchylosed with the lumbar vertebræ. So that the degree to which these middle dorsal vertebræ are anchylosed together is variable.

Seven is the normal number of true ribs, but in two skeletons of the Sultan fowl (in which the fourteenth cervical vertebra was not furnished with little ribs) there were eight pairs; the eighth pair seemed to be developed on a vertebra corresponding with the first lumbar in *G. bankiva*; the sternal portion of both the seventh and eighth ribs did not reach the sternum. In four skeletons in which ribs were developed on the fourteenth cervical vertebra, there were, when these cervical ribs are included, eight pairs; but in one Game-cock, in which the fourteenth cervical was furnished with ribs, there were only six pairs of true dorsal ribs; the sixth pair in this case did not have processes, and thus resembled the seventh pair in other skeletons; in this game-cock, as far as could be judged from the appearance of the lumbar vertebræ, a whole dorsal vertebra with its ribs was missing. We thus see that the ribs (whether or not the little pair attached to the fourteenth cervical vertebra be counted) vary from six to eight pair. The sixth pair is frequently not furnished with processes. The sternal portion of the seventh pair is extremely broad in Cochins, and is completely ossified. As previously stated, it is scarcely possible to count the lumbo-sacral vertebræ; but they certainly do not correspond in shape or number in the several skeletons. The caudal vertebræ are closely similar in all the skeletons, the only difference being, whether or not the basal one is anchylosed to the pelvis; they hardly vary even in length, not being shorter in Cochins, with their short tail-feathers, than in other breeds; in a Spanish cock, however, the caudal vertebræ were a little elongated. In three rumpless fowls the caudal vertebræ were few in number, and anchylosed together into a misformed mass.

In the individual vertebræ the differences in structure are very slight. In the atlas the cavity for the occipital condyle is either ossified into a ring, or is, as in Bankiva, open on its upper margin. The upper arc of the spinal canal is a little more arched in Cochins, in conformity with the shape of occipital foramen, than in *G. bankiva*. In several skeletons a difference, but not of much importance, may be observed, which commences at the fourth cervical vertebra, and is greatest at about the sixth, seventh, or eighth vertebra; this consists in the hæmal descending processes being united to the body of the vertebra by a sort of buttress. This structure may be observed in Cochins, Polish, some Hamburghs, and probably other breeds; but is absent, or barely developed, in Game, Dorking, Spanish, Bantam, and

Fig. 37.—Sixth Cervical Vertebra, of natural size, viewed laterally. A. Wild *Gallus bankiva*. B. Cochin Cock.

several other breeds examined by me. On the dorsal surface of the sixth cervical vertebra in Cochins three prominent points are more strongly developed than in the corresponding vertebra of the Game-fowl or *G. bankiva.*

Pelvis.—This differs in some few points in the several skeletons. The anterior margin of the ilium seems at first to vary much in outline, but this is chiefly due to the degree to which the margin in the middle part is ossified to the crest of the spine; the outline, however, does differ in being more truncated in Bantams, and more rounded in certain breeds, as in Cochins. The outline of the ischiadic foramen differs considerably, being nearly circular in Bantams, instead of egg-shaped as in the Bankiva, and more regularly oval in some skeletons, as in the Spanish. The obturator notch is also much less elongated in some skeletons than in others. The end of the pubic bone presents the greatest difference; being hardly enlarged in the Bankiva; considerably and gradually enlarged in Cochins, and in a lesser degree in some other breeds; and abruptly enlarged in Bantams. In one Bantam this bone extended very little beyond the extremity of the ischium. The whole pelvis in this latter bird differed widely in its proportions, being far broader proportionally to its length than in Bankiva.

Sternum.—This bone is generally so much deformed that it is scarcely possible to compare its form strictly in the several breeds. The shape of the triangular extremity of the lateral processes differs considerably, being either almost equilateral or much elongated. The front margin of the crest is more or less perpendicular and varies greatly, as does the curvature of the posterior end, and the flatness of the lower surface. The outline of the manubrial process also varies, being wedge-shaped in the Bankiva, and rounded in the Spanish breed. The *furcula* differs in being more or less arched, and greatly, as may be seen in the accompanying outlines, in the shape of the terminal plate; but the shape of this part differed a little in two skeletons of the wild Bankiva. The *coracoids* present no difference worth notice. The *scapula* varies in shape, being of nearly uniform breadth in Bankiva, much broader in the middle in the Polish fowl, and abruptly narrowed towards the apex in the two Sultan fowls.

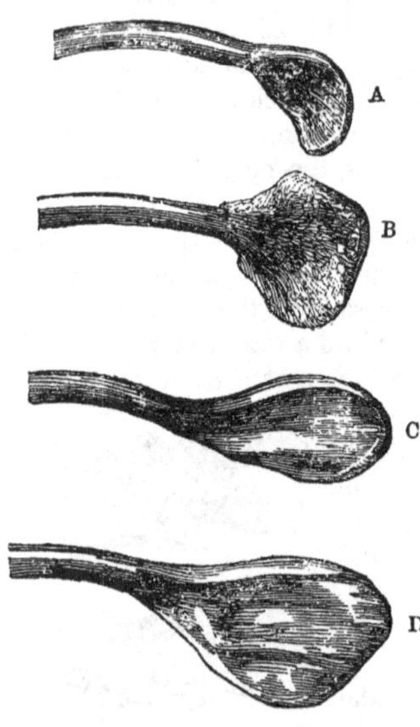

Fig. 38.—Extremity of the Furcula, of natural size, viewed laterally. A. Wild *Gallus bankiva.* B. Spangled Polish Fowl. C. Spanish Fowl. D. Dorking Fowl.

I carefully compared each separate bone of the leg and wing, relatively to the same bones in the wild Bankiva, in the following breeds, which I thought were the most likely to differ; namely, in Cochin, Dorking,

Spanish, Polish, Burmese Bantam, Frizzled Indian, and black-boned Silk fowls; and it was truly surprising to see how absolutely every process, articulation, and pore agreed, though the bones differed greatly in size. The agreement is far more absolute than in other parts of the skeleton. In stating this, I do not refer to the relative thickness and length of the several bones; for the tarsi varied considerably in both these respects. But the other limb-bones varied little even in relative length.

Finally, I have not examined a sufficient number of skeletons to say whether any of the foregoing differences, except in the skull, are characteristic of the several breeds. Apparently some differences are more common in certain breeds than in others,—as an additional rib to the fourteenth cervical vertebra in Hamburghs and Games, and the breadth of the end of the pubic bone in Cochins. Both skeletons of the Sultan fowl had eight dorsal vertebræ, and the end of the scapula in both was somewhat attenuated. In the skull, the deep medial furrow in the frontal bones and the vertically elongated occipital foramen seem to be characteristic of Cochins; as is the great breadth of the frontal bones in Dorkings; the separation and open spaces between the tips of the ascending branches of the premaxillaries and nasal bones, as well as the front part of the skull being but little depressed, characterise Hamburghs; the globular shape of the posterior part of the skull seems to be characteristic of laced Bantams; and lastly, the protuberance of the skull with the ascending branches of the premaxillaries partially aborted, together with the other differences before specified, are eminently characteristic of Polish and other Crested fowls.

But the most striking result of our examination of the skeleton is the great variability of all the bones except those of the extremities. To a certain extent we can understand why the skeleton fluctuates so much in structure; fowls have been exposed to unnatural conditions of life, and their whole organisation has thus been rendered variable; but the breeder is quite indifferent to, and never intentionally selects, any modifications in the skeleton. External characters, if not attended to by man,—such as the number of the tail and wing feathers and their relative lengths, which in wild birds are generally constant points,—fluctuate in our domestic fowls in the same manner as the several parts of the skeleton. An additional toe is a "point" in Dorkings, and has become a fixed character, but is variable in

Cochins and Silk-fowls. The colour of the plumage and the form of the comb are in most breeds, or even sub-breeds, eminently fixed characters; but in Dorkings these points have not been attended to, and are variable. When any modification in the skeleton is related to some external character which man values, it has been, unintentionally on his part, acted on by selection, and has become more or less fixed. We see this in the wonderful protuberance of the skull, which supports the crest of feathers in Polish fowls, and which by correlation has affected other parts of the skull. We see the same result in the two protuberances which support the horns in the horned fowl, and in the flattened shape of the front of the skull in Hamburghs consequent on their flattened and broad "rose-combs." We know not in the least whether additional ribs, or the changed outline of the occipital foramen, or the changed form of the scapula, or of the extremity of the furcula, are in any way correlated with other structures, or have arisen from the changed conditions and habits of life to which our fowls have been subjected; but there is no reason to doubt that these various modifications in the skeleton could be rendered, either by direct selection, or by the selection of correlated structures, as constant and as characteristic of each breed, as are the size and shape of the body, the colour of the plumage, and the form of the comb.

Effects of the Disuse of Parts.

Judging from the habits of our European gallinaceous birds, *Gallus bankiva* in its native haunts would use its legs and wings more than do our domestic fowls, which rarely fly except to their roosts. The Silk and the Frizzled fowls, from having imperfect wing-feathers, cannot fly at all; and there is reason to believe that both these breeds are ancient, so that their progenitors during many generations cannot have flown. The Cochins, also, from their short wings and heavy bodies, can hardly fly up to a low perch. Therefore in these breeds, especially in the two first, a considerable diminution in the wing-bones might have been expected, but this is not the case. In every specimen, after disarticulating and cleaning the bones, I carefully compared the relative length of the two main bones of the wing to each other, and of the two main bones of the leg to each other, with those of *G. bankiva*; and it was surprising to see (except in the case of the tarsi) how exactly the same relative length had been retained. This fact is curious, from showing how truly the proportions of an organ may be inherited, although not fully exercised during many generations. I then compared in several breeds the length of the femur and tibia with the humerus and ulna, and likewise these same bones with those of *G. bankiva;* the result was that the wing-bones in all the breeds (except the Burmese Jumper, which has unnaturally short legs) are slightly shortened relatively to the leg-bones; but the decrease is so slight that it may be due to the standard specimen of *G. bankiva* having accidentally had wings of slightly greater length than usual; so that the measurements are not worth giving. But it deserves notice that the Silk and Frizzled fowls, which are quite incapable of flight, had their wings *less* reduced relatively to their legs than in almost any other breed! We have seen with domesticated pigeons that the bones of the wings are somewhat reduced in length, whilst the primary feathers are rather increased in length, and it is just possible, though not probable, that in the Silk and Frizzled fowls any tendency to decrease in the length of the wing-bones from disuse may have been checked through the law of compensation, by the decreased growth of the wing-feathers, and consequent increased supply of nutriment. The wing-bones, however, in both these breeds, are found to be slightly reduced in length when judged by the standard of the length of the sternum or head, relatively to these same parts in *G. bankiva.*

The actual weight of the main bones of the leg and wing in twelve breeds is given in the two first columns in the following table. The calculated weight of the wing-bones relatively to the leg-bones, in comparison with the leg and wing-bones of *G. bankiva*, are given in the third column,—the weight of the wing-bones in *G. bankiva* being called a hundred.[73]

TABLE I.

	Names of Breeds.		Actual Weight of Femur and Tibia.	Actual Weight of Humerus and Ulna.	Weight of Wing-bones relatively to the Leg-bones, in comparison with these same bones in G. bankiva.
			Grains.	Grains.	
	Gallus bankiva	wild male	86	54	100
1	Cochin	male	311	162	83
2	Dorking	male	557	248	70
3	Spanish (Minorca) ..	male	386	183	75
4	Gold Spangled Polish	male	306	145	75
5	Game, black-breasted	male	293	143	77
6	Malay	female	231	116	80
7	Sultan	male	189	94	79
8	Indian Frizzled	male	206	88	67
9	Burmese Jumper ..	female	53	36	108
10	Hamburgh (pencilled)	male	157	104	106
11	Hamburgh (pencilled)	female	114	77	108
12	Silk (black-boned) ..	female	88	57	103

[73] It may be well to explain how the calculation has been made for the third column. In *G. bankiva* the leg-bones are to the wing-bones as 86 : 54, or as (neglecting decimals) 100 : 62;—in Cochins as 311 : 162, or as 100 : 52;—

In the eight first birds, belonging to distinct breeds, in this table, we see a decided reduction in the weight of the bones of the wing. In the Indian Frizzled fowl, which cannot fly, the reduction is carried to the greatest extent, namely, to thirty-three per cent. of their proper proportional weight. In the next four birds, including the Silk-hen, which is incapable of flight, we see that the wings, relatively to the legs, are slightly increased in weight; but it should be observed that, if in these birds the legs had become from any cause reduced in weight, this would give the false appearance of the wings having increased in relative weight. Now a reduction of this nature has certainly occurred with the Burmese Jumper, in which the legs are abnormally short, and in the two Hamburghs and Silk fowl, the legs, though not short, are formed of remarkably thin and light bones. I make these statements, not judging by mere eyesight, but after having calculated the weights of the leg-bones relatively to those of *G. bankiva*, according to the only two standards of comparison which I could use, namely, the relative lengths of the head and sternum; for I do not know the weight of the body in *G. bankiva*, which would have been a better standard. According to these standards, the leg-bones in these four fowls are in a marked manner far lighter than in any other breed. It may therefore be concluded that in all cases in which the legs have not been through some unknown cause much reduced in weight, the wing-bones have become reduced in weight relatively to the leg-bones, in comparison with those of *G. bankiva*. And this reduction of weight may, I apprehend, safely be attributed to disuse.

To make the foregoing table quite satisfactory, it ought to have been shown that in the eight first birds the leg-bones have not actually increased in weight out of due proportion with the rest of the body; this I cannot show, from not knowing, as already remarked, the weight of the wild Bankiva.[74] I am indeed inclined to suspect that the leg-bones in the Dorking, No. 2 in the table, are proportionally too heavy; but this bird was a very large one, weighing 7 lb. 2 oz., though very thin. Its leg-bones were more than ten times as heavy as those of the Burmese Jumper! I tried to ascertain the length both of the leg-bones and wing-bones relatively to other parts of the body and skeleton; but the whole organisation in these birds, which have been so long domesticated, has become so variable, that no certain conclusions could be reached. For instance, the legs of the above Dorking cock were nearly three-quarters of an inch too short relatively to the length of the sternum, and more than

in Dorkings as 557 : 248, or as 100 : 44; and so on for the other breeds. We thus get the series of 62, 52, 44 for the relative weights of the wing-bones in *G. bankiva*, Cochins, Dorkings, &c. And now taking 100, instead of 62, for the weight of the wing-bones in *G. bankiva*, we get, by another rule of three, 83 as the weight of the wing-bones in Cochins; 70 in the Dorkings; and so on for the remainder of the third column in the table.

[74] Mr. Blyth (in 'Annals and Mag. of Nat. Hist.,' 2nd series, vol. i., 1848, p. 456) gives 3¼ lb. as the weight of a full-grown male *G. bankiva*; but from what I have seen of the skins and skeletons of various breeds, I cannot believe that my two specimens of *G. bankiva* could have weighed so much.

three-quarters of an inch too long relatively to the length of the skull, in comparison with these same parts in *G. bankiva.*

In the following Table II. in the two first columns we see in inches and decimals the length of the sternum, and the extreme depth of its crest to which the pectoral muscles are attached. In the third column we have the calculated depth of the crest, relatively to the length of the sternum, in comparison with these same parts in *G. bankiva.*[75]

TABLE II.

	Names of Breeds.		Length of Sternum.	Depth of Crest of Sternum.	Depth of Crest, relatively to the length of the Sternum, in comparison with G. bankiva.
			Inches.	Inches.	
	Gallus bankiva	male	4·20	1·40	100
1	Cochin	male	5·83	1·55	78
2	Dorking	male	6·95	1·97	84
3	Spanish	male	6·10	1·83	90
4	Polish	male	5·07	1·50	87
5	Game	male	5·55	1·55	81
6	Malay	female	5·10	1·50	87
7	Sultan	male	4·47	1·36	90
8	Frizzled hen	male	4·25	1·20	84
9	Burmese Jumper	female	3·06	0·85	81
10	Hamburgh	male	5·08	1·40	81
11	Hamburgh	female	4·55	1·26	81
12	Silk fowl	female	4·49	1·01	66

By looking to the third column we see that in every case the depth of the crest relatively to the length of the sternum, in comparison with *G. bankiva*, is diminished, generally between 10 and 20 per cent. But the degree of reduction varies much, partly in consequence of the frequently deformed state of the sternum. In the Silk-fowl, which cannot fly, the crest is 34 per cent. less deep than what it ought to have been. This reduction of the crest in all the breeds probably accounts for the great variability, before referred to, in the curvature of the furcula, and in the shape of its sternal extremity. Medical men believe that the abnormal form of the spine so commonly observed in women of the higher ranks results from the attached muscles not being fully exercised. So it is with our domestic fowls, for they use their pectoral muscles but little, and, out of twenty-five sternums examined by me, three alone were perfectly symmetrical, ten were moderately crooked, and twelve were deformed to an extreme degree.

Finally, we may conclude with respect to the various breeds of the fowl, that the main bones of the wing have probably been shortened in a very slight degree; that they have cer-

[75] The third column is calculated on the same principle as explained in the previous foot-note, p. 271.

tainly become lighter relatively to the leg-bones in all the breeds in which these latter bones are not unnaturally short or delicate; and that the crest of the sternum, to which the pectoral muscles are attached, has invariably become less prominent, the whole sternum being also extremely liable to deformity. These results we may attribute to the lessened use of the wings.

Correlation of Growth.—I will here sum up the few facts which I have collected on this obscure, but important, subject. In Cochins and Game-fowls there is some relation between the colour of the plumage and the darkness of the egg-shell and even of the yolk. In Sultans the additional sickle-feathers in the tail are apparently related to the general redundancy of the plumage, as shown by the feathered legs, large crest, and beard. In two tailless fowls which I examined the oil-gland was aborted. A large crest of feathers, as Mr. Tegetmeier has remarked, seems always accompanied by a great diminution or almost entire absence of the comb. A large beard is similarly accompanied by diminished or absent wattles. These latter cases apparently come under the law of compensation or balancement of growth. A large beard beneath the lower jaw and a large top-knot on the skull often go together. The comb when of any peculiar shape, as with Horned, Spanish, and Hamburgh fowls, affects in a corresponding manner the underlying skull; and we have seen how wonderfully this is the case with Crested fowls when the crest is largely developed. With the protuberance of the frontal bones the shape of the internal surface of the skull and of the brain is greatly modified. The presence of a crest influences in some unknown way the development of the ascending branches of the premaxillary bone, and of the inner processes of the nasal bones; and likewise the shape of the external orifice of the nostrils. There is a plain and curious correlation between a crest of feathers and the imperfectly ossified condition of the skull. Not only does this hold good with nearly all crested fowls, but likewise with tufted ducks, and as Dr. Günther informs me with tufted geese in Germany.

Lastly, the feathers composing the crest in male Polish fowls resemble hackles, and differ greatly in shape from those in the crest of the female. The neck, wing-coverts, and loins in the male bird are properly covered with hackles, and it would appear that feathers of this shape have spread by correlation to the head of the male. This little fact is interesting; because, though both sexes of some wild gallinaceous birds have their heads similarly ornamented, yet there is often a difference in the size and shape of feathers forming their crests. Furthermore there is in some cases, as in the male Gold and in the male Amherst pheasants (*P. pictus* and *Amherstii*), a close relation in colour, as well as in structure, between the plumes on the head and on the loins. Hence it would appear that the same law has regulated the state of the feathers on the head and body, both with species living under their natural conditions, and with birds which have varied under domestication.

CHAPTER VIII.

DUCKS — GOOSE — PEACOCK — TURKEY — GUINEA-FOWL — CANARY-BIRD — GOLD-FISH — HIVE-BEES — SILK-MOTHS.

DUCKS, SEVERAL BREEDS OF — PROGRESS OF DOMESTICATION — ORIGIN OF, FROM THE COMMON WILD-DUCK — DIFFERENCES IN THE DIFFERENT BREEDS — OSTEOLOGICAL DIFFERENCES — EFFECTS OF USE AND DISUSE ON THE LIMB-BONES.
GOOSE, ANCIENTLY DOMESTICATED — LITTLE VARIATION OF — SEBASTOPOL BREED.
PEACOCK, ORIGIN OF BLACK-SHOULDERED BREED.
TURKEY, BREEDS OF — CROSSED WITH THE UNITED STATES SPECIES — EFFECTS OF CLIMATE ON.
GUINEA-FOWL, CANARY-BIRD, GOLD-FISH, HIVE-BEES.
SILK-MOTHS, SPECIES AND BREEDS OF — ANCIENTLY DOMESTICATED — CARE IN THEIR SELECTION—DIFFERENCES IN THE DIFFERENT RACES—IN THE EGG, CATERPILLAR, AND COCOON STATES — INHERITANCE OF CHARACTERS — IMPERFECT WINGS — LOST INSTINCTS — CORRELATED CHARACTERS.

I WILL, as in previous cases, first briefly describe the chief domestic breeds of the duck:—

BREED 1. *Common Domestic Duck.*—Varies much in colour and in proportions, and differs in instincts and disposition from the wild-duck. There are several sub-breeds:—(1) The Aylesbury, of great size, white, with pale-yellow beak and legs; abdominal sack largely developed. (2) The Rouen, of great size, coloured like the wild-duck, with green or mottled beak; abdominal sack largely developed. (3) Tufted Duck, with a large top-knot of fine downy feathers, supported on a fleshy mass, with the skull perforated beneath. The top-knot in a duck which I imported from Holland was two and a half inches in diameter. (4) Labrador (or Canadian, or Buenos Ayres, or East Indian); plumage entirely black; beak broader, relatively to its length, than in the wild-duck; eggs slightly tinted with black. This sub-breed perhaps ought to be ranked as a breed; it includes two sub-varieties, one as large as the common domestic duck, which I have kept alive, and the other smaller and often capable of flight.[1] I presume it is this latter sub-variety which has been described in France[2] as flying well, being rather wild, and when cooked having the flavour of the wild-duck; nevertheless this sub-variety is polygamous, like other domesticated ducks and unlike the wild duck. These black Labrador ducks breed true;

[1] 'Poultry Chronicle (1854), vol. ii. p. 91, and vol. i. p. 330.

[2] Dr. Turral, in 'Bull. Soc. d'Acclimat.,' tom. vii., 1860, p. 541.

but a case is given by Dr. Turral of the French sub-variety producing young with some white feathers on the head and neck, and with an ochre-coloured patch on the breast.

BREED 2. *Hook-billed Duck.*—This bird presents an extraordinary appearance from the downward curvature of the beak. The head is often tufted. The common colour is white, but some are coloured like wild-ducks. It is an ancient breed, having been noticed in 1676.[3] It shows its prolonged domestication by almost incessantly laying eggs, like the fowls which are called everlasting layers.[4]

BREED 3. *Call-Duck.*—Remarkable from its small size, and from the extraordinary loquacity of the female. Beak short. These birds are either white, or coloured like the wild-duck.

BREED 4. *Penguin Duck.*—This is the most remarkable of all the breeds, and seems to have originated in the Malayan archipelago. It walks with its body extremely erect, and with its thin neck stretched straight upwards. Beak rather short. Tail upturned, including only 18 feathers. Femur and meta-tarsi elongated.

Almost all naturalists admit that the several breeds are descended from the common wild duck (*Anas boschas*); most fanciers, on the other hand, take as usual a very different view.[5] Unless we deny that domestication, prolonged during centuries, can affect even such unimportant characters as colour, size, and in a slight degree proportional dimensions and mental disposition, there is no reason whatever to doubt that the domestic duck is descended from the common wild species, for the one differs from the other in no important character. We have some historical evidence with respect to the period and progress of the domestication of the duck. It was unknown[6] to the ancient Egyptians, to the Jews of the Old Testament, and to the Greeks of the Homeric period. About eighteen centuries ago Columella[7] and Varro speak of the necessity of keeping ducks in netted enclosures like other wild fowl, so that at this period there was danger of their flying away.

[3] Willughby's 'Ornithology,' by Ray, p. 381. This breed is also figured by Albin, in 1734, in his 'Nat. Hist. of Birds,' vol. ii. p. 86.

[4] F. Cuvier, in 'Annales du Muséum,' tom. ix. p. 128, says that moulting and incubation alone stop these ducks laying. Mr. B. P. Brent makes a similar remark in the 'Poultry Chronicle,' 1855, vol. iii. p. 512.

[5] Rev. E. S. Dixon, 'Ornamental and Domestic Poultry' (1848), p. 117. Mr. B. P. Brent, in 'Poultry Chronicle,' vol. iii., 1855, p. 512.

[6] Crawfurd on the 'Relation of Domesticated Animals to Civilisation,' read before the Brit. Assoc. at Oxford, 1860.

[7] Dureau de la Malle, in 'Annales des Sciences Nat.,' tom. xvii. p. 164; and tom. xxi. p. 55. Rev. E. S. Dixon, 'Ornamental Poultry,' p. 118. Tame ducks were not known in Aristotle's time, as remarked by Volz, in his 'Beiträge zur Kulturgeschichte,' 1852, s. 78.

Moreover, the plan recommended by Columella to those who might wish to increase their stock of ducks, namely, to collect the eggs of the wild bird and to place them under a hen, shows, as Mr. Dixon remarks, "that the duck had not at this time become a naturalised and prolific inmate of the Roman poultry-yard." The origin of the domestic duck from the wild species is recognised in nearly every language of Europe, as Aldrovandi long ago remarked, by the same name being applied to both. The wild duck has a wide range from the Himalayas to North America. It crosses readily with the domestic bird, and the crossed offspring are perfectly fertile.

Both in North America and Europe the wild duck has been found easy to tame and breed. In Sweden this experiment was carefully tried by Tiburtius; he succeeded in rearing wild ducks for three generations, but, though they were treated like common ducks, they did not vary even in a single feather. The young birds suffered from being allowed to swim about in cold water,[8] as is known to be the case, though the fact is a strange one, with the young of the common domestic duck. An accurate and well-known observer in England[9] has described in detail his often repeated and successful experiments in domesticating the wild duck. Young birds are easily reared from eggs hatched under a bantam; but to succeed it is indispensable not to place the eggs of both the wild and tame duck under the same hen, for in this case "the young wild ducks die off, leaving their more hardy brethren in undisturbed possession of their foster-mother's care. The difference of habit at the onset in the newly-hatched ducklings almost entails such a result to a certainty." The wild ducklings were from the first quite tame towards those who took care of them as long as they wore the same clothes, and likewise to the dogs and cats of the house. They would even snap with their beaks at the dogs, and drive them away from any spot which they coveted. But they were much alarmed at strange men and dogs. Differently from what

[8] I quote this account from 'Die Enten, Schwanen-zucht,' Ulm, 1828, s. 143. *See* Audubon's 'Ornithological Biography,' vol. iii. p. 168, on the taming of ducks on the Mississippi. For the same fact in England, *see* Mr. Waterton, in Loudon's 'Mag. of Nat. Hist.,' vol. viii., 1835, p. 542; and Mr. St. John, 'Wild Sports and Nat. Hist. of the Highlands,' 1846, p. 129.

[9] Mr. E. Hewitt, in 'Journal of Horticulture,' 1862, p. 773; and 1863, p. 39.

occurred in Sweden, Mr. Hewitt found that his young birds always changed and deteriorated in character in the course of two or three generations; notwithstanding that great care was taken to prevent any crossing with tame ducks. After the third generation his birds lost the elegant carriage of the wild species, and began to acquire the gait of the common duck. They increased in size in each generation, and their legs became less fine. The white collar round the neck of the mallard became broader and less regular, and some of the longer primary wing-feathers became more or less white. When this occurred, Mr. Hewitt always destroyed his old stock and procured fresh eggs from wild nests; so that he never bred the same family for more than five or six generations. His birds continued to pair together, and never became polygamous like the common domestic duck. I have given these details, because no other case, as far as I know, has been so carefully recorded by a competent observer of the progress of change in wild birds reared for several generations in a domestic condition.

From these considerations there can hardly be a doubt that the wild duck is the parent of the common domestic kind; nor need we look to distinct species for the parentage of the more distinct breeds, namely, Penguin, Call, Hook-billed, Tufted, and Labrador ducks. I will not repeat the arguments used in the previous chapters on the improbability of man having in ancient times domesticated several species since become unknown or extinct, though ducks are not readily exterminated in the wild state;—on some of the supposed parent-species having had abnormal characters in comparison with all the other species of the genus, as with hook-billed and penguin ducks;—on all the breeds, as far as is known, being fertile together;[10]—on all the breeds having the same general disposition, instinct, &c. But one fact bearing on this question may be noticed: in the great duck family, one species alone, namely, the male of

[10] I have met with several statements on the fertility of the several breeds when crossed. Mr. Yarrell assured me that Call and common ducks are perfectly fertile together. I crossed Hook-billed and common ducks, and a Penguin and Labrador, and the crossed ducks were quite fertile, though they were not bred *inter se*, so that the experiment was not fully tried. Some half-bred Penguins and Labradors were again crossed with Penguins, and subsequently bred by me *inter se*, and they were extremely fertile.

A. boschas, has its four middle tail-feathers curled upwardly; now in every one of the above-named domestic breeds these curled feathers exist, and on the supposition that they are descended from distinct species, we must assume that man formerly hit upon species all of which had this now unique character. Moreover, sub-varieties of each breed are coloured almost exactly like the wild duck, as I have seen with the largest and smallest breeds, namely Rouens and Call-ducks, and, as Mr. Brent states,[11] is the case with Hook-billed ducks. This gentleman, as he informs me, crossed a white Aylesbury drake and a black Labrador duck, and some of the ducklings as they grew up assumed the plumage of the wild duck.

With respect to Penguins, I have not seen many specimens, and none were coloured precisely like the wild duck; but Sir James Brooke sent me three skins from Lombok and Bali, in the Malayan archipelago; the two females were paler and more rufous than the wild duck, and the drake differed in having the whole under and upper surface (excepting the neck, tail-coverts, tail, and wings) silver-grey, finely pencilled with dark lines, closely like certain parts of the plumage of the wild mallard. But I found this drake to be identical in every feather with a variety of the common breed procured from a farm-yard in Kent, and I have occasionally elsewhere seen similar specimens. The occurrence of a duck bred under so peculiar a climate as that of the Malayan archipelago, where the wild species does not exist, with exactly the same plumage as may occasionally be seen in our farm-yards, is a fact worth notice. Nevertheless the climate of the Malayan archipelago apparently does tend to cause the duck to vary much, for Zollinger,[12] speaking of the Penguin breed, says that in Lombok "there is an unusual and very wonderful variety of ducks." One Penguin drake which I kept alive differed from those of which the skins were sent me from Lombok, in having its breast and back partially coloured with chesnut-brown, thus more closely resembling the Mallard.

From these several facts, more especially from the drakes of all the breeds having curled tail-feathers, and from certain sub-varieties in each breed occasionally resembling in general

[11] 'Poultry Chronicle,' 1855, vol. iii. p. 512.

[12] 'Journal of the Indian Archipelago,' vol. v. p. 334.

plumage the wild duck, we may conclude with confidence that all the breeds are descended from *A. boschas*.

I will now notice some of the peculiarities characteristic of the several breeds. The eggs vary in colour; some common ducks laying pale-greenish and others quite white eggs. The eggs which are first laid during each season by the black Labrador duck, are tinted black, as if rubbed with ink. So that with ducks, as with poultry, some degree of correlation exists between the colour of the plumage and the egg-shell. A good observer assured me that one year his Labrador ducks laid almost perfectly white eggs, but that the yolks were this same season dirty olive-green, instead of as usual of a golden yellow, so that the black tint appeared to have passed inwards. Another curious case shows what singular variations sometimes occur and are inherited; Mr. Hansell[13] relates that he had a common duck which always laid eggs with the yolk of a dark-brown colour like melted glue; and the young ducks, hatched from these eggs, laid the same kind of eggs, so that the breed had to be destroyed.

The hook-billed duck has a most remarkable appearance (see fig. of skull, woodcut No. 39); and its peculiar beak has been inherited at least since the year 1676. This structure is evidently analogous with that described in the Bagadotten carrier pigeon. Mr. Brent[14] says that, when hook-billed ducks are crossed with common ducks, "many young ones are produced with the upper mandible shorter than the lower, which not unfrequently causes the death of the bird." A tuft of feathers on the head is by no means a rare occurrence; namely, in the true tufted breed, the hook-billed, the common farmyard duck, and in a duck having no other peculiarity which was sent to me from the Malayan archipelago. The tuft is only so far interesting as it affects the skull, which is thus rendered slightly more globular, and is perforated by numerous apertures. Call-ducks are remarkable from their extraordinary loquacity: the drake only hisses like common drakes; nevertheless, when paired with the common duck, he transmits to his female offspring a strong quacking tendency. This loquacity seems at first a surprising character to have been acquired under domestication. But the voice varies in the different breeds; Mr. Brent[15] says that hook-billed ducks are very loquacious, and that Rouens utter a "dull, loud, and monotonous cry, easily distinguishable by an experienced ear." As the loquacity of the Call-duck is highly serviceable, these birds being used in decoys, this quality may have been increased by selection. For instance, Colonel Hawker says, if young wild-ducks cannot be got for a decoy, "by way of make-shift, *select* tame birds which are the most clamorous, even if their colour should not be like that of wild ones."[16] It has been

---

[13] 'The Zoologist,' vols. vii., viii. (1849–1850, p. 2353.

[14] 'Poultry Chronicle,' 1855, vol. iii. p. 512.

[15] 'Poultry Chronicle,' vol. iii., 1855, p. 312. With respect to Rouens, *see* ditto, vol. i., 1854, p. 167.

[16] Col. Hawker's 'Instructions to young Sportsmen,' quoted by Mr. Dixon in his 'Ornamental Poultry,' p. 125.

falsely asserted that Call-ducks hatch their eggs in less time than common ducks.[17]

The Penguin duck is the most remarkable of all the breeds; the thin neck and body are carried erect; the wings are small; the tail is upturned; and the thigh-bones and metatarsi are considerably lengthened in proportion with the same bones in the wild duck. In five specimens examined by me there were only eighteen tail-feathers instead of twenty as in the wild duck; but I have also found only eighteen and nineteen tail-feathers in two Labrador ducks. On the middle toe, in three specimens, there were twenty-seven or twenty-eight scutellæ, whereas in two wild ducks there were thirty-one and thirty-two. The Penguin when crossed transmits with much power its peculiar form of body and gait to its offspring; this was manifest with some hybrids raised in the Zoological Gardens between one of these birds and the Egyptian goose[18] (*Tadorna Ægyptiaca*), and likewise with some mongrels which I raised between the Penguin and Labrador duck. I am not much surprised that some writers have maintained that this breed must be descended from an unknown and distinct species; but from the reasons already assigned, it seems to me far more probable that it is the descendant, much modified by domestication under an unnatural climate, of *Anas boschas*.

Osteological Characters.—The skulls of the several breeds differ from each

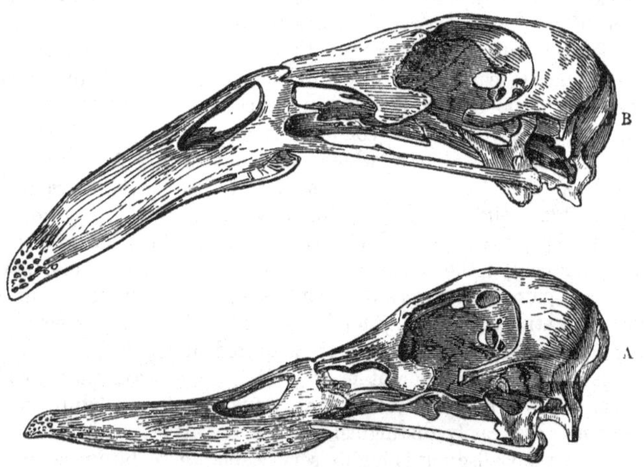

Fig. 39.—Skulls, viewed laterally, reduced to two-thirds of the natural size.
A. Wild Duck. B. Hook-billed Duck.

other and from the skull of the wild duck in very little except in the proportional length and curvature of the premaxillaries. These latter bones in the Call-duck are short, and a line drawn from their extremities to the summit of the skull is nearly straight, instead of being concave as in the

[17] 'Cottage Gardener,' April 9th, 1861.

[18] These hybrids have been described by M. Selys-Longchamps in the 'Bulletins (tom. xii. No. 10) Acad. Roy. de Bruxelles.'

common duck; so that the skull resembles that of a small goose. In the hook-billed duck (fig. 39) these same bones as well as the lower jaw curve downwards in a most remarkable manner, as represented. In the Labrador duck the premaxillaries are rather broader than in the wild duck; and in two skulls of this breed the vertical ridges on each side of the supra-occipital bone are very prominent. In the Penguin the premaxillaries are relatively shorter than in the wild duck; and the inferior points of the paramastoids more prominent. In a Dutch tufted duck, the skull under the enormous tuft was slightly more globular and was perforated by two large apertures; in this skull the lachrymal bones were produced much further backwards, so as to have a different shape and to nearly touch the post. lat. processes of the frontal bones, thus almost completing the bony orbit of the eye. As the quadrate and pterygoid bones are of such complex shape and stand in relation with so many other bones, I carefully compared them in all the principal breeds; but excepting in size they presented no difference.

Vertebræ and Ribs.—In one skeleton of the Labrador duck there were the usual fifteen cervical vertebræ and the usual nine dorsal vertebræ bearing ribs; in the other skeleton there were fifteen cervical and ten dorsal vertebræ with ribs; nor, as far as could be judged, was this owing merely to a rib having been developed on the first lumbar vertebra; for in both skeletons the lumbar vertebræ agreed perfectly in number, shape, and size with those of the wild duck. In two skeletons of the Call-duck there were fifteen cervical and nine dorsal vertebræ; in a third skeleton small ribs were attached to the so-called fifteenth cervical vertebra, making ten pairs of ribs; but these ten ribs do not correspond, or arise from the same vertebræ, with the ten in the above-mentioned Labrador duck. In the Call-duck, which had small ribs attached to the fifteenth cervical vertebra, the hæmal spines of the thirteenth and fourteenth (cervical) and of the seventeenth (dorsal) vertebræ corresponded with the spines on the fourteenth, fifteenth, and eighteenth vertebræ of the wild duck: so that each of these vertebræ had acquired a structure proper to one posterior to it in position. In the twelfth cervical vertebra of this same Call-duck (fig. 40, B), the two branches of the hæmal spine stand much closer together than in the wild duck (A), and the descending hæmal processes are much shortened. In the Penguin duck the neck from its thinness and erectness falsely appears (as ascertained by measurement) to be much elongated, but the cervical and dorsal vertebræ present no difference; the posterior dorsal vertebræ, however, are more completely anchylosed to

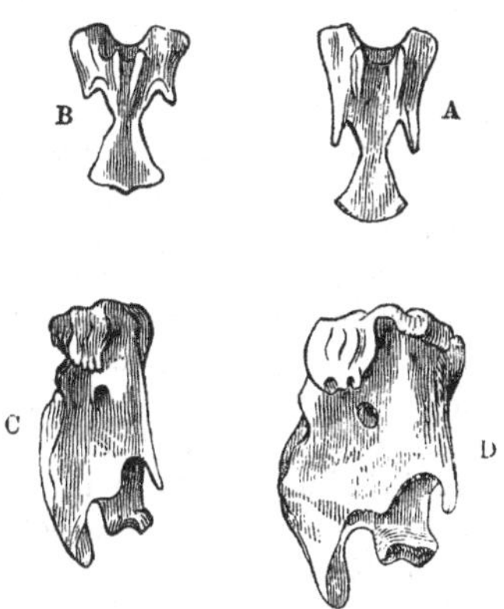

Fig. 40.—Cervical Vertebræ, of natural size. A. Eighth cervical vertebra of Wild Duck, viewed on hæmal surface. B Eighth cervical vertebra of Call Duck, viewed as above. C. Twelfth cervical vertebra of Wild Duck, viewed laterally. D. Twelfth cervical vertebra of Aylesbury Duck, viewed laterally.

the pelvis than in the wild duck. The Aylesbury duck has fifteen cervical and ten dorsal vertebræ furnished with ribs, but the same number of lumbar, sacral, and caudal vertebræ, as far as could be traced, as in the wild duck. The cervical vertebræ in this same duck (fig. 40, D) were much broader and thicker relatively to their length than in the wild (C); so much so, that I have thought it worth while to give a sketch of the eighth cervical vertebra in these two birds. From the foregoing statements we see that the fifteenth cervical vertebra occasionally becomes modified into a dorsal vertebra, and when this occurs all the adjoining vertebræ are modified. We also see that an additional dorsal vertebra bearing a rib is occasionally developed, the number of the cervical and lumbar vertebræ apparently remaining the same as usual.

I examined the bony enlargement of the trachea in the males of the Penguin, Call, Hook-billed, Labrador, and Aylesbury breeds; and in all it was identical in shape.

The *Pelvis* is remarkably uniform; but in the skeleton of the Hook-billed duck the anterior part is much bowed inwards; in the Aylesbury and some other breeds the ischiadic foramen is less elongated. In the sternum, furcula, coracoids, and scapula, the differences are so slight and so variable as not to be worth notice, except that in two skeletons of the Penguin duck the terminal portion of the scapula was much attenuated.

In the bones of the leg and wing no modification in shape could be observed. But in Penguin and Hook-billed ducks, the terminal phalanges of the wing are a little shortened. In the former, the femur and metatarsus (but not the tibia) are considerably lengthened, relatively to the same bones in the wild duck, and to the wing-bones in both birds. This elongation of the leg-bones could be seen whilst the bird was alive, and is no doubt connected with its peculiar upright manner of walking. In a large Aylesbury duck, on the other hand, the tibia was the only bone of the leg which relatively to the other bones was slightly lengthened.

On the effects of the increased and decreased Use of the Limbs.—In all the breeds the bones of the wing (measured separately after having been cleaned) relatively to those of the leg have become slightly shortened, in comparison with the same bones in the wild duck, as may be seen in the following table:—

Name of Breed.	Length of Femur, Tibia, and Metatarsus together.	Length of Humerus, Radius, and Metacarpus together.	Or as
	Inches.	Inches.	
Wild mallard	7·14	9·28	100 : 129
Aylesbury	8·64	10·43	100 : 120
Tufted (Dutch)	8·25	9·83	100 : 119
Penguin	7·12	8·78	100 : 123
Call	6·20	7·77	100 : 125
	Length of same Bones.	Length of all the Bones of Wing.	
	Inches.	Inches.	
Wild duck (another specimen)	6·85	10·07	100 : 147
Common domestic duck ..	8·15	11 26	100 : 138

In the foregoing table we see that, in comparison with the wild duck, the reduction in the length of the bones of the wing, relatively to those of the legs, though slight, is universal. The reduction is least in the Call-duck, which has the power and the habit of frequently flying.

In weight there is a greater relative difference between the bones of the leg and wing, as may be seen in the following table:—

Name of Breed.	Weight of Femur, Tibia, and Metatarsus.	Weight of Humerus, Radius, and Metacarpus.	Or as
	Grains.	Grains.	
Wild mallard	54	97	100 : 179
Aylesbury	164	204	100 : 124
Hooked-bill	107	160	100 : 149
Tufted (Dutch)	111	148	100 : 133
Penguin	75	90·5	100 : 120
Labrador	141	165	100 : 117
Call	57	93	100 : 163
	Weight of all the Bones of the Leg and Foot.	Weight of all the Bones of the Wing.	
	Grains.	Grains.	
Wild (another specimen)	66	115	100 : 173
Common domestic duck	127	158	100 : 124

In these domesticated birds, the considerably lessened weight of the bones of the wing (*i. e.* on an average, twenty-five per cent. of their proper proportional weight), as well as their slightly lessened length, relatively to the leg-bones, might follow, not from any actual decrease in the wing-bones, but from the increased weight and length of the bones of the legs. The first of the two tables on the next page shows that the leg-bones relatively to the weight of the entire skeleton have really increased in weight; but the second table shows that according to the same standard the wing-bones have also really decreased in weight; so that the relative disproportion shown in the foregoing tables between the wing and leg bones, in comparison with those of the wild duck, is partly due to the increase in weight and length of the leg-bones, and partly to the decrease in weight and length of the wing-bones.

With respect to the two following tables, I may first state that I tested them by taking another skeleton of a wild duck and of a common domestic duck, and by comparing the weight of *all* the bones of the leg with *all* those of the wings, and the result was the same. In the first of these tables we see that the leg-bones in each case have increased in actual weight. It might have been expected that, with the increased or decreased weight of the entire skeleton, the leg-bones would have become proportionally heavier or lighter; but their greater weight in all the breeds relatively to the other bones can be accounted for only by these domestic birds having used their legs in walking and standing much more than the wild, for they never fly, and the more artificial breeds rarely swim. In the second

Name of Breed.	Weight of entire Skeleton. (N.B. One Metatarsus and Foot was removed from each skeleton, as it had been accidentally lost in two cases.)	Weight of Femur, Tibia, and Metatarsus.	Or as
	Grains.	Grains.	
Wild mallard	839	54	1000 : 64
Aylesbury	1925	164	1000 : 85
Tufted (Dutch)	1404	111	1000 : 79
Penguin	871	75	1000 : 86
Call (from Mr. Fox)	717	57	1000 : 79
	Weight of Skeleton as above.	Weight of Humerus, Radius and Ulna, and Metacarpus.	
	Grains.	Grains.	
Wild mallard	839	97	1000 : 115
Aylesbury	1925	204	1000 : 105
Tufted (Dutch)	1404	148	1000 : 105
Penguin	871	90	1000 : 103
Call (from Mr. Baker)	914	100	1000 : 109
Call (from Mr. Fox)	713	92	1000 : 129

table we see, with the exception of one case, a plain reduction in the weight of the bones of the wing, and this no doubt has resulted from their lessened use. The one exceptional case, namely, in one of the Call-ducks, is in truth no exception, for this bird was constantly in the habit of flying about; and I have seen it day after day rise from my grounds, and fly for a long time in circles of more than a mile in diameter. In this Call-duck there is not only no decrease, but an actual increase in the weight of the wing-bones relatively to those of the wild duck; and this probably is consequent on the remarkable lightness and thinness of all the bones of the skeleton.

Lastly, I weighed the furcula, coracoids, and scapula of a wild duck and of a common domestic duck, and I found that their weight, relatively to that of the whole skeleton, was as one hundred in the former to eighty-nine in the latter; this shows that these bones in the domestic duck have been reduced eleven per cent. of their due proportional weight. The prominence of the crest of the sternum, relatively to its length, is also much reduced in all the domestic breeds. These changes have evidently been caused by the lessened use of the wings.

It is well known that several birds, belonging to different Orders, and inhabiting oceanic islands, have their wings greatly reduced in size and are incapable of flight. I suggested in my 'Origin of Species' that, as these birds are not persecuted by any enemies, the reduction of their wings has probably been caused by gradual disuse. Hence, during the earlier stages of the process of reduction, such birds might be expected to resemble in the state of their organs of flight our domesticated ducks. This is the case with the water-hen (*Gallinula nesiotis*) of Tristan d'Acunha, which "can flutter a little, but obviously uses its legs, and not its wings, as a mode of escape." Now Mr. Sclater[19] finds in this bird that the wings, sternum, and coracoids, are all reduced in length, and the crest of the sternum in depth, in comparison with the same bones in the European water-hen (*G. chloropus*). On the other hand, the thigh-bones and pelvis are increased in length, the former by four lines, relatively to the same bones in the common water-hen. Hence in the skeleton of this natural species nearly the same changes have occurred, only carried a little further, as with our domestic ducks, and in this latter case I presume no one will dispute that they have resulted from the lessened use of the wings and the increased use of the legs.

THE GOOSE.

THIS bird deserves some notice, as hardly any other anciently domesticated bird or quadruped has varied so little. That geese were anciently domesticated we know from certain verses in Homer; and from these birds having been kept (388 B.C.) in the Capitol at Rome as sacred to Juno, which sacredness implies great antiquity.[20] That the goose has varied in some degree, we may infer from naturalists not being unanimous with respect to its wild parent-form; though the difficulty is chiefly due to the existence of three or four closely allied wild European species.[21] A large majority of capable judges are convinced that our geese are descended from the wild Grey-lag goose (*A. ferus*); the young of which can easily be tamed,[22] and are domesticated by the Laplanders. This species, when crossed with the domestic goose, produced in the Zoological Gardens, as I was assured in

[19] 'Proc. Zoolog. Soc.,' 1861, p. 261.

[20] 'Ceylon,' by Sir J. E. Tennent, 1859, vol. i. p. 485; also J. Crawfurd on the 'Relation of Domest. Animals to Civilisation,' read before Brit. Assoc., 1860. *See* also 'Ornamental Poultry,' by Rev. E. S. Dixon, 1848, p. 132. The goose figured on the Egyptian monuments seems to have been the Red goose of Egypt.

[21] Macgillivray's 'British Birds,' vol. iv. p. 593.

[22] Mr. A. Strickland ('Annals and Mag. of Nat. Hist.,' 3rd Series, vol. iii. 1859, p. 122) reared some young wild geese, and found them in habits and in all characters identical with the domestic goose.

1849, perfectly fertile offspring.[23] Yarrell[24] has observed that the lower part of the trachea of the domestic goose is sometimes flattened, and that a ring of white feathers sometimes surrounds the base of the beak. These characters seem at first good indications of a cross at some former period with the white-fronted goose (*A. albifrons*); but the white ring is variable in this latter species, and we must not overlook the law of analogous variation; that is, of one species assuming some of the characters of allied species.

As the goose has proved so inflexible in its organization under long-continued domestication, the amount of variation which can be detected is worth giving. It has increased in size and in productiveness;[25] and varies from white to a dusky colour. Several observers[26] have stated that the gander is more frequently white than the goose, and that when old it almost invariably becomes white; but this is not the case with the parent-form, the *A. ferus*. Here, again, the law of analogous variation may have come into play, as the snow-white male of the Rock-Goose (*Bernicla antarctica*) standing on the sea-shore by his dusky partner is a sight well known to all those who have traversed the sounds of Tierra del Fuego and the Falkland Islands. Some geese have topknots; and the skull beneath, as before stated, is perforated. A sub-breed has lately been formed with the feathers reversed at the back of the head and neck.[27] The beak varies a little in size, and is of a yellower tint than in the wild species; but its colour and that of the legs are both slightly variable.[28] This latter fact deserves attention, because the colour of the legs and beak is highly serviceable in discriminating the several closely allied wild forms.[29] At our

[23] *See* also Hunter's 'Essays,' edited by Owen, vol. ii. p. 322.

[24] Yarrell's 'British Birds,' vol. iii. p. 142. He refers to the Laplanders domesticating the goose.

[25] L. Lloyd, 'Scandinavian Adventures,' 1854, vol. ii. p. 413, says that the wild goose lays from five to eight eggs, which is a much fewer number than that laid by our domestic goose.

[26] The Rev. L. Jenyns seems first to have made this observation in his 'British Animals.' *See* also Yarrell, and Dixon in his 'Ornamental Poultry' (p. 139), and 'Gardener's Chronicle,' 1857, p. 45.

[27] Mr. Bartlett exhibited the head and neck of a bird thus characterised at the Zoological Soc., Feb. 1860.

[28] W. Thompson, 'Natural Hist. of Ireland,' 1851, vol. iii. p. 31. The Rev. E. S. Dixon gave me some information on the varying colour of the beak and legs.

[29] Mr. A. Strickland, in 'Annals and Mag. of Nat. Hist.,' 3rd series, vol. iii., 1859, p. 122.

Shows two breeds are exhibited; viz. the Embden and Toulouse; but they differ in nothing except colour.[30] Recently a smaller and singular variety has been imported from Sebastopol,[31] with the scapular feathers (as I hear from Mr. Tegetmeier, who sent me specimens) greatly elongated, curled, and even spirally twisted. The margins of these feathers are rendered plumose by the divergence of the barbs and barbules, so that they resemble in some degree those on the back of the black Australian swan. These feathers are likewise remarkable from the central shaft, which is excessively thin and transparent, being split into fine filaments, which, after running for a space free, sometimes coalesce again. It is a curious fact that these filaments are regularly clothed on each side with fine down or barbules, precisely like those on the proper barbs of the feather. This structure of the feathers is transmitted to half-bred birds. In *Gallus sonneratii* the barbs and barbules blend together, and form thin horny plates of the same nature with the shaft: in this variety of the goose, the shaft divides into filaments which acquire barbules, and thus resemble true barbs.

Although the domestic goose certainly differs somewhat from any known wild species, yet the amount of variation which it has undergone, as compared with most domesticated animals, is singularly small. This fact can be partially accounted for by selection not having come largely into play. Birds of all kinds which present many distinct races are valued as pets or ornaments; no one makes a pet of the goose; the name, indeed, in more languages than one, is a term of reproach. The goose is valued for its size and flavour, for the whiteness of its feathers which adds to their value, and for its prolificness and tameness. In all these points the goose differs from the wild parent-form; and these are the points which have been selected. Even in ancient times the Roman gourmands valued the liver of the *white* goose; and Pierre Belon[32] in 1555 speaks of two varieties, one of which was larger, more fecund, and of a better colour than the other; and he expressly states that good managers

[30] 'Poultry Chronicle,' vol. i., 1854, p. 498; vol. iii. p. 210.

[31] 'The Cottage Gardener,' Sept. 4th, 1860, p. 348.

[32] 'L'Hist. de la Nature des Oiseaux,' par P. Belon, 1555, p. 156. With respect to the livers of white geese being preferred by the Romans, *see* Isid. Geoffroy St. Hilaire, 'Hist. Nat. Gén.,' tom. iii. p. 58.

attended to the colour of their goslings, so that they might know which to preserve and select for breeding.

THE PEACOCK.

THIS is another bird which has hardly varied under domestication, except in sometimes being white or piebald. Mr. Waterhouse carefully compared, as he informs me, skins of the wild Indian and domestic bird, and they were identical in every respect, except that the plumage of the latter was perhaps rather thicker. Whether our birds are descended from those introduced into Europe in the time of Alexander, or have been subsequently imported, is doubtful. They do not breed very freely with us, and are seldom kept in large numbers,—circumstances which would greatly interfere with the gradual selection and formation of new breeds.

There is one strange fact with respect to the peacock, namely, the occasional appearance in England of the "japanned" or "black-shouldered" kind. This form has lately been named on the high authority of Mr. Sclater as a distinct species, viz. *Pavo nigripennis*, which he believes will hereafter be found wild in some country, but not in India, where it is certainly unknown. These japanned birds differ conspicuously from the common peacock in the colour of their secondary wing-feathers, scapulars, wing-coverts, and thighs; the females are much paler, and the young, as I hear from Mr. Bartlett, likewise differ. They can be propagated perfectly true. Although they do not resemble the hybrids which have been raised between *P. cristatus* and *muticus*, nevertheless they are in some respects intermediate in character between these two species; and this fact favours, as Mr. Sclater believes, the view that they form a distinct and natural species.[33]

On the other hand, Sir R. Heron states[34] that this breed suddenly appeared within his memory in Lord Brownlow's large stock of pied, white, and common peacocks. The same thing occurred in Sir J. Trevelyan's flock composed entirely of the

[33] Mr. Sclater on the black-shouldered peacock of Latham, 'Proc. Zoolog. Soc.,' April 24th, 1860.

[34] 'Proc. Zoolog. Soc.,' April 14th, 1835.

common kind, and in Mr. Thornton's stock of common and pied peacocks. It is remarkable that in these two latter instances the black-shouldered kind increased, "to the extinction of the previously existing breed." I have also received through Mr. Sclater a statement from Mr. Hudson Gurney that he reared many years ago a pair of black-shouldered peacocks from the common kind; and another ornithologist, Prof. A. Newton, states that, five or six years ago, a female bird, in all respects similar to the female of the black-shouldered kind, was produced from a stock of common peacocks in his possession, which during more than twenty years had not been crossed with birds of any other strain. Here we have five distinct cases of japanned birds suddenly appearing in flocks of the common kind kept in England. Better evidence of the first appearance of a new variety could hardly be desired. If we reject this evidence, and believe that the japanned peacock is a distinct species, we must suppose in all these cases that the common breed had at some former period been crossed with the supposed *P. nigripennis*, but had lost every trace of the cross, yet that the birds occasionally produced offspring which suddenly and completely reacquired through reversion the characters of *P. nigripennis*. I have heard of no other such case in the animal or vegetable kingdom. To perceive the full improbability of such an occurrence, we may suppose that a breed of dogs had been crossed at some former period with a wolf, but had lost every trace of the wolf-like character, yet that the breed gave birth in five instances in the same country, within no great length of time, to a wolf perfect in every character; and we must further suppose that in two of the cases the newly produced wolves afterwards spontaneously increased to such an extent as to lead to the extinction of the parent-breed of dogs. So remarkable a form as the *P. nigripennis*, when first imported, would have realized a large price; it is therefore improbable that it should have been silently introduced and its history subsequently lost. On the whole the evidence seems to me, as it did to Sir R. Heron, to preponderate strongly in favour of the black-shouldered breed being a variation, induced either by the climate of England, or by some unknown cause, such as reversion to a primordial and extinct condition of the species. On the view that the black-shouldered

U 2

peacock is a variety, the case is the most remarkable ever recorded of the abrupt appearance of a new form, which so closely resembles a true species that it has deceived one of the most experienced of living ornithologists.

THE TURKEY.

IT seems fairly well established by Mr. Gould,[35] that the turkey, in accordance with the history of its first introduction, is descended from a wild Mexican species (*Meleagris Mexicana*) which had been already domesticated by the natives before the discovery of America, and which differs specifically, as it is generally thought, from the common wild species of the United States. Some naturalists, however, think that these two forms should be ranked only as well-marked geographical races. However this may be, the case deserves notice because in the United States wild male turkeys sometimes court the domestic hens, which are descended from the Mexican form, "and are generally received by them with great pleasure."[36] Several accounts have likewise been published of young birds, reared in the United States from the eggs of the wild species, crossing and commingling with the common breed. In England, also, this same species has been kept in several parks; from two of which the Rev. W. D. Fox procured birds, and they crossed freely with the common domestic kind, and during many years afterwards, as he informs me, the turkeys in his neighbourhood clearly showed traces of their crossed parentage. We here have an instance of a domestic race being modified by a cross with a distinct species or wild race. F. Michaux[37] suspected in 1802 that the common domestic turkey was not descended from the United States species alone, but likewise from a southern form, and he went so far as to believe that English and French

[35] 'Proc. Zoolog. Soc.,' April 8th, 1856, p. 61. Prof. Baird believes (as quoted in Tegetmeier's 'Poultry Book,' 1866, p. 269) that our turkeys are descended from a West Indian species now extinct. But besides the improbability of a bird having long ago become extinct in these large and luxuriant islands, it appears (as we shall presently see) that the turkey degenerates in India, and this fact indicates that it was not aboriginally an inhabitant of the lowlands of the tropics.

[36] Audubon's 'Ornithological Biograph.,' vol. i., 1831, pp. 4–13; and 'Naturalist's Library,' vol. xiv., Birds, p. 138.

[37] F. Michaux, 'Travels in N. America,' 1802, Eng. translat., p. 217.

turkeys differed from having different proportions of the blood of the two parent-forms.

English turkeys are smaller than either wild form. They have not varied in any great degree; but there are some breeds which can be distinguished—as Norfolks, Suffolks, Whites, and Copper-coloured (or Cambridge), all of which, if precluded from crossing with other breeds, propagate their kind truly. Of these kinds, the most distinct is the small, hardy, dull-black Norfolk turkey, of which the chickens are black, with occasionally white patches about the head. The other breeds scarcely differ except in colour, and their chickens are generally mottled all over with brownish-grey.[38] The tuft of hair on the breast, which is proper to the male alone, occasionally appears on the breast of the domesticated female.[39] The inferior tail-coverts vary in number, and according to a German superstition the hen lays as many eggs as the cock has feathers of this kind.[40] In Holland there was formerly, according to Temminck, a beautiful buff-yellow breed, furnished with an ample white topknot. Mr. Wilmot has described[41] a white turkey-cock with a crest formed of "feathers about four inches long, with bare quills, and a tuft of soft white down growing at the end." Many of the young birds whilst young inherited this kind of crest, but afterwards it either fell off or was pecked out by the other birds. This is an interesting case, as with care a new breed might probably have been formed; and a topknot of this nature would have been to a certain extent analogous to that borne by the males in several allied genera, such as Euplocomus, Lophophorus, and Pavo.

Wild turkeys, believed in every instance to have been imported from the United States, have been kept in the parks of Lords Powis, Leicester, Hill, and Derby. The Rev. W. D. Fox procured birds from the two first-named parks, and he informs me that they certainly differed a little from each other in the shape of their bodies and in the barred plumage on their wings. These birds likewise differed from Lord Hill's stock. Some of the latter kept at Oulton by Sir P. Egerton, though precluded from

[38] 'Ornamental Poultry,' by the Rev. E. S. Dixon, 1848, p. 34.

[39] Rev. E. S. Dixon, id., p. 35.

[40] Bechstein, 'Naturgesch. Deutschlands,' B. iii., 1793, s. 309.

[41] 'Gardener's Chronicle,' 1852, p. 699.

crossing with common turkeys, occasionally produced much paler-coloured birds, and one that was almost white, but not an albino. These half-wild turkeys in thus slightly differing from each other present an analogous case with the wild cattle kept in the several British parks. We must suppose that the differences have resulted from the prevention of free intercrossing between birds ranging over a wide area, and from the changed conditions to which they have been exposed in England. In India the climate has apparently wrought a still greater change in the turkey, for it is described by Mr. Blyth[42] as being much degenerated in size, "utterly incapable of rising on the wing," of a black colour, and "with the long pendulous appendages over the beak enormously developed."

THE GUINEA FOWL.

THE domesticated guinea-fowl is now believed by naturalists to be descended from the *Numida ptilorhynca*, which inhabits very hot, and, in parts, extremely arid districts in Eastern Africa; consequently it has been exposed in this country to extremely different conditions of life. Nevertheless it has hardly varied at all, except in the plumage being either paler or darker-coloured. It is a singular fact that this bird varies more in colour in the West Indies and on the Spanish Main, under a hot though humid climate, than in Europe.[43] The guinea-fowl has become thoroughly feral in Jamaica and in St. Domingo,[44] and has diminished in size; the legs are black, whereas the legs of the aboriginal African bird are said to be grey. This small change is worth notice on account of the often-repeated statement that all feral animals invariably revert in every character to their original type.

[42] E. Blyth, in 'Annals and Mag. of Nat. Hist.,' 1847, vol. xx. p. 391.

[43] Roulin makes this remark in 'Mém. de divers Savans, l'Acad. des Sciences,' tom. vi., 1835, p. 349. Mr. Hill, of Spanish Town, in a letter to me, describes five varieties of the guinea-fowl in Jamaica. I have seen singular pale-coloured varieties imported from Barbadoes and Demerara.

[44] For St. Domingo, *see* M. A. Salle, in 'Proc. Soc. Zoolog.,' 1857, p. 236. Mr. Hill remarks to me, in his letter, on the colour of the legs of the feral birds in Jamaica.

THE CANARY BIRD.

As this bird has been recently domesticated, namely, within the last 350 years, its variability deserves notice. It has been crossed with nine or ten other species of Fringillidæ, and some of the hybrids are almost completely fertile; but we have no evidence that any distinct breed has originated from such crosses. Notwithstanding the modern domestication of the canary, many varieties have been produced; even before the year 1718 a list of twenty-seven varieties was published in France,[45] and in 1779 a long schedule of the desired qualities was printed by the London Canary Society, so that methodical selection has been practised during a considerable period. The greater number of the varieties differ only in colour and in the markings of their plumage. Some breeds, however, differ in shape, such as the hooped or bowed canaries, and the Belgian canaries with their much elongated bodies. Mr. Brent[46] measured one of the latter and found it eight inches in length, whilst the wild canary is only five and a quarter inches long. There are topknotted canaries, and it is a singular fact, that, if two topknotted birds are matched, the young, instead of having very fine topknots, are generally bald, or even have a wound on their heads.[47] It would appear as if the topknot were due to some morbid condition which is increased to an injurious degree when two birds in this state are paired. There is a feather-footed breed, and another with a kind of frill running down the breast. One other character deserves notice from being confined to one period of life and from being strictly inherited at the same period: namely, the wing and tail feathers in prize canaries being black, "but this colour is retained only until the first moult; once moulted, the peculiarity ceases."[48] Canaries differ much in disposition and character, and in some small degree in song. They produce eggs three or four times during the year.

[45] Mr. B. P. Brent, 'The Canary, British Finches,' &c., pp. 21, 30.

[46] 'Cottage Gardener,' Dec. 11th, 1855, p. 184. An account is here given of all the varieties. For many measurements of the wild birds, *see* Mr. E. Vernon Harcourt, id., Dec. 25th, 1855, p. 223.

[47] Bechstein, 'Naturgesch. der Stubenvögel,' 1840, s. 243; *see* s. 252, on the inherited song of Canary-birds. With respect to their baldness, *see* also W. Kidd's 'Treatise on Song-Birds.'

[48] W. Kidd's 'Treatise on Song-Birds, p. 18.

GOLD-FISH.

BESIDES mammals and birds, few animals belonging to the other great classes have been domesticated; but to show that it is an almost universal law that animals, when removed from their natural conditions of life, vary, and that races can be formed when selection is applied, it is necessary to say a few words on gold-fish, bees, and silk-moths.

Gold-fish (*Cyprinus auratus*) were introduced into Europe only two or three centuries ago; but it is believed that they have been kept in confinement from an ancient period in China. Mr. Blyth[49] suspects from the analogous variation of other fishes that golden-coloured fish do not occur in a state of nature. These fishes frequently live under the most unnatural conditions, and their variability in colour, size, and in some important points of structure is very great. M. Sauvigny has described and given coloured drawings of no less than eighty-nine varieties.[50] Many of the varieties, however, such as triple tail-fins, &c., ought to be called monstrosities; but it is difficult to draw any distinct line between a variation and a monstrosity. As gold-fish are kept for ornament or curiosity, and as "the Chinese are just the people to have secluded a chance variety of any kind, and to have matched and paired from it,"[51] we may feel nearly confident that selection has been largely practised in the formation of new breeds. It is however a singular fact that some of the monstrosities or variations are not inherited; for Sir R. Heron[52] kept many of these fishes, and placed all the deformed fishes, namely those destitute of dorsal fins, and those furnished with a double anal fin, or triple tail, in a pond by themselves; but they did "not produce a greater proportion of deformed offspring than the perfect fishes."

Passing over an almost infinite diversity of colour, we meet with the most extraordinary modifications of structure. Thus, out of about two dozen specimens bought in London, Mr. Yarrell observed some with the dorsal fin extending along more than

[49] The 'Indian Field,' 1858, p. 255.

[50] Yarrell's 'British Fishes,' vol. i. p. 319.

[51] Mr. Blyth, in the 'Indian Field,' 1858, p. 255.

[52] 'Proc. Zoolog. Soc.,' May 25th. 1842.

half the length of the back; others with this fin reduced to only five or six rays; and one with no dorsal fin. The anal fins are sometimes double, and the tail is often triple. This latter deviation of structure seems generally to occur "at the expense of the whole or part of some other fin;"[53] but Bory de Saint Vincent[54] saw at Madrid gold-fish furnished with a dorsal fin and a triple tail. One variety is characterized by a hump on its back near the head; and the Rev. L. Jenyns[55] has described a most singular variety, imported from China, almost globular in form like a Diodon, with "the fleshy part of the tail as if entirely cut away; the caudal fin being set on a little behind the dorsal and immediately above the anal." In this fish the anal and caudal fins were double; the anal fin being attached to the body in a vertical line: the eyes also were enormously large and protuberant.

HIVE-BEES.

BEES have been domesticated from an ancient period; if indeed their state can be considered one of domestication, for they search for their own food, with the exception of a little generally given to them during the winter. Their habitation is a hive instead of a hole in a tree. Bees, however, have been transported into almost every quarter of the world, so that climate ought to have produced whatever direct effect it is capable of producing. It is frequently asserted that the bees in different parts of Great Britain differ in size, colour, and temper; and Godron[56] says that they are generally larger in the south than in other parts of France; it has also been asserted that the little brown bees of High Burgundy, when transported to La Bresse, become large and yellow in the second generation. But these statements require confirmation. As far as size is concerned, it is known that bees produced in very old combs are smaller, owing to the cells having become smaller from the

[53] Yarrell's 'British Fishes,' vol. i. p. 319.

[54] 'Dict. Class. d'Hist. Nat.,' tom. v. p. 276.

[55] 'Observations in Nat. Hist.,' 1846, p. 211. Dr. Gray has described, in 'Annals and Mag. of Nat. Hist.,' 1860, p. 151, a nearly similar variety, but destitute of a dorsal fin.

[56] 'De l'Espèce,' 1859, p. 459. With respect to the bees of Burgundy, *see* M. Gérard, art. 'Espèce,' in 'Dict. Univers. d'Hist. Nat.'

successive old cocoons. The best authorities[57] concur that, with the exception of the Ligurian race or species, presently to be mentioned, distinct breeds do not exist in Britain or on the Continent. There is, however, even in the same stock, some variability in colour. Thus Mr. Woodbury states[58] that he has several times seen queen bees of the common kind annulated with yellow like Ligurian queens, and the latter dark-coloured like common bees. He has also observed variations in the colour of the drones, without any corresponding difference in the queens or workers of the same hive. The great apiarian Dzierzon, in answer to my queries on this subject, says[59] that in Germany bees of some stocks are decidedly dark, whilst others are remarkable for their yellow colour. Bees also seem to differ in habits in different districts, for Dzierzon adds, "If many stocks with their offspring are more inclined to swarm, whilst others are richer in honey, so that some bee-keepers even distinguish between swarming and honey-gathering bees, this is a habit which has become second nature, caused by the customary mode of keeping the bees and the pasturage of the district. For example; what a difference in this respect one may perceive to exist between the bees of the Lüneburg heath and those of this country!" ...... "Removing an old queen and substituting a young one of the current year is here an infallible mode of keeping the strongest stock from swarming and preventing drone-breeding; whilst the same means if adopted in Hanover would certainly be of no avail." I procured a hive full of dead bees from Jamaica, where they have long been naturalised, and, on carefully comparing them under the microscope with my own bees, I could detect not a trace of difference.

This remarkable uniformity in the hive-bee, wherever kept, may probably be accounted for by the great difficulty, or rather impossibility, of bringing selection into play by pairing particular queens and drones, for these insects unite only during

[57] *See* a discussion on this subject, in answer to a question of mine, in 'Journal of Horticulture,' 1862, pp. 225-242; also Mr. Bevan Fox, in ditto, 1862, p. 284.

[58] This excellent observer may be implicitly trusted; *see* 'Journal of Horticulture,' July 14th, 1863, p. 39.

[59] 'Journal of Horticulture,' Sept. 9th, 1862, p. 463; *see* also Herr Kleine on same subject (Nov. 11th, p. 643), who sums up, that, though there is some variability in colour, no constant or perceptible differences can be detected in the bees of Germany.

flight. Nor is there any record, with a single partial exception, of any person having separated and bred from a hive in which the workers presented some appreciable difference. In order to form a new breed, seclusion from other bees would, as we now know, be indispensable; for since the introduction of the Ligurian bee into Germany and England, it has been found that the drones wander at least two miles from their own hives, and often cross with the queens of the common bee.[60] The Ligurian bee, although perfectly fertile when crossed with the common kind, is ranked by most naturalists as a distinct species, whilst by others it is ranked as a natural variety: but this form need not here be noticed, as there is no reason to believe that it is the product of domestication. The Egyptian and some other bees are likewise ranked by Dr. Gerstäcker,[61] but not by other highly competent judges, as geographical races; and he grounds his conclusion in chief part on the fact that in certain districts, as in the Crimea and Rhodes, the hive-bee varies so much in colour, that the several geographical races can be closely connected by intermediate forms.

I have alluded to a single instance of the separation and preservation of a particular stock of bees. Mr. Lowe[62] procured some bees from a cottager a few miles from Edinburgh, and perceived that they differed from the common bee in the hairs on the head and thorax being lighter coloured and more profuse in quantity. From the date of the introduction of the Ligurian bee into Great Britain we may feel sure that these bees had not been crossed with this form. Mr. Lowe propagated this variety, but unfortunately did not separate the stock from his other bees, and after three generations the new character was almost completely lost. Nevertheless, as he adds, "a great number of the bees still retain traces, though faint, of the original colony." This case shows us what could probably be effected by careful and long-continued selection applied exclusively to the workers, for, as we have seen, queens and drones cannot be selected and paired.

[60] Mr. Woodbury has published several such accounts in 'Journal of Horticulture,' 1861 and 1862.

[61] 'Annals and Mag. of Nat. Hist.,' 3rd series, vol. xi. p. 339.

[62] 'The Cottage Gardener,' May, 1860, p. 110; and ditto in 'Journal of Hort.,' 1862, p. 242.

SILK-MOTHS.

THESE insects are in several respects interesting to us, more especially because they have varied largely at early periods of life, and the variations have been inherited at corresponding periods. As the value of the silk-moth depends entirely on the cocoon, every change in its structure and qualities has been carefully attended to, and races differing much in the cocoon, but hardly at all in the adult state, have been produced. With the races of most other domestic animals, the young resemble each other closely, whilst the adults differ much.

It would be useless, even if it were possible, to describe all the many kinds of silk-worms. Several distinct species exist in India and China which produce useful silk, and some of these are capable of freely crossing with the common silk-moth, as has been recently ascertained in France. Captain Hutton[63] states that throughout the world at least six species have been domesticated; and he believes that the silk-moths reared in Europe belong to two or three species. This, however, is not the opinion of several capable judges who have particularly attended to the cultivation of this insect in France; and hardly accords with some facts presently to be given.

The common silk-moth (*Bombyx mori*) was brought to Constantinople in the sixth century, whence it was carried into Italy, and in 1494 into France.[64] Everything has been favourable for the variation of this insect. It is believed to have been domesticated in China as long ago as 2700 B.C. It has been kept under unnatural and diversified conditions of life, and has been transported into many countries. There is reason to believe that the nature of the food given to the caterpillar influences to a certain extent the character of the breed.[65] Disuse has apparently aided in checking the development of the wings. But the most important element in the production of the many now existing, much modified races, no doubt has

[63] 'Transact. Entomolog. Soc.,' 3rd series, vol. iii. pp. 143-173, and pp. 295-331.

[64] Godron, 'De l'Espèce,' 1859, tom. i. p. 460. The antiquity of the silk-worm in China is given on the authority of Stanislas Julien.

[65] *See* the remarks of Prof. Westwood, General Hearsey, and others, at the meeting of the Entomolog. Soc. of London, July, 1861.

been the close attention which has long been applied in many countries to every promising variation. The care taken in Europe in the selection of the best cocoons and moths for breeding is notorious,[66] and the production of eggs is followed as a distinct trade in parts of France. I have made inquiries through Dr. Falconer, and am assured that in India the natives are equally careful in the process of selection. In China the production of eggs is confined to certain favourable districts, and the raisers are precluded by law from producing silk, so that their whole attention may be necessarily given up to this one object.[67]

The following details on the differences between the several breeds are taken, when not stated to the contrary, from M. Robinet's excellent work,[68] which bears every sign of care and large experience. The *eggs* in the different races vary in colour, in shape (being round, elliptic, or oval), and in size. The eggs laid in June in the south of France, and in July in the central provinces, do not hatch until the following spring; and it is in vain, says M. Robinet, to expose them to a temperature gradually raised, in order that the caterpillar may be quickly developed. Yet occasionally, without any known cause, batches of eggs are produced, which immediately begin to undergo the proper changes, and are hatched in from twenty to thirty days. From these and some other analogous facts it may be concluded that the Trevoltini silkworms of Italy, of which the caterpillars are hatched in from fifteen to twenty days, do not necessarily form, as has been maintained, a distinct species. Although the breeds which live in temperate countries produce eggs which cannot be immediately hatched by artificial heat, yet when they are removed to and reared in a hot country they gradually acquire the character of quick development, as in the Trevoltini races.[69]

Caterpillars.—These vary greatly in size and colour. The skin is generally white, sometimes mottled with black or grey, and occasionally quite black. The colour, however, as M. Robinet asserts, is not constant, even in perfectly pure breeds; except in the *race tigrée*, so called from being marked with transverse black stripes. As the general colour of the caterpillar is not correlated with that of the silk,[70] this character is disregarded

[66] *See*, for instance, M. A. de Quatrefage's 'Etudes sur les Maladies actuelles du Ver à Soie,' 1859, p. 101.

[67] My authorities for these statements will be given in the chapter on Selection.

[68] 'Manuel de l'Educateur de Vers à Soie,' 1848.

[69] Robinet, idem, pp. 12, 318. I may add that the eggs of N. American silkworms taken to the Sandwich Islands were very irregularly developed; and the moths thus raised produced eggs which were even worse in this respect. Some were hatched in ten days, and others not until after the lapse of many months. No doubt a regular early character would ultimately have been acquired. *See* review in 'Athenæum,' 1844, p. 329, of J. Jarves' 'Scenes in the Sandwich Islands.'

[70] 'The Art of rearing Silk-worms,' translated from Count Dandolo, 1825, p. 23.

by cultivators, and has not been fixed by selection. Captain Hutton, in the paper before referred to, has argued with much force that the dark tiger-like marks, which so frequently appear during the later moults in the caterpillars of various breeds, are due to reversion; for the caterpillars of several allied wild species of Bombyx are marked and coloured in this manner. He separated some caterpillars with the tiger-like marks, and in the succeeding spring (pp. 149, 298) nearly all the caterpillars reared from them were dark-brindled, and the tints became still darker in the third generation. The moths reared from these caterpillars[71] also became darker, and resembled in colouring the wild *B. Huttoni*. On this view of the tiger-like marks being due to reversion, the persistency with which they are transmitted is intelligible.

Several years ago Mrs. Whitby took great pains in breeding silkworms on a large scale, and she informed me that some of her caterpillars had dark eyebrows. This is probably the first step in reversion towards the tiger-like marks, and I was curious to know whether so trifling a character would be inherited; at my request she separated in 1848 twenty of these caterpillars, and having kept the moths separate, bred from them. Of the many caterpillars thus reared, "every one without exception had eyebrows, some darker and more decidedly marked than the others, but *all* had eyebrows more or less plainly visible." Black caterpillars occasionally appear amongst those of the common kind, but in so variable a manner, that according to M. Robinet the same race will one year exclusively produce white caterpillars, and the next year many black ones; nevertheless, I have been informed by M. A. Bossi of Geneva, that, if these black caterpillars are separately bred from, they reproduce the same colour; but the cocoons and moths reared from them do not present any difference.

The caterpillar in Europe ordinarily moults four times before passing into the cocoon stage; but there are races "à trois mues," and the Trevoltini race likewise moults only thrice. It might have been thought that so important a physiological difference would not have arisen under domestication; but M. Robinet[72] states that, on the one hand, ordinary caterpillars occasionally spin their cocoons after only three moults, and, on the other hand, "presque toutes les races à trois mues, que nous avons expérimentées, ont fait quatre mues à la seconde ou à la troisième année, ce qui semble prouver qu'il a suffi de les placer dans des conditions favorables pour leur rendre une faculté qu'elles avaient perdue sous des influences moins favorables."

Cocoons.—The caterpillar in changing into the cocoon loses about 50 per cent. of its weight; but the amount of loss differs in different breeds, and this is of importance to the cultivator. The cocoon in the different races presents characteristic differences; being large or small;—nearly spherical with no constriction, as in the *Race de Loriol*, or cylindrical with either a deep or slight constriction in the middle;—with the two ends, or with one end alone, more or less pointed. The silk varies in fineness and quality, and in being nearly white, of two tints, or yellow. Generally the colour of

---

[71] 'Transact. Ent. Soc.,' ut supra, pp. 153, 308.

[72] Robinet, idem, p. 317.

the silk is not strictly inherited: but in the chapter on Selection I shall give a curious account how, in the course of sixty-five generations, the number of yellow cocoons in one breed has been reduced in France from one hundred to thirty-five in the thousand. According to Robinet, the white race, called Sina, by careful selection during the last seventy-five years, "est arrivée à un tel état de pureté, qu'on ne voit pas un seul cocon jaune dans des millions de cocons blancs."[73] Cocoons are sometimes formed, as is well known, entirely destitute of silk, which yet produce moths; unfortunately Mrs. Whitby was prevented by an accident from ascertaining whether this character would prove hereditary.

Adult stage.—I can find no account of any constant difference in the moths of the most distinct races. Mrs. Whitby assured me that there was none in the several kinds bred by her; and I have received a similar statement from the eminent naturalist M. de Quatrefages. Captain Hutton also says[74] that the moths of all kinds vary much in colour, but in nearly the same inconstant manner. Considering how much the cocoons in the several races differ, this fact is of interest, and may probably be accounted for on the same principle as the fluctuating variability of colour in the caterpillar, namely, that there has been no motive for selecting and perpetuating any particular variation.

The males of the wild Bombycidæ "fly swiftly in the day-time and evening, but the females are usually very sluggish and inactive."[75] In several moths of this family the females have abortive wings, but no instance is known of the males being incapable of flight, for in this case the species could hardly have been perpetuated. In the silk-moth both sexes have imperfect, crumpled wings, and are incapable of flight; but still there is a trace of the characteristic difference in the two sexes; for though, on comparing a number of males and females, I could detect no difference in the development of their wings, yet I was assured by Mrs. Whitby that the males of the moths bred by her used their wings more than the females, and could flutter downwards, though never upwards. She also states that, when the females first emerge from the cocoon, their wings are less expanded than those of the male. The degree of imperfection, however, in the wings varies much in different races and under different circumstances; M. Quatrefages[76] says that he has seen a number of moths with their wings reduced to a third, fourth, or tenth part of their normal dimensions, and even to mere short straight stumps: "il me semble qu'il y a là un véritable arrêt de développement partiel." On the other hand, he describes the female moths of the André Jean breed as having "leurs ailes larges et étalées. Un seul présente quelques courbures irrégulières et des plis anomaux." As moths and butterflies of all kinds reared from wild caterpillars under confinement often have crippled wings, the same cause, whatever it may be, has probably acted on silk-

---

[73] Robinet, idem, pp. 306-317.

[74] 'Transact. Ent. Soc.,' ut supra, p. 317.

[75] Stephens' Illustrations, 'Haustellala,' vol. ii. p. 35. *See* also Capt. Hutton, 'Transact. Ent. Soc.' idem, p. 152.

[76] 'Etudes sur les Maladies du Ver à Soie,' 1859, pp. 304, 209.

moths, but the disuse of their wings during so many generations has, it may be suspected, likewise come into play.

The moths of many breeds fail to glue their eggs to the surface on which they are laid,[77] but this proceeds, according to Capt. Hutton,[78] merely from the glands of the ovipositor being weakened.

As with other long-domesticated animals, the instincts of the silk-moth have suffered. The caterpillars, when placed on a mulberry-tree, often commit the strange mistake of devouring the base of the leaf on which they are feeding, and consequently fall down; but they are capable, according to M. Robinet,[79] of again crawling up the trunk. Even this capacity sometimes fails, for M. Martins[80] placed some caterpillars on a tree, and those which fell were not able to remount and perished of hunger; they were even incapable of passing from leaf to leaf.

Some of the modifications which the silk-moth has undergone stand in correlation with each other. Thus the eggs of the moths which produce white cocoons and of those which produce yellow cocoons differ slightly in tint. The abdominal feet also of the caterpillars which yield white cocoons are always white, whilst those which give yellow cocoons are invariably yellow.[81] We have seen that the caterpillars with dark tiger-like stripes produce moths which are more darkly shaded than other moths. It seems well established[82] that in France the caterpillars of the races which produce white silk, and certain black caterpillars, have resisted, better than other races, the disease which has recently devastated the silk-districts. Lastly, the races differ constitutionally, for some do not succeed so well under a temperate climate as others; and a damp soil does not equally injure all the races.[83]

From these various facts we learn that silk-moths, like the higher animals, vary greatly under long-continued domestication. We learn also the more important fact that variations may occur at various periods of life, and be inherited at corresponding periods. And finally we see that insects are amenable to the great principle of Selection.

77 Quatrefages, 'Etudes,' &c., p. 214.

78 'Transact. Ent. Soc.,' ut supra, p. 151.

79 'Manuel de l'Educateur,' &c., p. 26.

80 Godron, 'De l'Espèce,' p. 462.

81 Quatrefages, 'Etudes,' &c., pp. 12, 209, 214.

82 Robinet, 'Manuel,' &c., p. 303.

83 Robinet, idem, p. 15.

CHAPTER IX.

CULTIVATED PLANTS: CEREAL AND CULINARY PLANTS.

PRELIMINARY REMARKS ON THE NUMBER AND PARENTAGE OF CULTIVATED PLANTS — FIRST STEPS IN CULTIVATION — GEOGRAPHICAL DISTRIBUTION OF CULTIVATED PLANTS.

CEREALIA. — DOUBTS ON THE NUMBER OF SPECIES. —— WHEAT: VARIETIES OF — INDIVIDUAL VARIABILITY — CHANGED HABITS — SELECTION — ANCIENT HISTORY OF THE VARIETIES. —— MAIZE: GREAT VARIATION OF — DIRECT ACTION OF CLIMATE ON.

CULINARY PLANTS. — CABBAGES: VARIETIES OF, IN FOLIAGE AND STEMS, BUT NOT IN OTHER PARTS — PARENTAGE OF — OTHER SPECIES OF BRASSICA. —— PEAS: AMOUNT OF DIFFERENCE IN THE SEVERAL KINDS, CHIEFLY IN THE PODS AND SEED — SOME VARIETIES CONSTANT, SOME HIGHLY VARIABLE — DO NOT INTERCROSS. —— BEANS. —— POTATOES: NUMEROUS VARIETIES OF — DIFFERING LITTLE, EXCEPT IN THE TUBERS — CHARACTERS INHERITED.

I SHALL not enter into so much detail on the variability of cultivated plants, as in the case of domesticated animals. The subject is involved in much difficulty. Botanists have generally neglected cultivated varieties, as beneath their notice. In several cases the wild prototype is unknown or doubtfully known; and in other cases it is hardly possible to distinguish between escaped seedlings and truly wild plants, so that there is no safe standard of comparison by which to judge of any supposed amount of change. Not a few botanists believe that several of our anciently cultivated plants have become so profoundly modified that it is not possible now to recognise their aboriginal parent-forms. Equally perplexing are the doubts whether some of them are descended from one species, or from several inextricably commingled by crossing and variation. Variations often pass into, and cannot be distinguished from, monstrosities; and monstrosities are of little significance for our purpose. Many varieties are propagated solely by grafts, buds, layers, bulbs, &c., and frequently it is not known how far their peculiarities can be transmitted by seminal generation. Nevertheless some facts of value can be gleaned; and other facts will hereafter be incidentally given. One chief object in the two following chapters is to show how generally almost every character in our cultivated plants has become variable.

Before entering on details a few general remarks on the origin of cultivated plants may be introduced. M. Alph. de Candolle [1] in an admirable discussion on this subject, in which he displays a wonderful amount of knowledge, gives a list of 157 of the most useful cultivated plants. Of these he believes that 85 are almost certainly known in their wild state; but on this head other competent judges [2] entertain great doubts. Of 40 of them, the origin is admitted by M. De Candolle to be doubtful, either from a certain amount of dissimilarity which they present when compared with their nearest allies in a wild state, or from the probability of the latter not being truly wild plants, but seedlings escaped from culture. Of the entire 157, 32 alone are ranked by M. De Candolle as quite unknown in their aboriginal condition. But it should be observed that he does not include in his list several plants which present ill-defined characters, namely, the various forms of pumpkins, millet, sorghum, kidney-bean, dolichos, capsicum, and indigo. Nor does he include flowers; and several of the more anciently cultivated flowers, such as certain roses, the common Imperial lily, the tuberose, and even the lilac, are said [3] not to be known in the wild state.

From the relative numbers above given, and from other arguments of much weight, M. De Candolle concludes that plants have rarely been so much modified by culture that they cannot be identified with their wild prototypes. But on this view, considering that savages probably would not have chosen rare plants for cultivation, that useful plants are generally conspicuous, and that they could not have been the inhabitants of deserts or of remote and recently discovered islands, it appears strange to me that so many of our cultivated plants should be still unknown or only doubtfully known in the wild state. If, on the other hand, many of these plants have been profoundly modified by culture, the difficulty disappears. Their

[1] 'Géographie Botanique Raisonnée,' 1855, pp. 810 to 991.

[2] Review by Mr. Bentham in 'Hort. Journal,' vol. ix. 1855, p. 133, entitled 'Historical Notes on cultivated Plants,' by Dr. A. Targioni-Tozzetti. *See* also 'Edinburgh Review,' 1866, p. 510.

[3] 'Hist. Notes,' as above, by Targioni-Tozzetti.

extermination during the progress of civilisation would likewise remove the difficulty; but M. De Candolle has shown that this probably has seldom occurred. As soon as a plant became cultivated in any country, the half-civilised inhabitants would no longer have need to search the whole surface of the land for it, and thus lead to its extirpation; and even if this did occur during a famine, dormant seeds would be left in the ground. In tropical countries the wild luxuriance of nature, as was long ago remarked by Humboldt, overpowers the feeble efforts of man. In anciently civilised temperate countries, where the whole face of the land has been greatly changed, it can hardly be doubted that some plants have been exterminated; nevertheless De Candolle has shown that all the plants historically known to have been first cultivated in Europe still exist here in the wild state.

MM. Loiseleur-Deslongchamps[4] and De Candolle have remarked that our cultivated plants, more especially the cereals, must originally have existed in nearly their present state; for otherwise they would not have been noticed and valued as objects of food. But these authors apparently have not considered the many accounts given by travellers of the wretched food collected by savages. I have read an account of the savages of Australia cooking, during a dearth, many vegetables in various ways, in the hopes of rendering them innocuous and more nutritious. Dr. Hooker found the half-starved inhabitants of a village in Sikhim suffering greatly from having eaten arum-roots,[5] which they had pounded and left for several days to ferment, so as partially to destroy their poisonous nature; and he adds that they cooked and ate many other deleterious plants. Sir Andrew Smith informs me that in South Africa a large number of fruits and succulent leaves, and especially roots, are used in times of scarcity. The natives, indeed, know the properties of a long catalogue of plants, some having

[4] 'Considérations sur les Céréales,' 1842, p. 37. 'Géographie Bot.,' 1855, p. 930. "Plus on suppose l'agriculture ancienne et remontant à une époque d'ignorance, plus il est probable que les cultivateurs avaient choisi des espèces offrant à l'origine même un avantage incontestable."

[5] Dr. Hooker has given me this information. *See*, also, his 'Himalayan Journals,' 1854, vol. ii. p. 49.

been found during famines to be eatable, others injurious to health, or even destructive to life. He met a party of Baquanas who, having been expelled by the conquering Zulus, had lived for years on any roots or leaves which afforded some little nutriment, and distended their stomachs, so as to relieve the pangs of hunger. They looked like walking skeletons, and suffered fearfully from constipation. Sir Andrew Smith also informs me that on such occasions the natives observe as a guide for themselves, what the wild animals, especially baboons and monkeys, eat.

From innumerable experiments made through dire necessity by the savages of every land, with the results handed down by tradition, the nutritious, stimulating, and medicinal properties of the most unpromising plants were probably first discovered. It appears, for instance, at first an inexplicable fact that untutored man, in three distant quarters of the world, should have discovered amongst a host of native plants that the leaves of the tea-plant and mattee, and the berries of the coffee, all included a stimulating and nutritious essence, now known to be chemically the same. We can also see that savages suffering from severe constipation would naturally observe whether any of the roots which they devoured acted as aperients. We probably owe our knowledge of the uses of almost all plants to man having originally existed in a barbarous state, and having been often compelled by severe want to try as food almost everything which he could chew and swallow.

From what we know of the habits of savages in many quarters of the world, there is no reason to suppose that our cereal plants originally existed in their present state so valuable to man. Let us look to one continent alone, namely, Africa: Barth[6] states that the slaves over a large part of the central region regularly collect the seeds of a wild grass, the *Pennisetum distichum;* in another district he saw women collecting the seeds of a Poa by swinging a sort of basket through the rich meadow-land. Near Tete Livingstone observed the natives collecting the seeds

[6] 'Travels in Central Africa,' Eng. translat., vol. i. pp. 529 and 390; vol. ii. pp. 29, 265, 270. Livingstone's 'Travels,' p. 551.

of a wild grass; and farther south, as Andersson informs me, the natives largely use the seeds of a grass of about the size of canary-seed, which they boil in water. They eat also the roots of certain reeds, and every one has read of the Bushmen prowling about and digging up with a fire-hardened stake various roots. Similar facts with respect to the collection of seeds of wild grasses in other parts of the world could be given.[7]

Accustomed as we are to our excellent vegetables and luscious fruits, we can hardly persuade ourselves that the stringy roots of the wild carrot and parsnip, or the little shoots of the wild asparagus, or crabs, sloes, &c., should ever have been valued; yet, from what we know of the habits of Australian and South African savages, we need feel no doubt on this head. The inhabitants of Switzerland during the Stone-period largely collected wild crabs, sloes, bullaces, hips of roses, elderberries, beech-mast, and other wild berries and fruit.[8] Jemmy Button, a Fuegian on board the *Beagle*, remarked to me that the poor and acid black-currants of Tierra del Fuego were too sweet for his taste.

The savage inhabitants of each land, having found out by many and hard trials what plants were useful, or could be rendered useful by various cooking processes, would after a time take the first step in cultivation by planting them near their usual abodes. Livingstone[9] states that the savage Batokas sometimes left wild fruit-trees standing in their gardens, and occasionally even planted them, "a practice seen nowhere else amongst the natives." But Du Chaillu saw a palm and some other wild fruit-trees which had been planted; and these trees were considered private property. The next step in cultivation, and this would require but little forethought, would be to sow

[7] As in both North and South America. Mr. Edgeworth ('Journal Proc. Linn. Soc.,' vol. vi. Bot., 1862, p. 181) states that in the deserts of the Punjab poor women sweep up, "by a whisk into straw baskets," the seeds of four genera of grasses, namely, of Agrostis, Panicum, Cenchrus, and Pennisetum, as well as the seeds of four other genera belonging to distinct families.

[8] Prof. O. Heer, 'Die Pflanzen der Pfahlbauten, 1865, aus dem Neujahr. Naturforsc. Gesellschaft,' 1866; and Dr. H. Christ, in Rütimeyer's 'Die Fauna der Pfahlbauten,' 1861, s. 226.

[9] 'Travels,' p. 535. Du Chaillu, 'Adventures in Equatorial Africa,' 1861, p. 445.

the seeds of useful plants; and as the soil near the hovels of the natives[10] would often be in some degree manured, improved varieties would sooner or later arise. Or a wild and unusually good variety of a native plant might attract the attention of some wise old savage; and he would transplant it, or sow its seed. That superior varieties of wild fruit-trees occasionally are found is certain, as in the case of the American species of hawthorns, plums, cherries, grapes, and hickories, specified by Professor Asa Gray.[11] Downing also refers to certain wild varieties of the hickory, as being "of much larger size and finer flavour than the common species." I have referred to American fruit-trees, because we are not in this case troubled with doubts whether or not the varieties are seedlings which have escaped from cultivation. Transplanting any superior variety, or sowing its seeds, hardly implies more forethought than might be expected at an early and rude period of civilisation. Even the Australian barbarians "have a law that no plant bearing seeds is to be dug up after it has flowered;" and Sir G. Grey[12] never saw this law, evidently framed for the preservation of the plant, violated. We see the same spirit in the superstitious belief of the Fuegians, that killing water-fowl whilst very young will be followed by "much rain, snow, blow much."[13] I may add, as showing forethought in the lowest barbarians, that the Fuegians when they find a stranded whale bury large portions in the sand, and during the often-recurrent famines travel from great distances for the remnants of the half-putrid mass.

It has often been remarked[14] that we do not owe a single useful plant to Australia or the Cape of Good Hope,—countries abounding to an unparalleled degree with endemic species,—or to New Zealand, or to America south of the Plata; and, according to some authors, not to America northward of Mexico. I do not believe that any edible or valuable plant, except the canary-

[10] In Tierra del Fuego the spot where wigwams had formerly stood could be distinguished at a great distance by the bright green tint of the native vegetation.

[11] 'American Acad. of Arts and Sciences,' April 10th, 1860, p. 413. Downing, 'The Fruits of America,' 1845, p. 261.

[12] 'Journals of Expeditions in Australia,' 1841, vol. ii. p. 292.

[13] Darwin's 'Journal of Researches,' 1845, p. 215.

[14] De Candolle has tabulated the facts in the most interesting manner in his 'Géographie Bot.,' p. 986.

grass, has been derived from an oceanic or uninhabited island. If nearly all our useful plants, natives of Europe, Asia, and South America, had originally existed in their present condition, the complete absence of similarly useful plants in the great countries just named would indeed be a surprising fact. But if these plants have been so greatly modified and improved by culture as no longer closely to resemble any natural species, we can understand why the above-named countries have given us no useful plants, for they were either inhabited by men who did not cultivate the ground at all, as in Australia and the Cape of Good Hope, or who cultivated it very imperfectly, as in some parts of America. These countries do yield plants which are useful to savage man; and Dr. Hooker[15] enumerates no less than 107 such species in Australia alone; but these plants have not been improved, and consequently cannot compete with those which have been cultivated and improved during thousands of years in the civilised world.

The case of New Zealand, to which fine island we as yet owe no widely cultivated plant, may seem opposed to this view; for, when first discovered, the natives cultivated several plants; but all inquirers believe, in accordance with the traditions of the natives, that the early Polynesian colonists brought with them seeds and roots, as well as the dog, which had all been wisely preserved during their long voyage. The Polynesians are so frequently lost on the ocean, that this degree of prudence would occur to any wandering party: hence the early colonists of New Zealand, like the later European colonists, would not have had any strong inducement to cultivate the aboriginal plants. According to De Candolle we owe thirty-three useful plants to Mexico, Peru, and Chile; nor is this surprising when we remember the civilized state of the inhabitants, as shown by the fact of their having practised artificial irrigation and made tunnels through hard rocks without the use of iron or gunpowder, and who, as we shall see in a future chapter, fully recognised, as far as animals were concerned, and therefore probably in the case of plants, the important principle of selection. We owe some plants to Brazil; and the early voyagers, namely Vespucius and Cabral, describe the country as thickly peopled

[15] 'Flora of Australia,' Introduction, p. cx.

and cultivated. In North America[16] the natives cultivated maize, pumpkins, gourds, beans, and peas, "all different from ours," and tobacco; and we are hardly justified in assuming that none of our present plants are descended from these North American forms. Had North America been civilized for as long a period, and as thickly peopled, as Asia or Europe, it is probable that the native vines, walnuts, mulberries, crabs, and plums, would have given rise, after a long course of cultivation, to a multitude of varieties, some extremely different from their parent-stocks; and escaped seedlings would have caused in the New, as in the Old World, much perplexity with respect to their specific distinctness and parentage.[17]

Cerealia.—I will now enter on details. The cereals cultivated in Europe consist of four genera—wheat, rye, barley, and oats. Of wheat the best modern authorities[18] make four or five, or even seven distinct species; of rye, one; of barley, three; and of oats, two, three, or four species. So that altogether our cereals are ranked by different authors under from ten to fifteen distinct species. These have given rise to a multitude of varieties. It is a remarkable fact that botanists are not universally agreed on the aboriginal parent-form of any one cereal plant. For instance, a high authority writes in 1855,[19] "We ourselves have no hesitation in stating our conviction, as the result of all the most reliable evidence, that none of these Cerealia exist, or have existed, truly wild in their present state, but that all are cultivated varieties of species now growing in great abundance in S. Europe or W. Asia." On the other hand, Alph. De Candolle[20] has adduced abundant evidence that common wheat (*Triticum vulgare*) has been found wild in various parts of Asia, where it is not likely to have escaped from cultivation; and there is

---

16 For Canada, *see* J. Cartier's Voyage in 1534; for Florida, *see* Narvaez and Ferdinand de Soto's Voyages. As I have consulted these and other old Voyages in more than one general collection of Voyages, I do not give precise references to the pages. *See* also, for several references, Asa Gray, in the 'American Journal of Science,' vol. xxiv., Nov. 1857, p. 441. For the traditions of the natives of New Zealand, *see* Crawfurd's 'Grammar and Dict. of the Malay Language,' 1852, p. cclx.

17 *See*, for example, M. Hewett C. Watson's remarks on our wild plums and cherries and crabs: 'Cybele Britannica,' vol. i. pp. 330, 334, &c. Van Mons (in his 'Arbres Fruitiers,' 1835, tom. i. p. 444) declares that he has found the types of all our cultivated varieties in wild seedlings, but then he looks on these seedlings as so many aboriginal stocks.

18 *See* A. De Candolle, 'Géograph. Bot.,' 1855, p. 928 *et seq.* Godron, 'De l'Espèce,' 1859, tom. ii. p. 70; and Metzger, 'Die Getreidearten,' &c., 1841.

19 Mr. Bentham, in his review, entitled 'Hist. Notes on cultivated Plants,' by Dr. A. Targioni-Tozzetti, in 'Journal of Hort. Soc.,' vol. ix. (1855), p. 133.

20 'Géograph. Bot.,' p. 928. The whole subject is discussed with admirable fullness and knowledge.

force in M. Godron's remark, that, supposing these plants to be escaped seedlings,[21] if they have propagated themselves in a wild state for several generations, their continued resemblance to cultivated wheat renders it probable that the latter has retained its aboriginal character. M. De Candolle insists strongly on the frequent occurrence in the Austrian dominions of rye and of one kind of oats in an apparently wild condition. With the exception of these two cases, which however are rather doubtful, and with the exception of two forms of wheat and one of barley, which he believes to have been found truly wild, M. De Candolle does not seem fully satisfied with the other reported discoveries of the parent-forms of our other cereals. With respect to oats, according to Mr. Buckman,[22] the wild English *Avena fatua* can be converted by a few years of careful cultivation and selection into forms almost identical with two very distinct cultivated races. The whole subject of the origin and specific distinctness of the various cereal plants is a most difficult one; but we shall perhaps be able to judge a little better after considering the amount of variation which wheat has undergone.

Metzger describes seven species of wheat, Godron refers to five, and De Candolle to only four. It is not improbable that, besides the kinds known in Europe, other strongly characterised forms exist in the more distant parts of the world; for Loiseleur-Deslongchamps[23] speaks of three new species or varieties, sent to Europe in 1822 from Chinese Mongolia, which he considers as being there indigenous. Moorcroft[24] also speaks of Hasora wheat in Ladakh as very peculiar. If those botanists are right who believe that at least seven species of wheat originally existed, then the amount of variation in any important character which wheat has undergone under cultivation has been slight; but if only four or a lesser number of species originally existed, then it is evident that varieties so strongly marked have arisen, that they have been considered by capable judges as specifically distinct. But the impossibility of deciding which forms ought to be ranked as species and which as varieties, makes it useless to specify in detail the differences between the various kinds of wheat. Speaking generally, the organs of vegetation differ little;[25] but some kinds grow close and upright, whilst others spread and trail along the ground. The straw differs in being more or less hollow, and in quality. The ears[26]

[21] Godron, 'De l'Espèce,' tom. ii. p. 72. A few years ago the excellent, though misinterpreted, observations of M. Fabre led many persons to believe that wheat was a modified descendant of Ægilops; but M. Godron (tom. i. p. 165) has shown by careful experiments that the first step in the series, viz. *Ægilops triticoides*, is a hybrid between wheat and *Æ. ovata*. The frequency with which these hybrids spontaneously arise, and the gradual manner in which the *Æ. triticoides* becomes converted into true wheat, alone leave any doubt on the subject.

[22] Report to British Association for 1857, p. 207.

[23] 'Considérations sur les Céréales, 1842–43, p. 29.

[24] 'Travels in the Himalayan Provinces,' &c., 1841, vol. i. p. 224.

[25] Col. J. Le Couteur on the 'Varieties of Wheat,' pp. 23, 79.

[26] Loiseleur-Deslongchamps, 'Consid. sur les Céréales,' p. 11.

differ in colour and in shape, being quadrangular, compressed, or nearly cylindrical; and the florets differ in their approximation to each other, in their pubescence, and in being more or less elongated. The presence or absence of barbs is a conspicuous difference, and in certain Gramineæ serves even as a generic character;[27] although, as remarked by Godron,[28] the presence of barbs is variable in certain wild grasses, and especially in those, such as *Bromus secalinus* and *Lolium temulentum*, which habitually grow mingled with our cereal crops, and which have thus unintentionally been exposed to culture. The grains differ in size, weight, and colour; in being more or less downy at one end, in being smooth or wrinkled, in being either nearly globular, oval, or elongated; and finally in internal texture, being tender or hard, or even almost horny, and in the proportion of gluten which they contain.

Nearly all the races or species of wheat vary, as Godron[29] has remarked, in an exactly parallel manner,—in the seed being downy or glabrous, and in colour,—and in the florets being barbed or not barbed, &c. Those who believe that all the kinds are descended from a single wild species may account for this parallel variation by the inheritance of a similar constitution, and a consequent tendency to vary in the same manner; and those who believe in the general theory of descent with modification may extend this view to the several species of wheat, if such ever existed in a state of nature.

Although few of the varieties of wheat present any conspicuous difference, their number is great. Dalbret cultivated during thirty years from 150 to 160 kinds, and excepting in the quality of the grain they all kept true: Colonel Le Couteur possessed upwards of 150, and Philippar 322 varieties.[30] As wheat is an annual, we thus see how strictly many trifling differences in character are inherited through many generations. Colonel Le Couteur insists strongly on this same fact: in his persevering and successful attempts to raise new varieties by selection, he began by choosing the best ears, but soon found that the grains in the same ear differed so that he was compelled to select them separately; and each grain generally transmitted its own character. The great amount of variability in the plants of the same variety is another interesting point, which would never have been detected except by an eye long practised to the work; thus Colonel Le Couteur relates[31] that in a field of his own wheat, which he considered at least as pure as that of any of his neighbours, Professor La Gasca found twenty-three sorts; and Professor Henslow has observed similar facts. Besides such individual variations, forms sufficiently well marked to be valued and to become widely cultivated

---

[27] *See* an excellent review in Hooker's Journ. of Botany,' vol. viii. p. 82, note.

[28] 'De l'Espèce,' tom. ii. p. 73.

[29] Idem, tom. ii. p. 75.

[30] For Dalbret and Philippar, *see* Loiseleur-Deslongchamps, 'Consid. sur les Céréales,' pp. 45, 70. Le Couteur on Wheat, p. 6.

[31] 'Varieties of Wheat,' Introduction, p. vi. Marshall, in his 'Rural Economy of Yorkshire,' vol. ii. p. 9, remarks that "in every field of corn there is as much variety as in a herd of cattle."

sometimes suddenly appear: thus Mr. Sheriff has had the good fortune to raise in his lifetime seven new varieties, which are now extensively grown in many parts of Britain.[32]

As in the case of many other plants, some varieties, both old and new, are far more constant in character than others. Colonel Le Couteur was forced to reject some of his new sub-varieties, which he suspected had been produced from a cross, as incorrigibly sportive. With respect to the tendency to vary, Metzger[33] gives from his own experience some interesting facts: he describes three Spanish sub-varieties, more especially one known to be constant in Spain, which in Germany assumed their proper character only during hot summers; another variety kept true only in good land, but after having been cultivated for twenty-five years became more constant. He mentions two other sub-varieties which were at first inconstant, but subsequently became, apparently without any selection, accustomed to their new homes, and retained their proper character. These facts show what small changes in the conditions of life cause variability, and they further show that a variety may become habituated to new conditions. One is at first inclined to conclude with Loiseleur-Deslongchamps, that wheat cultivated in the same country is exposed to remarkably uniform conditions; but manures differ, seed is taken from one soil to another, and what is far more important the plants are exposed as little as possible to struggle with other plants, and are thus enabled to exist under diversified conditions. In a state of nature each plant is confined to that particular station and kind of nutriment which it can seize from the other plants by which it is surrounded.

Wheat quickly assumes new habits of life. The summer and winter kinds were classed by Linnæus as distinct species; but M. Monnier[34] has proved that the difference between them is only temporary. He sowed winter-wheat in spring, and out of one hundred plants four alone produced ripe seeds; these were sown and resown, and in three years plants were reared which ripened all their seed. Conversely, nearly all the plants raised from summer-wheat, which was sown in autumn, perished from frost; but a few were saved and produced seed, and in three years this summer-variety was converted into a winter-variety. Hence it is not surprising that wheat soon becomes to a certain extent acclimatised, and that seed brought from distant countries and sown in Europe vegetates at first, or even for a considerable period,[35] differently from our European varieties. In Canada the first settlers, according to Kalm,[36] found their winters too severe for winter-wheat brought from France, and their summers often too short for summer-wheat; and until they procured summer-wheat from the northern parts of Europe, which succeeded well, they thought that their

---

32 'Gardener's Chron. and Agricult. Gazette,' 1862, p. 963.

33 'Getreidearten,' 1841, s. 66, 91, 92, 116, 117.

34 Quoted by Godron, 'De l'Espèce,' vol. ii. p. 74. So it is, according to Metzger ('Getreidearten,' s. 18), with summer and winter barley.

35 Loiseleur-Deslongchamps, 'Céréales,' part ii. p. 224. Le Couteur, p. 70. Many other accounts could be added.

36 'Travels in North America,' 1753-1761, Eng. translat., vol. iii. p. 165.

country was useless for corn crops. It is notorious that the proportion of gluten differs much under different climates. The weight of the grain is also quickly affected by climate: Loiseleur-Deslongchamps[37] sowed near Paris 54 varieties, obtained from the South of France and from the Black Sea, and 52 of these yielded seed from 10 to 40 per cent. heavier than the parent-seed. He then sent these heavier grains back to the South of France, but there they immediately yielded lighter seed.

All those who have closely attended to the subject insist on the close adaptation of numerous varieties of wheat to various soils and climates even within the same country; thus Colonel Le Couteur[38] says, "It is the suitableness of each sort to each soil that will enable the farmer to pay his rent by sowing one variety, where he would be unable to do so by attempting to grow another of a seemingly better sort." This may be in part due to each kind becoming habituated to its conditions of life, as Metzger has shown certainly occurs, but it is probably in main part due to innate differences between the several varieties.

Much has been written on the deterioration of wheat; that the quality of the flour, size of grain, time of flowering, and hardiness may be modified by climate and soil, seems nearly certain; but that the whole body of any one sub-variety ever becomes changed into another and distinct sub-variety, there is no reason to believe. What apparently does take place, according to Le Couteur,[39] is, that some one sub-variety out of the many which may always be detected in the same field is more prolific than the others, and gradually supplants the variety which was first sown.

With respect to the natural crossing of distinct varieties the evidence is conflicting, but preponderates against its frequent occurrence. Many authors maintain that impregnation takes place in the closed flower, but I am sure from my own observations that this is not the case, at least with those varieties to which I have attended. But as I shall have to discuss this subject in another work, it may be here passed over.

In conclusion, all authors admit that numerous varieties of wheat have arisen; but their differences are unimportant, unless, indeed, some of the so-called species are ranked as varieties. Those who believe that from four to seven wild species of Triticum originally existed in nearly the same condition as at present, rest their belief chiefly on the great antiquity of the several forms.[40] It is an important fact, which we have recently learnt from the admirable researches

[37] 'Céréales,' part ii. pp. 179–183.

[38] 'On the Varieties of Wheat,' Introduct., p. vii. *See* Marshall, 'Rural Econ. of Yorkshire,' vol. ii. p. 9. With respect to similar cases of adaptation in the varieties of oats, *see* some interesting papers in the 'Gardener's Chron. and Agricult. Gazette,' 1850, pp. 204, 219.

[39] 'On the Varieties of Wheat,' p. 59. Mr. Sheriff, and a higher authority cannot be given ('Gard. Chron. and Agricult. Gazette,' 1862, p. 963), says, "I have never seen grain which has either been improved or degenerated by cultivation, so as to convey the change to the succeeding crop."

[40] Alph. De Candolle, 'Géograph. Bot.,' p. 930.

of Heer,[41] that the inhabitants of Switzerland, even so early as the Neolithic period, cultivated no less than ten cereal plants, namely, five kinds of wheat, of which at least four are commonly looked at as distinct species, three kinds of barley, a panicum, and a setaria. If it could be shown that at the earliest dawn of agriculture five kinds of wheat and three of barley had been cultivated, we should of course be compelled to look at these forms as distinct species. But, as Heer has remarked, agriculture even at the period of the lake-habitations had already made considerable progress; for, besides the ten cereals, peas, poppies, flax, and apparently apples, were cultivated. It may also be inferred, from one variety of wheat being the so-called Egyptian, and from what is known of the native country of the panicum and setaria, as well as from the nature of the weeds which then grew mingled with the crops, that the lake-inhabitants either still kept up commercial intercourse with some southern people or had originally proceeded as colonists from the South.

Loiseleur-Deslongchamps[42] has argued that, if our cereal plants had been greatly modified by cultivation, the weeds which habitually grow mingled with them would have been equally modified. But this argument shows how completely the principle of selection has been overlooked. That such weeds have not varied, or at least do not vary now in any extreme degree, is the opinion of Mr. H. C. Watson and Professor Asa Gray, as they inform me; but who will pretend to say that they do not vary as much as the individual plants of the same sub-variety of wheat? We have already seen that pure varieties of wheat, cultivated in the same field, offer many slight variations, which can be selected and separately propagated; and that occasionally more strongly pronounced variations appear, which, as Mr. Sheriff has proved, are well worthy of extensive cultivation. Not until equal attention be paid to the variability and selection of weeds, can the argument from their constancy under unintentional culture be of any value. In accordance with the principles of selection we can understand how it is that in the several cultivated varieties of wheat the organs of vegetation differ so little; for if a plant

41 'Pflanzen der Pfahlbauten,' 1866.

42 'Les Céréales,' p. 94.

with peculiar leaves appeared, it would be neglected unless the grains of corn were at the same time superior in quality or size. The selection of seed-corn was strongly recommended[43] in ancient times by Columella and Celsus; and as Virgil says,—

> "I've seen the largest seeds, tho' view'd with care,
> Degenerate, unless th' industrious hand
> Did yearly cull the largest."

But whether in ancient times selection was methodically pursued we may well doubt, when we hear how laborious the work was found by Le Couteur. Although the principle of selection is so important, yet the little which man has effected, by incessant efforts[44] during thousands of years, in rendering the plants more productive or the grains more nutritious than they were in the time of the old Egyptians, would seem to speak strongly against its efficacy. But we must not forget that at each successive period the state of agriculture and the quantity of manure supplied to the land will have determined the maximum degree of productiveness; for it would be impossible to cultivate a highly productive variety, unless the land contained a sufficient supply of the necessary chemical elements.

We now know that man was sufficiently civilized to cultivate the ground at an immensely remote period; so that wheat might have been improved long ago up to that standard of excellence which was possible under the then existing state of agriculture. One small class of facts supports this view of the slow and gradual improvement of our cereals. In the most ancient lake-habitations of Switzerland, when men employed only flint-tools, the most extensively cultivated wheat was a peculiar kind, with remarkably small ears and grains.[45] "Whilst the grains of the modern forms are in section from seven to eight millimètres in length, the larger grains from the lake-habitations are six, seldom seven, and the smaller ones only four. The ear is thus much narrower, and the spikelets stand out more horizontally, than in our present forms." So again with barley, the most ancient and most extensively cultivated kind had small ears, and the grains

[43] Quoted by Le Couteur, p. 16.

[44] A. De Candolle, 'Géograph. Bot.,' p. 932.

[45] O. Heer, 'Die Pflanzen der Pfahlbauten,' 1866. The following passage is quoted from Dr. Christ, in 'Die Fauna der Pfahlbauten von Dr. Rütimeyer,' 1861, s. 225.

were "smaller, shorter, and nearer to each other, than in that now grown; without the husk they were $2\frac{1}{2}$ lines long, and scarcely $1\frac{1}{2}$ broad, whilst those now grown have a length of three lines, and almost the same in breadth."[46] These small-grained varieties of wheat and barley are believed by Heer to be the parent-forms of certain existing allied varieties, which have supplanted their early progenitors.

Heer gives an interesting account of the first appearance and final disappearance of the several plants which were cultivated in greater or less abundance in Switzerland during former successive periods, and which generally differed more or less from our existing varieties. The peculiar small-eared and small-grained wheat, already alluded to, was the commonest kind during the Stone period; it lasted down to the Helvetico-Roman age, and then became extinct. A second kind was rare at first, but afterwards became more frequent. A third, the Egyptian wheat (*T. turgidum*), does not agree exactly with any existing variety, and was rare during the Stone period. A fourth kind (*T. dicoccum*) differs from all known varieties of this form. A fifth kind (*T. monococcum*) is known to have existed during the Stone period only by the presence of a single ear. A sixth kind, the common *T. spelta*, was not introduced into Switzerland until the Bronze age. Of barley, besides the short-eared and small-grained kind, two others were cultivated, one of which was very scarce, and resembled our present common *H. distichum*. During the Bronze age rye and oats were introduced; the oat-grains being somewhat smaller than those produced by our existing varieties. The poppy was largely cultivated during the Stone period, probably for its oil; but the variety which then existed is not now known. A peculiar pea with small seeds lasted from the Stone to the Bronze age, and then became extinct; whilst a peculiar bean, likewise having small seeds, came in at the Bronze period and lasted to the time of the Romans. These details sound like the description given by a palæontologist of the mutations in form, of the first appearance, the increasing rarity, and final extinction of fossil species, embedded in the successive stages of a geological formation.

[46] Heer, as quoted by Carl Vogt, 'Lectures on Man,' Eng. translat., p. 355.

Finally, every one must judge for himself whether it is more probable that the several forms of wheat, barley, rye, and oats are descended from between ten and fifteen species, most of which are now either unknown or extinct, or whether they are descended from between four and eight species, which may have either closely resembled our present cultivated forms, or have been so widely different as to escape identification. In this latter case, we must conclude that man cultivated the cereals at an enormously remote period, and that he formerly practised some degree of selection, which in itself is not improbable. We may, perhaps, further believe that, when wheat was first cultivated, the ears and grains increased quickly in size, in the same manner as the roots of the wild carrot and parsnip are known to increase quickly in bulk under cultivation.

Maize: Zea Mays.—Botanists are nearly unanimous that all the cultivated kinds belong to the same species. It is undoubtedly[47] of American origin, and was grown by the aborigines throughout the continent from New England to Chili. Its cultivation must have been extremely ancient, for Tschudi[48] describes two kinds, now extinct or not known in Peru, which were taken from tombs apparently prior to the dynasty of the Incas. But there is even stronger evidence of antiquity, for I found on the coast of Peru[49] heads of maize, together with eighteen species of recent sea-shell, embedded in a beach which had been upraised at least 85 feet above the level of the sea. In accordance with this ancient cultivation, numerous American varieties have arisen. The aboriginal form has not as yet been discovered in the wild state. A peculiar kind,[50] in which the grains, instead of being naked, are concealed by husks as much as eleven lines in length, has been stated on insufficient evidence to grow wild in Brazil. It is almost certain that the aboriginal form would have had its grains thus protected;[51] but the seeds of the Brazilian variety produce, as I hear from Professor Asa Gray, and as is stated in two published accounts, either common or husked maize; and it is not cre-

[47] *See* Alph. De Candolle's long discussion in his 'Géograph. Bot.,' p. 942. With respect to New England, *see* Silliman's 'American Journal,' vol. xliv. p. 99.

[48] 'Travels in Peru,' Eng. translat., p. 177.

[49] 'Geolog. Observ. on S. America,' 1846, p. 49.

[50] This maize is figured in Bonafous' magnificent work, 'Hist. Nat. du Mais,' 1836, Pl. v. bis, and in the 'Journal of Hort. Soc.,' vol. i., 1846, p. 115, where an account is given of the result of sowing the seed. A young Guarany Indian, on seeing this kind of maize, told Auguste St. Hilaire (*see* De Candolle, 'Géograph. Bot.,' p. 951) that it grew wild in the humid forests of his native land. Mr. Teschemacher, in 'Proc. Boston Soc. Nat. Hist.,' Oct. 19th, 1842, gives an account of sowing the seed.

[51] Moquin-Tandon, 'Éléments de Tératologie,' 1841, p. 126.

dible that a wild species, when first cultivated, should vary so quickly and in so great a degree.

Maize has varied in an extraordinary and conspicuous manner. Metzger,[52] who paid particular attention to the cultivation of this plant, makes twelve races (unter-art) with numerous sub-varieties; of the latter some are tolerably constant, others quite inconstant. The different races vary in height from 15-18 feet to only 16-18 inches, as in a dwarf variety described by Bonafous. The whole ear is variable in shape, being long and narrow, or short and thick, or branched. The ear in one variety is more than four times as long as in a dwarf kind. The seeds are arranged in the ear in from six to even twenty rows, or are placed irregularly. The seeds are coloured—white, pale-yellow, orange, red, violet, or elegantly streaked with black;[53] and in the same ear there are sometimes seeds of two colours. In a small collection I found that a single grain of one variety nearly equalled in weight seven grains of another variety. The shape of the seed varies greatly, being very flat, or nearly globular, or oval; broader than long, or longer than broad; without any point, or produced into a sharp tooth, and this tooth is sometimes recurved. One variety (the rugosa of Bonafous) has its seeds curiously wrinkled, giving to the whole ear a singular appearance. Another variety (the cymosa of Bon.) carries its ears so crowded together that it is called *maïs à bouquet*. The seeds of some varieties contain much glucose instead of starch. Male flowers sometimes appear amongst the female flowers, and Mr. J. Scott has lately observed the rarer case of female flowers on a true male panicle, and likewise hermaphrodite flowers.[54] Azara describes[55] a variety in Paraguay the grains of which are very tender, and he states that several varieties are fitted for being cooked in various ways. The varieties also differ greatly in precocity, and have different powers of resisting dryness and the action of violent wind.[56] Some of the foregoing differences would certainly be considered of specific value with plants in a state of nature.

Le Comte Ré states that the grains of all the varieties which he cultivated ultimately assumed a yellow colour. But Bonafous[57] found that most of those which he sowed for ten consecutive years kept true to their proper tints; and he adds that in the valleys of the Pyrenees and on the plains of Piedmont a white maize has been cultivated for more than a century, and has undergone no change.

The tall kinds grown in southern latitudes, and therefore exposed to great heat, require from six to seven months to ripen their seed; whereas the dwarf kinds, grown in northern and colder climates, require only from

[52] 'Die Getreidearten,' 1841, s. 208. I have modified a few of Metzger's statements in accordance with those made by Bonafous in his great work, 'Hist. Nat. du Maïs,' 1836.

[53] Godron, 'De l'Espèce,' tom. ii. p. 80; Al. De Candolle, idem, p. 951.

[54] 'Transact. Bot. Soc. of Edinburgh,' vol. viii. p. 60.

[55] 'Voyages dans l'Amérique Méridionale,' tom. i. p. 147.

[56] Bonafous' 'Hist. Nat. du Maïs,' p. 31.

[57] Idem, p. 31.

three to four months.[58] Peter Kalm,[59] who particularly attended to this plant, says, that in the United States, in proceeding from south to north, the plants steadily diminish in bulk. Seeds brought from lat. 37° in Virginia, and sown in lat. 43°-44° in New England, produce plants which will not ripen their seed, or ripen them with the utmost difficulty. So it is with seed carried from New England to lat. 45°-47° in Canada. By taking great care at first, the southern kinds after some years' culture ripen their seed perfectly in their northern homes, so that this is an analogous case with that of the conversion of summer into winter wheat, and conversely. When tall and dwarf maize are planted together, the dwarf kinds are in full flower before the others have produced a single flower; and in Pennsylvania they ripen their seed six weeks earlier than the tall maize. Metzger also mentions a European maize which ripens its seed four weeks earlier than another European kind. With these facts, so plainly showing inherited acclimatisation, we may readily believe Kalm, who states that in North America maize and some other plants have gradually been cultivated further and further northward. All writers agree that to keep the varieties of maize pure they must be planted separately so that they shall not cross.

The effects of the climate of Europe on the American varieties is highly remarkable. Metzger obtained seed from various parts of America, and cultivated several kinds in Germany. I will give an abstract of the changes observed[60] in one case, namely, with a tall kind (Breit-korniger mays, Zea altissima) brought from the warmer parts of America. During the first year the plants were twelve feet high, and few seeds were perfected; the lower seeds in the ear kept true to their proper form, but the upper seeds became slightly changed. In the second generation the plants were from nine to ten feet in height, and ripened their seed better; the depression on the outer side of the seed had almost disappeared, and the original beautiful white colour had become duskier. Some of the seeds had even become yellow, and in their now rounded form they approached common European maize. In the third generation nearly all resemblance to the original and very distinct American parent-form was lost. In the sixth generation this maize perfectly resembled a European variety, described as the second sub-variety of the fifth race. When Metzger published his book, this variety was still cultivated near Heidelberg, and could be distinguished from the common kind only by a somewhat more vigorous growth. Analogous results were obtained by the cultivation of another American race, the "white-tooth corn," in which the tooth nearly disappeared even in the second generation. A third race, the "chicken-corn," did not undergo so great a change, but the seeds became less polished and pellucid.

These facts afford the most remarkable instance known to me of the direct and prompt action of climate on a plant. It might

58 Metzger, 'Getreidearten,' s. 206.

59 'Description of Maize,' by P. Kalm, 1752, in 'Swedish Acts,' vol. iv. I have consulted an old English MS. translation.

60 'Getreidearten,' s. 208.

have been expected that the tallness of the stem, the period of vegetation, and the ripening of the seed, would have been thus affected; but it is a much more surprising fact that the seeds should have undergone so rapid and great a change. As, however, flowers, with their product the seed, are formed by the metamorphosis of the stem and leaves, any modification in these latter organs would be apt to extend, through correlation, to the organs of fructification.

Cabbage (*Brassica oleracea*).—Every one knows how greatly the various kinds of cabbage differ in appearance. In the island of Jersey, from the effects of particular culture and of climate, a stalk has grown to the height of sixteen feet, and "had its spring shoots at the top occupied by a magpie's nest:" the woody stems are not unfrequently from ten to twelve feet in height, and are there used as rafters[61] and as walking-sticks. We are thus reminded that in certain countries plants belonging to the generally herbaceous order of the Cruciferæ are developed into trees. Every one can appreciate the difference between green or red cabbages with great single heads; Brussel-sprouts with numerous little heads; broccolis and cauliflowers with the greater number of their flowers in an aborted condition, incapable of producing seed, and borne in a dense corymb instead of an open panicle; savoys with their blistered and wrinkled leaves; and borecoles and kales, which come nearest to the wild parent-form. There are also various frizzled and laciniated kinds, some of such beautiful colours that Vilmorin in his Catalogue of 1851 enumerates ten varieties, valued solely for ornament, which are propagated by seed. Some kinds are less commonly known, such as the Portuguese Couve Tronchuda, with the ribs of its leaves greatly thickened; and the Kohlrabi or choux-raves, with their stems enlarged into great turnip-like masses above the ground; and the recently formed new race[62] of choux-raves, already including nine sub-varieties, in which the enlarged part lies beneath the ground like a turnip.

Although we see such great differences in the shape, size, colour, arrangement, and manner of growth of the leaves and stem, and of the flower-stems in the broccoli and cauliflower, it is remarkable that the flowers themselves, the seed-pods, and seeds, present extremely slight differences or none at all.[63] I compared the flowers of all the principal kinds; those of the Couve Tronchuda are white and rather smaller than in common cabbages; those of the Portsmouth broccoli have narrower sepals, and smaller, less elongated petals; and in no other cabbage could any difference be detected. With respect to the seed-pods, in the purple Kohlrabi alone,

[61] 'Cabbage Timber,' 'Gardener's Chron., 1856, p. 744, quoted from Hooker's 'Journal of Botany.' A walking-stick made from a cabbage-stalk is exhibited in the Museum at Kew.

[62] 'Journal de la Soc. Imp. d'Horticulture,' 1855, p. 254, quoted from 'Gartenflora,' Ap. 1855.

[63] Godron, 'De l'Espèce,' tom. ii. p. 52; Metzger, 'Syst. Beschreibung der Kult. Kohlarten,' 1833, s. 6.

do they differ, being a little longer and narrower than usual. I made a collection of the seeds of twenty-eight different kinds, and most of them were undistinguishable; when there was any difference it was excessively slight; thus, the seeds of various broccolis and cauliflowers, when seen in mass, are a little redder; those of the early green Ulm savoy are rather smaller; and those of the Breda kail slightly larger than usual, but not larger than the seeds of the wild cabbage from the coast of Wales. What a contrast in the amount of difference is presented if, on the one hand, we compare the leaves and stems of the various kinds of cabbage with their flowers, pods, and seeds, and on the other hand the corresponding parts in the varieties of maize and wheat! The explanation is obvious; the seeds alone are valued in our cereals, and their variations have been selected; whereas the seeds, seed-pods, and flowers have been utterly neglected in the cabbage, whilst many useful variations in their leaves and stems have been noticed and preserved from an extremely remote period, for cabbages were cultivated by the old Celts.[64]

It would be useless to give a classified description [65] of the numerous races, sub-races, and varieties of the cabbage; but it may be mentioned that Dr. Lindley has lately proposed [66] a system founded on the state of development of the terminal and lateral leaf-buds, and of the flower-buds. Thus, I. All the leaf-buds active and open, as in the wild-cabbage, kail, &c. II. All the leaf-buds active, but forming heads, as in Brussel-sprouts, &c. III. Terminal leaf-bud alone active, forming a head as in common cabbages, savoys, &c. IV. Terminal leaf-bud alone active and open, with most of the flowers abortive and succulent, as in the cauliflower and broccoli. V. All the leaf-buds active and open, with most of the flowers abortive and succulent, as in the sprouting-broccoli. This latter variety is a new one, and bears the same relation to common broccoli, as Brussel-sprouts do to common cabbages; it suddenly appeared in a bed of common broccoli, and was found faithfully to transmit its newly-acquired and remarkable characters.

The principal kinds of cabbage existed at least as early as the sixteenth century,[67] so that numerous modifications of structure have been inherited for a long period. This fact is the more remarkable as great care must be taken to prevent the crossing of the different kinds. To give one proof of this: I raised 233 seedlings from cabbages of different kinds, which had purposely been planted near each other, and of the seedlings no less than 155 were plainly deteriorated and mongrelized; nor were the remaining 78 all perfectly true. It may be doubted whether many permanent varieties have been formed by intentional or accidental crosses; for such crossed plants are found to be very inconstant. One kind, however, called "Cottager's Kale," has lately been produced by crossing common kale and Brussel-sprouts, recrossed with purple broccoli,[68] and is said to be true, but plants

[64] Regnier, 'De l'Économie Publique des Celtes,' 1818, p. 438.

[65] *See* the elder De Candolle, in 'Transact. of Hort. Soc.,' vol. v.; and Metzger 'Kohlarten,' &c.

[66] 'Gardener's Chronicle,' 1859, p. 992.

[67] Alph. De Candolle, 'Géograph. Bot.,' pp. 842 and 989.

[68] 'Gardener's Chron.,' Feb. 1858, p. 128.

raised by me were not nearly so constant in character as any common cabbage.

Although most of the kinds keep true if carefully preserved from crossing, yet the seed-beds must be yearly examined, and a few seedlings are generally found false; but even in this case the force of inheritance is shown, for, as Metzger has remarked[69] when speaking of Brussel-sprouts, the variations generally keep to their "unter art," or main race. But in order that any kind may be truly propagated there must be no great change in the conditions of life; thus cabbages will not form heads in hot countries, and the same thing has been observed with an English variety grown during an extremely warm and damp autumn near Paris.[70] Extremely poor soil also affects the characters of certain varieties.

Most authors believe that all the races are descended from the wild cabbage found on the western shores of Europe; but Alph. De Candolle[71] forcibly argues on historical and other grounds that it is more probable that two or three closely allied forms, generally ranked as distinct species, still living in the Mediterranean region, are the parents, now all commingled together, of the various cultivated kinds. In the same manner as we have often seen with domesticated animals, the supposed multiple origin of the cabbage throws no light on the characteristic differences between the cultivated forms. If our cabbages are the descendants of three or four distinct species, every trace of any sterility which may originally have existed between them is now lost, for none of the varieties can be kept distinct without scrupulous care to prevent intercrossing.

The other cultivated forms of the genus Brassica are descended, according to the view adopted by Godron and Metzger,[72] from two species, *B. napus* and *rapa*; but according to other botanists from three species; whilst others again strongly suspect that all these forms, both wild and cultivated, ought to be ranked as a single species. *Brassica napus* has given rise to two large groups, namely, Swedish turnips (by some believed to be of hybrid origin)[73] and Colzas, the seeds of which yield oil. *Brassica rapa* (of Koch) has also given rise to two races, namely, common turnips and the oil-giving rape. The evidence is unusually clear that these latter plants, though so different in external appearance, belong to the same species; for the turnip has been observed by Koch and Godron to lose its thick roots in uncultivated soil, and when rape and turnips are sown together they cross to such a degree that scarcely a single plant comes true.[74] Metzger by culture converted the biennial or winter rape into the annual or summer rape,—varieties which have been thought by some authors to be specifically distinct.[75]

In the production of large, fleshy, turnip-like stems, we have a case

[69] 'Kohlarten,' s. 22.

[70] Godron, 'De l'Espèce,' tom. ii. p. 52; Metzger, 'Kohlarten,' s. 22.

[71] 'Géograph, Bot.,' p. 840.

[72] Godron, 'De l'Espèce,' tom. ii. p. 54; Metzger, 'Kohlarten,' s. 10.

[73] 'Gardener's Chron. and Agricult. Gazette,' 1856, p. 729.

[74] 'Gardener's Chron. and Agricult. Gazette,' 1855, p. 730.

[75] Metzger, 'Kohlarten,' s. 51.

of analogous variation in three forms which are generally considered as distinct species. But scarcely any modification seems so easily acquired as a succulent enlargement of the stem or root—that is a store of nutriment laid up for the plant's own future use. We see this in our radishes, beet, and in the less generally known "turnip-rooted" celery, and in the finocchio or Italian variety of the common fennel. Mr. Buckman has lately proved by his interesting experiments how quickly the roots of the wild parsnip can be enlarged, as Vilmorin formerly proved in the case of the carrot.[76] This latter plant, in its cultivated state, differs in scarcely any character from the wild English species, except in general luxuriance and in the size and quality of its roots; but in the root ten varieties, differing in colour, shape, and quality, are cultivated [77] in England, and come true by seed. Hence, with the carrot, as in so many other cases, for instance with the numerous varieties and sub-varieties of the radish, that part of the plant which is valued by man, falsely appears alone to have varied. The truth is that variations in this part alone have been selected; and the seedlings inheriting a tendency to vary in the same way, analogous modifications have been again and again selected, until at last a great amount of change has been effected.

Pea (*Pisum sativum*).—Most botanists look at the garden-pea as specifically distinct from the field-pea (*P. arvense*). The latter exists in a wild state in Southern Europe; but the aboriginal parent of the garden-pea has been found by one collector alone, as he states, in the Crimea.[78] Andrew Knight crossed, as I am informed by the Rev. A. Fitch, the field pea with a well-known garden variety, the Prussian pea, and the cross seems to have been perfectly fertile. Dr. Alefeld has recently studied [79] the genus with care, and, after having cultivated about fifty varieties, concludes that they all certainly belong to the same species. It is an interesting fact already alluded to, that, according to O. Heer,[80] the peas found in the lake-habitations of Switzerland of the Stone and Bronze ages, belong to an extinct variety, with exceedingly small seeds, allied to *P. arvense*, or field-pea. The varieties of the common garden-pea are numerous, and differ considerably from each other. For comparison I planted at the same time forty-one English and French varieties, and in this one case I will describe minutely their differences. The varieties

[76] These experiments by Vilmorin have been quoted by many writers. An eminent botanist, Prof. Decaisne, has lately expressed doubts on the subject from his own negative results, but these cannot be valued equally with positive results. On the other hand, M. Carrière has lately stated ('Gard. Chronicle,' 1865, p. 1154) that he took seed from a wild carrot, growing far from any cultivated land, and even in the first generation the roots of his seedlings differed in being spindle-shaped, longer, softer and less fibrous than those of the wild plant. From these seedlings he raised several distinct varieties.

[77] Loudon's 'Encyclop. of Gardening,' p. 835.

[78] Alph. De Candolle 'Géograph. Bot.,' 960. Mr. Bentham ('Hort. Journal,' vol. ix. (1855), p. 141) believes that garden and field peas belong to the same species, and in this respect he differs from Dr. Targioni.

[79] 'Botanische Zeitung,' 1860, s. 204.

[80] 'Die Pflanzen der Pfahlbauten,' 1866, s. 23.

differ greatly in height,—namely from between 6 and 12 inches to 8 feet,[81] —in manner of growth, and in period of maturity. Some varieties differ in general aspect even while only two or three inches in height. The stems of the *Prussian* pea are much branched. The tall kinds have larger leaves than the dwarf kinds, but not in strict proportion to their height:—*Hairs' Dwarf Monmouth* has very large leaves, and the *Pois nain hatif*, and the moderately tall *Blue Prussian*, have leaves about two-thirds of the size of the tallest kind. In the *Danecroft* the leaflets are rather small and a little pointed; in the *Queen of Dwarfs* rather rounded; and in the *Queen of England* broad and large. In these three peas the slight differences in the shape of the leaves are accompanied by slight differences in colour. In the *Pois géant sans parchemin*, which bears purple flowers, the leaflets in the young plant are edged with red; and in all the peas with purple flowers the stipules are marked with red.

In the different varieties, one or two, or several flowers in a small cluster, are borne on the same peduncle; and this is a difference which with some of the Leguminosæ is considered of specific value. In all the varieties the flowers closely resemble each other except in colour and size. They are generally white, sometimes purple, but the colour is inconstant even in the same variety. In *Warner's Emperor*, which is a tall kind, the flowers are nearly double the size of those of the *Pois nain hatif*, but *Hairs' Dwarf Monmouth*, which has large leaves, likewise has large flowers. The calyx in the *Victoria Marrow* is large, and in *Bishop's Long Pod* the sepals are rather narrow. In no other kind is there any difference in the flower.

The pods and seeds, which with natural species afford such constant characters, differ greatly in the cultivated varieties of the pea; and these are the valuable, and consequently the selected parts. *Sugar peas*, or *Pois sans parchemin*, are remarkable from their thin pods, which, whilst young, are cooked and eaten whole; and in this group, which, according to Mr. Gordon includes eleven sub-varieties, it is the pod which differs most: thus *Lewis's Negro-podded pea* has a straight, broad, smooth, and dark-purple pod, with the husk not so thin as in the other kinds; the pod of another variety is extremely bowed; that of the *Pois géant* is much pointed at the extremity; and in the variety "*à grands cosses*" the peas are seen through the husk in so conspicuous a manner that the pod, especially when dry, can hardly at first be recognised as that of a pea.

In the ordinary varieties the pods also differ much in size;—in colour, that of *Woodford's Green Marrow* being bright-green when dry, instead of pale brown, and that of the purple-podded pea being expressed by its name;—in smoothness, that of *Danecroft* being remarkably glossy, whereas that of the *Ne plus ultra* is rugged;—in being either nearly cylindrical, or broad and flat;—in being pointed at the end as in *Thurston's Reliance*, or much truncated as in the *American Dwarf*. In the *Auvergne pea* the whole end of

[81] A variety called the Rouncival attains this height, as is stated by Mr. Gordon in 'Transact. Hort. Soc.' (2nd series), vol. i., 1835, p. 374, from which paper I have taken some facts.

the pod is bowed upwards. In the *Queen of the Dwarfs* and in *Scimitar péas* the pod is almost elliptic in shape. I here give drawings of the four most distinct pods produced by the plants cultivated by me.

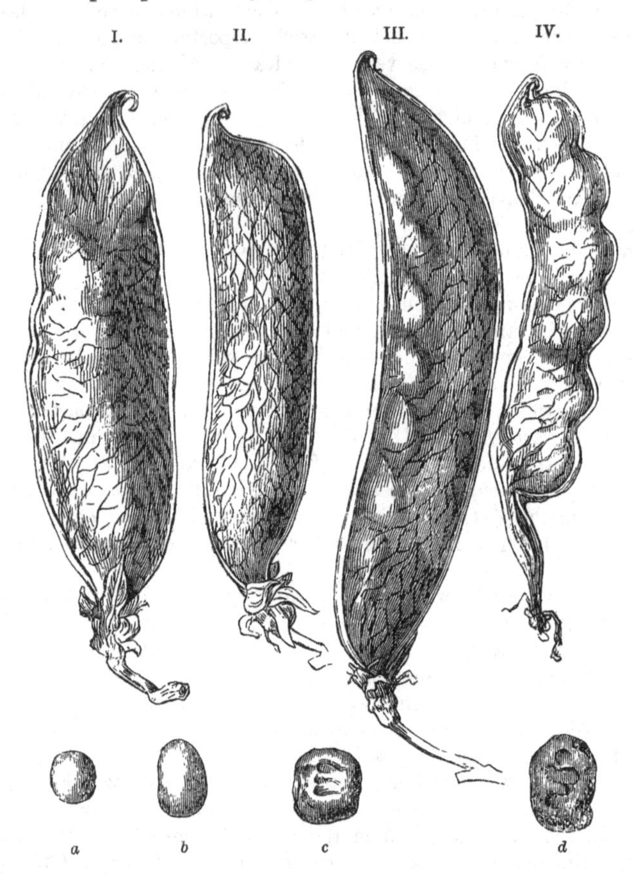

Fig. 41.—Pods and Peas. I. Queen of Dwarfs. II. American Dwarf. III. Thurston's Reliance. IV. Pois Géant sans parchemin. *a.* Dan O'Rourke Pea. *b.* Queen of Dwarfs Pea. *c.* Knight's Tall White Marrow. *d.* Lewis's Negro Pea.

In the pea itself we have every tint between almost pure white, brown, yellow, and intense green; in the varieties of the *sugar peas* we have these same tints, together with red passing through fine purple into a dark chocolate tint. These colours are either uniform or distributed in dots, striæ, or moss-like marks; they depend in some cases on the colour of the cotyledons seen through the skin, and in other cases on the outer coats of the pea itself. In the different varieties the pods contain, according to Mr. Gordon, from eleven or twelve to only four or five peas. The largest peas are nearly twice as much in diameter as the smallest; and the latter are not always borne by the most dwarfed kinds. Peas differ much in shape, being smooth and spherical, smooth and oblong, nearly oval in the *Queen of Dwarfs*, and nearly cubical and crumpled in many of the larger kinds.

With respect to the value of the differences between the chief varieties, it cannot be doubted that, if one of the tall *Sugar-peas*, with purple flowers, thin-skinned pods of an extraordinary shape, including large, dark-purple peas, grew wild by the side of the lowly *Queen of the Dwarfs*, with white flowers, greyish-green, rounded leaves, scimitar-like pods, containing oblong, smooth, pale-coloured peas, which became mature at a different season; or by the side of one of the gigantic sorts, like the *Champion of England*, with leaves of great size, pointed pods, and large, green, crumpled, almost cubical peas,—all three kinds would be ranked as indisputably distinct species.

Andrew Knight[82] has observed that the varieties of peas keep very true, because they are not crossed by insects. As far as the fact of keeping true is concerned, I hear from Mr. Masters of Canterbury, well known as the originator of several new kinds, that certain varieties have remained constant for a considerable time,—for instance, *Knight's Blue Dwarf*, which came out about the year 1820.[83] But the greater number of varieties have a singularly short existence: thus Loudon remarks[84] that "sorts which were highly approved in 1821, are now, in 1833, nowhere to be found;" and on comparing the lists of 1833 with those of 1855, I find that nearly all the varieties have changed. Mr. Masters informs me that the nature of the soil causes some varieties to lose their character. As with other plants, certain varieties can be propagated truly, whilst others show a determined tendency to vary; thus two peas differing in shape, one round and the other wrinkled, were found by Mr. Masters within the same pod, but the plants raised from the wrinkled kind always evinced a strong tendency to produce round peas. Mr. Masters also raised from a plant of another variety four distinct sub-varieties, which bore blue and round, white and round, blue and wrinkled, and white and wrinkled peas; and although he sowed these four varieties separately during several successive years, each kind always reproduced all four kinds mixed together!

With respect to the varieties not naturally intercrossing, I have ascertained that the pea, which in this respect differs from some other Leguminosæ, is perfectly fertile without the aid of insects. Yet I have seen humble-bees whilst sucking the nectar depress the keel-petals, and become so thickly dusted with pollen, that some could hardly fail to be left on the stigma of the next flower which was visited. I have made inquiries from several great raisers of seed-peas, and I find that but few sow them separately; the majority take no precaution; and it is certain, as I have myself found, that true seed may be saved during at least several generations from distinct varieties growing close together.[85] Under these circumstances, Mr. Fitch raised, as he informs me, one variety for twenty

82 'Phil. Transact.,' 1799, p. 196.

83 'Gardener's Magazine,' vol. i., 1826, p. 153.

84 'Encyclopædia of Gardening,' p. 823.

85 *See* Dr. Anderson to the same effect in the 'Bath Soc. Agricultural Papers,' vol. iv. p. 87.

years, which always came true. From the analogy of kidney-beans I should have expected [86] that occasionally, perhaps at long intervals of time, when some slight degree of sterility had supervened from long-continued self-fertilisation, varieties thus growing near each other would have crossed; and I shall give in the eleventh chapter two cases of distinct varieties which spontaneously intercrossed, as shown (in a manner hereafter to be explained) by the pollen of the one variety having acted directly on the seeds of the other. Whether the incessant supply of new varieties is partly due to such occasional and accidental crosses, and their fleeting existence to changes of fashion; or again, whether the varieties which arise after a long course of continued self-fertilisation are weakly and soon perish, I cannot even conjecture. It may, however, be noticed that several of Andrew Knight's varieties, which have endured longer than most kinds, were raised towards the close of the last century by artificial crosses; some of them, I believe, were still, in 1860, vigorous; but now, in 1865, a writer, speaking [87] of Knight's four kinds of marrows, says, they have acquired a famous history, but their glory has departed.

With respect to Beans (*Faba vulgaris*), I will say but little. Dr. Alefeld has given [88] short diagnostic characters of forty varieties. Every one who has seen a collection must have been struck with the great difference in shape, thickness, proportional length and breadth, colour, and size which beans present. What a contrast between a Windsor and Horse-bean! As in the case of the pea, our existing varieties were preceded during the Bronze age in Switzerland by a peculiar and now extinct variety producing very small beans.[89]

Potato (*Solanum tuberosum*).—There is little doubt about the parentage of this plant; for the cultivated varieties differ extremely little in general appearance from the wild species, which can be recognised in its native land at the first glance.[90] The varieties cultivated in Britain are numerous; thus Lawson [91] gives a description of 175 kinds. I planted eighteen kinds in adjoining rows; their stems and leaves differed but little, and in several cases there was as great an amount of difference between the individuals of the same variety as between the different varieties. The flowers vary in size, and in colour between white and purple, but in no other respect, except that in one kind the sepals were somewhat elongated. One strange variety has been described which always produces two sorts of flowers, the first double and sterile, the second single and fertile.[92] The fruit or berries also differ, but only in a slight degree.[93]

---

[86] I have published full details of experiments on this subject in the 'Gardener's Chronicle,' 1857, Oct. 25th.

[87] 'Gardener's Chronicle,' 1865, p. 387.

[88] 'Bonplandia,' x., 1862, s. 348.

[89] O. Heer, 'Die Pflanzen der Pfahlbauten,' 1866, s. 22.

[90] Darwin, 'Journal of Researches,' 1845, p. 285.

[91] Synopsis of the vegetable products of Scotland, quoted in Wilson's 'British Farming,' p. 317.

[92] Sir G. Mackenzie, in 'Gardener's Chronicle,' 1845, p. 790.

[93] 'Putsche und Vertuch, Versuch einer Monographie der Kartoffeln,' 1819, s. 9, 15. *See* also Dr. Anderson's 'Recreations in Agriculture,' vol. iv. p. 325.

The tubers, on the other hand, present a wonderful amount of diversity. This fact accords with the principle that the valuable and selected parts of all cultivated productions present the greatest amount of modification. They differ much in size and shape, being globular, oval, flattened, kidney-like, or cylindrical. One variety from Peru is described [94] as being quite straight, and at least six inches in length, though no thicker than a man's finger. The eyes or buds differ in form, position, and colour. The manner in which the tubers are arranged on the so-called roots is different; thus in the *gurken-kartoffeln* they form a pyramid with the apex downwards, and in another variety they bury themselves deep in the ground. The roots themselves run either near the surface or deep in the ground. The tubers also differ in smoothness and colour, being externally white, red, purple, or almost black, and internally white, yellow, or almost black. They differ in flavour and quality, being either waxy or mealy; in their period of maturity, and in their capacity for long preservation.

As with many other plants which have been long propagated by bulbs, tubers, cuttings, &c., by which means the same individual is exposed during a length of time to diversified conditions, seedling potatoes generally display innumerable slight differences. Several varieties, even when propagated by tubers, are far from constant, as will be seen in the chapter on Bud-variation. Dr. Anderson [95] procured seed from an Irish purple potato, which grew far from any other kind, so that it could not at least in this generation have been crossed, yet the many seedlings varied in almost every possible respect, so that "scarcely two plants were exactly alike." Some of the plants which closely resembled each other above ground, produced extremely dissimilar tubers; and some tubers which externally could hardly be distinguished, differed widely in quality when cooked. Even in this case of extreme variability, the parent-stock had some influence on the progeny, for the greater number of the seedlings resembled in some degree the parent Irish potato. Kidney potatoes must be ranked amongst the most highly cultivated and artificial races; yet their peculiarities can often be strictly propagated by seed. A great authority, Mr. Rivers,[96] states that "seedlings from the ash-leaved kidney always bear a strong resemblance to their parent. Seedlings from the fluke-kidney are still more remarkable for their adherence to their parent-stock, for, on closely observing a great number during two seasons, I have not been able to observe the least difference either in earliness, productiveness, or in the size or shape of their tubers."

---

[94] 'Gardener's Chronicle,' 1862, p. 1052.

[95] 'Bath Society Agricult. Papers,' vol. v. p. 127. And 'Recreations in Agriculture,' vol. v. p. 86.

[96] 'Gardener's Chronicle,' 1863, p. 643.

CHAPTER X.

PLANTS *continued* — FRUITS — ORNAMENTAL TREES — FLOWERS.

FRUITS. — GRAPES — VARY IN ODD AND TRIFLING PARTICULARS. —— MULBERRY. —— THE ORANGE GROUP — SINGULAR RESULTS FROM CROSSING. —— PEACH AND NECTARINE — BUD-VARIATION — ANALOGOUS VARIATION — RELATION TO THE ALMOND. —— APRICOT. —— PLUMS — VARIATION IN THEIR STONES. —— CHERRIES — SINGULAR VARIETIES OF. —— APPLE. —— PEAR. —— STRAWBERRY — INTERBLENDING OF THE ORIGINAL FORMS. —— GOOSEBERRY — STEADY INCREASE IN SIZE OF THE FRUIT — VARIETIES OF. —— WALNUT. —— NUT. —— CUCURBITACEOUS PLANTS — WONDERFUL VARIATION OF.

ORNAMENTAL TREES — THEIR VARIATION IN DEGREE AND KIND — ASH-TREE — SCOTCH-FIR — HAWTHORN.

FLOWERS — MULTIPLE ORIGIN OF MANY KINDS — VARIATION IN CONSTITUTIONAL PECULIARITIES — KIND OF VARIATION. —— ROSES — SEVERAL SPECIES CULTIVATED. —— PANSY. —— DAHLIA. —— HYACINTH, HISTORY AND VARIATION OF.

The Vine (*Vitis vinifera*).—THE best authorities consider all our grapes as the descendants of one species which now grows wild in western Asia, which grew during the Bronze-age wild in Italy,[1] and which has recently been found fossil in a tufaceous deposit in the south of France.[2] Some authors, however, entertain much doubt about the single parentage of our cultivated varieties, owing to the number of semi-wild forms found in Southern Europe, especially as described by Clemente,[3] in a forest in Spain; but as the grape sows itself freely in Southern Europe, and as several of the chief kinds transmit their characters by seed,[4] whilst others are extremely variable, the existence of many different escaped forms could hardly fail to occur in countries where this plant has been cultivated from the remotest antiquity. That the vine varies much when propagated by seed, we may infer from the largely increased number of varieties since the earlier historical records. New hot-house varieties are produced almost every year; for instance,[5] a golden-coloured variety has been recently raised in England from a black grape without the aid of a cross.

---

[1] Heer, 'Pflanzen der Pfahlbauten,' 1866, s. 28.

[2] Alph. De Candolle, 'Géograph. Bot.,' p. 872; Dr. A. Targioni-Tozzetti, in 'Jour. Hort. Soc.,' vol. ix. p. 133. For the fossil vine found by Dr. G. Planchon, *see* 'Nat. Hist. Review,' 1865, April, p. 224.

[3] Godron, 'De l'Espèce,' tom. ii. p. 100.

[4] *See* an account of M. Vibert's experiments, by Alex. Jordan, in 'Mém. de l'Acad. de Lyon,' tom. ii., 1852, p. 108.

[5] 'Gardener's Chronicle,' 1864, p. 488.

Van Mons[6] reared a multitude of varieties from the seed of one vine, which was completely separated from all others, so that there could not, at least in this generation, have been any crossing, and the seedlings presented "les analogues de toutes les sortes," and differed in almost every possible character both in the fruit and foliage.

The cultivated varieties are extremely numerous; Count Odart says that he will not deny that there may exist throughout the world 700 or 800, perhaps even 1000 varieties, but not a third of these have any value. In the Catalogue of fruit cultivated in the Horticultural Gardens of London, published in 1842, 99 varieties are enumerated. Wherever the grape is grown many varieties occur: Pallas describes 24 in the Crimea, and Burnes mentions 10 in Cabool. The classification of the varieties has much perplexed writers, and Count Odart is reduced to a geographical system; but I will not enter on this subject, nor on the many and great differences between the varieties. I will merely specify a few curious and trifling peculiarities, all taken from Odart's highly esteemed work,[7] for the sake of showing the diversified variability of this plant. Simon has classed grapes into two main divisions, those with downy leaves and those with smooth leaves, but he admits that in one variety, namely the Rebazo, the leaves are either smooth or downy; and Odart (p. 70) states that some varieties have the nerves alone, and other varieties their young leaves, downy, whilst the old ones are smooth. The Pedro-Ximenes grape (Odart, p. 397) presents a peculiarity by which it can be at once recognised amongst a host of other varieties, namely, that when the fruit is nearly ripe the nerves of the leaves or even the whole surface becomes yellow. The Barbera d'Asti is well marked by several characters (p. 426), amongst others, "by some of the leaves, and it is always the lowest on the branches, suddenly becoming of a dark red colour." Several authors in classifying grapes have founded their main divisions on the berries being either round or oblong; and Odart admits the value of this character; yet there is one variety, the Maccabeo (p. 71), which often produces small round, and large oblong, berries in the same bunch. Certain grapes called Nebbiolo (p. 429) present a constant character, sufficient for their recognition, namely, "the slight adherence of that part of the pulp which surrounds the seeds to the rest of the berry, when cut through transversely." A Rhenish variety is mentioned (p. 228) which likes a dry soil; the fruit ripens well, but at the moment of maturity, if much rain falls, the berries are apt to rot; on the other hand, the fruit of a Swiss variety (p. 243) is valued for well sustaining prolonged humidity. This latter variety sprouts late in the spring, yet matures its fruit early; other varieties (p. 362) have the fault of being too much excited by the April sun, and in consequence suffer from frost. A Styrian variety (p. 254) has brittle foot-stalks, so that the clusters of fruit are often blown off; this variety is said to be particularly attractive to wasps and bees. Other varieties have tough stalks, which resist the wind. Many other variable characters could be given, but the foregoing facts are sufficient to show in how many small structural and

---

[6] 'Arbres Fruitiers,' 1836, tom. ii. p. 290.

[7] Odart, 'Ampélographie Universelle,' 1849.

constitutional details the vine varies. During the vine disease in France certain whole groups of varieties[8] have suffered far more from mildew than others. Thus "the group of the Chasselas, so rich in varieties, did not afford a single fortunate exception;" certain other groups suffered much less; the true old Burgundy, for instance, was comparatively free from disease, and the Carminat likewise resisted the attack. The American vines, which belong to a distinct species, entirely escaped the disease in France; and we thus see that those European varieties which best resist the disease must have acquired in a slight degree the same constitutional peculiarities as the American species.

White Mulberry (*Morus alba*).—I mention this plant because it has varied in certain characters, namely, in the texture and quality of the leaves, fitting them to serve as food for the domesticated silkworm, in a manner not observed with other plants; but this has arisen simply from such variations in the mulberry having been attended to, selected, and rendered more or less constant. M. de Quatrefages[9] briefly describes six kinds cultivated in one valley in France: of these the *amourouso* produces excellent leaves, but is rapidly being abandoned because it produces much fruit mingled with the leaves: the *antofino* yields deeply cut leaves of the finest quality, but not in great quantity: the *claro* is much sought for because the leaves can be easily collected: lastly, the *roso* bears strong hardy leaves, produced in large quantity, but with the one inconvenience, that they are best adapted for the worms after their fourth moult. MM. Jacquemet-Bonnefont, of Lyon, however, remark in their catalogue (1862) that two sub-varieties have been confounded under the name of the *roso*, one having leaves too thick for the caterpillars, the other being valuable because the leaves can easily be gathered from the branches without the bark being torn.

In India the mulberry has also given rise to many varieties. The Indian form is thought by many botanists to be a distinct species; but as Royle remarks,[10] "so many varieties have been produced by cultivation that it is difficult to ascertain whether they all belong to one species;" they are, as he adds, nearly as numerous as those of the silkworm.

The Orange Group.—We here meet with great confusion in the specific distinction and parentage of the several kinds. Gallesio,[11] who almost devoted his life-time to the subject, considers that there are four species, namely, sweet and bitter oranges, lemons, and citrons, each of which has given rise to whole groups of varieties, monsters, and supposed hybrids. One high authority[12] believes that these four reputed species are all

---

[8] M. Bouchardat, in 'Comptes Rendus,' Dec. 1st, 1851, quoted in 'Gardener's Chron.,' 1852, p. 435.

[9] 'Études sur les Maladies actuelles du Ver à Soie,' 1859, p. 321.

[10] 'Productive Resources of India,' p. 130.

[11] 'Traité du Citrus,' 1811. 'Teoria della Riproduzione Vegetale,' 1816. I quote chiefly from this second work. In 1839 Gallesio published in folio 'Gli Agrumi dei Giard. Bot. di Firenze,' in which he gives a curious diagram of the supposed relationship of all the forms.

[12] Mr. Bentham, Review of Dr. A. Targioni-Tozzetti, 'Journal of Hort. Soc.,' vol. ix. p. 133.

varieties of the wild *Citrus medica*, but that the shaddock (*Citrus decumana*), which is not known in a wild state, is a distinct species; though its distinctness is doubted by another writer "of great authority on such matters," namely, Dr. Buchanan Hamilton. Alph. De Candolle,[13] on the other hand—and there cannot be a more capable judge—advances what he considers sufficient evidence of the orange (he doubts whether the bitter and sweet kinds are specifically distinct), the lemon, and citron, having been found wild, and consequently that they are distinct. He mentions two other forms cultivated in Japan and Java, which he ranks as undoubted species; he speaks rather more doubtfully about the shaddock, which varies much, and has not been found wild; and finally he considers some forms, such as Adam's apple and the bergamotte, as probably hybrids.

I have briefly abstracted these opinions for the sake of showing those who have never attended to such subjects, how perplexed with doubt they are. It would, therefore, be useless for my purpose to give a sketch of the conspicuous differences between the several forms. Besides the ever-recurrent difficulty of determining whether forms found wild are truly aboriginal or are escaped seedlings, many of the forms, which must be ranked as varieties, transmit their characters almost perfectly by seed. Sweet and bitter oranges differ in no important respect except in the flavour of their fruit, but Gallesio[14] is most emphatic that both kinds can be propagated by seed with absolute certainty. Consequently, in accordance with his simple rule, he classes them as distinct species; as he does sweet and bitter almonds, the peach and nectarine, &c. He admits, however, that the soft-shelled pine-tree produces not only soft-shelled but some hard-shelled seedlings, so that a little greater force in the power of inheritance would, according to this rule, raise the soft-shelled pine-tree into the dignity of an aboriginally created species. The positive assertion made by Macfayden[15] that the pips of sweet oranges produce in Jamaica, according to the nature of the soil in which they are sown, either sweet or bitter oranges, is probably an error; for M. Alph. De Candolle informs me that since the publication of his great work he has received accounts from Guiana, the Antilles, and Mauritius, that in these countries sweet oranges faithfully transmit their character. Gallesio found that the willow-leafed and the Little China oranges reproduced their proper leaves and fruit; but the seedlings were not quite equal in merit to their parents. The red-fleshed orange, on the other hand, fails to reproduce itself. Gallesio also observed that the seeds of several other singular varieties all reproduced trees having a peculiar physiognomy, but partly resembling their parent-forms. I can adduce another case: the myrtle-leaved orange is ranked by all authors as a variety, but is very distinct in general aspect: in my father's greenhouse, during many years, it rarely yielded any seed, but at last produced one; and a tree thus raised was identical with the parent-form.

Another and more serious difficulty in determining the rank of the several forms is that, according to Gallesio,[16] they largely intercross without

---

13 'Géograph. Bot.,' p. 863.

14 'Teoria della Riproduzione,' pp. 52–57.

15 Hooker's 'Bot. Misc.,' vol. i. p. 302; vol. ii. p. 111.

16 'Teoria della Riproduzione,' p. 53.

artificial aid; thus he positively states that seeds taken from lemon-trees (*C. lemonum*) growing mingled with the citron (*C. medica*), which is generally considered as a distinct species, produced a graduated series of varieties between these two forms. Again, an Adam's apple was produced from the seed of a sweet orange, which grew close to lemons and citrons. But such facts hardly aid us in determining whether to rank these forms as species or varieties; for it is now known that undoubted species of Verbascum, Cistus, Primula, Salix, &c., frequently cross in a state of nature. If indeed it were proved that plants of the orange tribe raised from these crosses were even partially sterile, it would be a strong argument in favour of their rank as species. Gallesio asserts that this is the case; but he does not distinguish between sterility from hybridism and from the effects of culture; and he almost destroys the force of this statement by another,[17] namely, that when he impregnated the flowers of the common orange with the pollen taken from undoubted *varieties* of the orange, monstrous fruits were produced, which included "little pulp, and had no seeds, or imperfect seeds."

In this tribe of plants we meet with instances of two highly remarkable facts in vegetable physiology: Gallesio[18] impregnated an orange with pollen from a lemon, and the fruit borne on the mother tree had a raised stripe of peel like that of a lemon both in colour and taste, but the pulp was like that of an orange and included only imperfect seeds. The possibility of pollen from one variety or species directly affecting the fruit produced by another variety or species, is a subject which I shall fully discuss in the following chapter.

The second remarkable fact is that two supposed hybrids[19] (for their hybrid nature was not ascertained) between an orange and either a lemon or citron produced, on the same tree, leaves, flowers, and fruit of both pure parent-forms, as well as of a mixed or crossed nature. A bud taken from any one of the branches and grafted on another tree produces either one of the pure kinds or a capricious tree reproducing the three kinds. Whether the sweet lemon, which includes within the same fruit segments of differently flavoured pulp,[20] is an analogous case, I know not. But to this subject I shall have to recur.

I will conclude by giving from A. Risso[21] a short account of a very singular variety of the common orange. It is the "*citrus aurantium fructu variabili*," which on the young shoots produces rounded-oval leaves spotted with yellow, borne on petioles with heart-shaped wings; when these leaves fall off, they are succeeded by longer and narrower leaves, with undulated margins, of a pale-green colour embroidered with yellow, borne on foot-stalks without wings. The fruit whilst young is pear-shaped, yellow, longitudinally striated, and sweet; but as it ripens, it becomes spherical, of a reddish-yellow, and bitter.

Peach and Nectarine (*Amygdalus Persica*). The best authorities are

---

[17] Gallesio, 'Teoria della Riproduzione,' p. 69.

[18] Gallesio, idem, p. 67.

[19] Gallesio, idem, pp. 75, 76.

[20] 'Gardener's Chronicle,' 1841, p. 613.

[21] 'Annales du Muséum,' tom. xx. p. 188.

nearly unanimous that the peach has never been found wild. It was introduced from Persia into Europe a little before the Christian era, and at this period few varieties existed. Alph. De Candolle,[22] from the fact of the peach not having spread from Persia at an earlier period, and from its not having

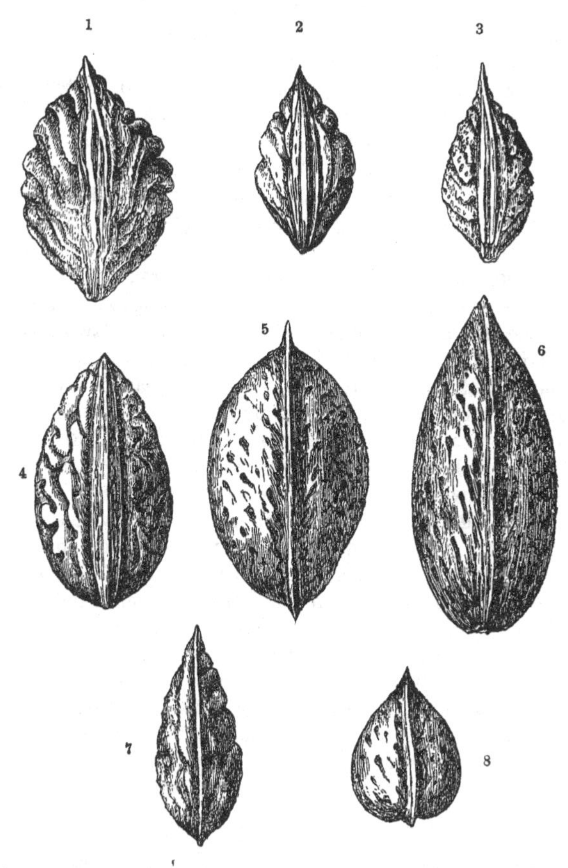

Fig. 42.—Peach and Almond Stones, of natural size, viewed edgeways. 1. Common English Peach. 2. Double, crimson-flowered, Chinese Peach. 3. Chinese Honey Peach. 4. English Almond. 5. Barcelona Almond. 6. Malaga Almond. 7. Soft-shelled French Almond. 8. Smyrna Almond.

pure Sanscrit or Hebrew names, believes that it is not an aboriginal of Western Asia, but came from the *terra incognita* of China. The supposition, however, that the peach is a modified almond which acquired its present character at a comparatively late period, would, I presume, account for these facts; on the same principle that the nectarine, the offspring of the peach, has few native names, and became known in Europe at a still later period.

22 'Géograph. Bot.,' p. 882.

Andrew Knight,[23] from finding that a seedling-tree, raised from a sweet almond fertilised by the pollen of a peach, yielded fruit quite like that of a peach, suspected that the peach-tree is a modified almond; and in this he has been followed by various authors.[24] A first-rate peach, almost globular in shape, formed of soft and sweet pulp, surrounding a hard, much furrowed, and slightly-flattened stone, certainly differs greatly from an almond, with its soft, slightly furrowed, much flattened, and elongated stone, protected by a tough, greenish layer of bitter flesh. Mr. Bentham[25] has particularly called attention to the stone of the almond being so much more flattened than that of the peach. But in the several varieties of the almond, the stone differs greatly in the degree to which it is compressed, in size, shape, strength, and in the depth of the furrows, as may be seen in the accompanying drawings (Nos. 4 to 8) of such kinds as I have been able to collect. With peach-stones also (Nos. 1 to 3) the degree of compression and elongation is seen to vary; so that the stone of the Chinese Honey-peach (fig. 3) is much more elongated and compressed than that of the (No. 8) Smyrna almond. Mr. Rivers of Sawbridgeworth, to whom I am indebted for some of the specimens above figured, and who has had such great horticultural experience, has called my attention to several varieties which connect the almond and the peach. In France there is a variety called the Peach-almond, which Mr. Rivers formerly cultivated, and which is correctly described in a French catalogue as being oval and swollen, with the aspect of a peach, including a hard stone surrounded by a fleshy covering, which is sometimes eatable.[26] A remarkable statement by M. Luizet has recently appeared in the 'Revue Horticole,'[27] namely, that a Peach-almond, grafted on a peach, bore during 1863 and 1864 almonds alone, but in 1865 bore six peaches and no almonds. M. Carrière, in commenting on this fact, cites the case of a double-flowered almond which, after producing during several years almonds, suddenly bore for two years in succession spherical fleshy peach-like fruits, but in 1865 reverted to its former state and produced large almonds.

Again, as I hear from Mr. Rivers, the double-flowering Chinese peaches resemble almonds in their manner of growth and in their flowers; the fruit is much elongated and flattened, with the flesh both bitter and sweet, but

---

[23] 'Transactions of Hort. Soc.,' vol. iii. p. 1, and vol. iv. p. 369, and note to p. 370. A coloured drawing is given of this hybrid.

[24] 'Gardener's Chronicle,' 1856, p. 532. A writer, it may be presumed Dr. Lindley, remarks on the perfect series which may be formed between the almond and the peach. Another high authority, Mr. Rivers, who has had such wide experience, strongly suspects ('Gardener's Chronicle,' 1863, p. 27) that peaches, if left to a state of nature, would in the course of time retrograde into thick-fleshed almonds.

[25] 'Journal of Hort. Soc.,' vol. ix. p. 168.

[26] Whether this is the same variety as one lately mentioned ('Gard. Chron.' 1865, p. 1154) by M. Carrière under the name of *Persica intermedia*, I know not: this var. is said to be intermediate in nearly all its characters between the almond and peach; it produces during successive years very different kinds of fruit.

[27] Quoted in 'Gard. Chron.' 1866, p. 800.

not uneatable, and it is said to be of better quality in China. From this stage one small step leads us to such inferior peaches as are occasionally raised from seed. For instance, Mr. Rivers sowed a number of peach-stones imported from the United States, where they are collected for raising stocks, and some of the trees raised by him produced peaches which were very like almonds in appearance, being small and hard, with the pulp not softening till very late in the autumn. Van Mons[28] also states that he once raised from a peach-stone a peach having the aspect of a wild tree, with fruit like that of the almond. From inferior peaches, such as these just described, we may pass by small transitions, through clingstones of poor quality, to our best and most melting kinds. From this gradation, from the cases of sudden variation above recorded, and from the fact that the peach has not been found wild, it seems to me by far the most probable view, that the peach is the descendant of the almond, improved and modified in a marvellous manner.

One fact, however, is opposed to this conclusion. A hybrid, raised by Knight from the sweet almond by the pollen of the peach, produced flowers with little or no pollen, yet bore fruit, having been apparently fertilised by a neighbouring nectarine. Another hybrid from a sweet almond by the pollen of a nectarine produced during the first three years imperfect blossoms, but afterwards perfect flowers with an abundance of pollen. If this slight degree of sterility cannot be accounted for by the youth of the trees (and this often causes lessened fertility), or by the monstrous state of the flowers, or by the conditions to which the trees were exposed, these two cases would afford a strong argument against the peach being the descendant of the almond.

Whether or not the peach has proceeded from the almond, it has certainly given rise to nectarines, or smooth peaches, as they are called by the French. Most of the varieties both of the peach and nectarine reproduce themselves truly by seed. Gallesio[29] says he has verified this with respect to eight races of the peach. Mr. Rivers[30] has given some striking instances from his own experience, and it is notorious that good peaches are constantly raised in North America from seed. Many of the American subvarieties come true or nearly true to their kind, such as the white-blossom, several of the yellow-fruited freestone peaches, the blood clingstone, the heath, and the lemon-clingstone. On the other hand, a clingstone peach has been known to give rise to a freestone.[31] In England it has been noticed that seedlings inherit from their parents flowers of the same size and colour. Some characters, however, contrary to what might have been expected, often are not inherited; such as the presence and form of the glands on the leaves.[32] With respect to nectarines, both cling and free-

[28] Quoted in 'Journal de la Soc. Imp. d'Horticulture,' 1855, p. 238.

[29] 'Teoria della Riproduzione Vegetale,' 1816, p. 86.

[30] 'Gardener's Chronicle,' 1862, p. 1195.

[31] Mr. Rivers, 'Gardener's Chron., 1859, p. 774.

[32] Downing, 'The Fruits of America,' 1845, pp. 475, 489, 492, 494, 496. *See* also F. Michaux, 'Travels in N. America' (Eng. translat.), p. 228. For similar cases in France *see* Godron, 'De l'Espèce,' tom. ii. p. 97.

stones are known in North America to reproduce themselves by seed.[33] In England the new white nectarine was a seedling of the old white, and Mr. Rivers[34] has recorded several similar cases. From this strong tendency to inheritance, which both peach and nectarine trees exhibit,—from certain slight constitutional differences[35] in their nature,—and from the great difference in their fruit both in appearance and flavour, it is not surprising, notwithstanding that the trees differ in no other respects and cannot even be distinguished, as I am informed by Mr. Rivers, whilst young, that they have been ranked by some authors as specifically distinct. Gallesio does not doubt that they are distinct; even Alph. De Candolle does not appear perfectly assured of their specific identity; and an eminent botanist has quite recently[36] maintained that the nectarine "probably constitutes a distinct species."

Hence it may be worth while to give all the evidence on the origin of the nectarine. The facts in themselves are curious, and will hereafter have to be referred to when the important subject of bud-variation is discussed. It is asserted[37] that the Boston nectarine was produced from a peach-stone, and this nectarine reproduced itself by seed.[38] Mr. Rivers states[39] that from stones of three distinct varieties of the peach he raised three varieties of nectarine; and in one of these cases no nectarine grew near the parent peach-tree. In another instance Mr. Rivers raised a nectarine from a peach, and in the succeeding generation another nectarine from this nectarine.[40] Other such instances have been communicated to me, but they need not be given. Of the converse case, namely, of nectarine-stones yielding peach-trees (both free and cling-stones), we have six undoubted instances recorded by Mr. Rivers; and in two of these instances the parent nectarines had been seedlings from other nectarines.[41]

With respect to the more curious case of full-grown peach-trees suddenly producing nectarines by bud-variation (or sports as they are called by gardeners), the evidence is superabundant; there is also good evidence of the same tree producing both peaches and nectarines, or half and half fruit;—by this term I mean a fruit with the one-half a perfect peach, and the other half a perfect nectarine.

Peter Collinson in 1741 recorded the first case of a peach-tree producing a nectarine,[42] and in 1766 he added two other instances. In the same work, the editor, Sir J. E. Smith, describes the more remarkable case of a tree in Norfolk, which usually bore both perfect nectarines and perfect peaches; but during two seasons some of the fruit were half-and-half in nature.

---

[33] Brickell's 'Nat. Hist. of N. Carolina,' p. 102, and Downing's Fruit Trees,' p. 505.

[34] 'Gardener's Chronicle,' 1862, p. 1196.

[35] The peach and nectarine do not succeed equally well in the same soil: *see* Lindley's 'Horticulture,' p. 351.

[36] Godron, 'De l'Espèce,' tom. ii. 1859, p. 97.

[37] 'Transact. Hort. Soc.,' vol. vi. p. 394.

[38] Downing's 'Fruit Trees,' p. 502.

[39] 'Gardener's Chronicle,' 1862, p. 1195.

[40] 'Journal of Horticulture,' Feb. 6th, 1866, p. 102.

[41] Mr. Rivers, in 'Gardener's Chron.,' 1859, p. 774; 1862, p. 1195; 1865, p. 1059; and 'Journal of Hort.,' 1866, p. 102.

[42] 'Correspondence of Linnæus,' 1821, pp. 7, 8, 70.

Mr. Salisbury in 1808[43] records six other cases of peach-trees producing nectarines. Three of the varieties are named; viz., the Alberge, Belle Chevreuse, and Royal George. This latter tree seldom failed to produce both kinds of fruit. He gives another case of a half-and-half fruit.

At Radford in Devonshire[44] a clingstone peach, purchased as the Chancellor, was planted in 1815, and in 1824, after having previously produced peaches alone, bore on one branch twelve nectarines; in 1825 the same branch yielded twenty-six nectarines, and in 1826 thirty-six nectarines together with eighteen peaches. One of the peaches was almost as smooth on one side as a nectarine. The nectarines were as dark as, but smaller than, the Elruge.

At Beccles a Royal George peach[45] produced a fruit, "three parts of it being peach and one part nectarine, quite distinct in appearance as well as in flavour." The lines of division were longitudinal, as represented in the engraving. A nectarine-tree grew five yards from this tree.

Professor Chapman states[46] that he has often seen in Virginia very old peach-trees bearing nectarines.

A writer in the 'Gardener's Chronicle' says that a peach-tree planted fifteen years previously[47] produced this year a nectarine between two peaches; a nectarine-tree grew close by.

In 1844[48] a Vanguard peach-tree produced, in the midst of its ordinary fruit, a single red Roman nectarine.

Mr. Calver is stated[49] to have raised in the United States a seedling peach which produced a mixed crop of both peaches and nectarines.

Near Dorking[50] a branch of the Têton de Venus peach, which reproduces itself truly by seed,[51] bore its own fruit "so remarkable for its prominent point, and a nectarine rather smaller but well formed and quite round."

The previous cases all refer to peaches suddenly producing nectarines, but at Carclew[52] the unique case occurred, of a nectarine-tree, raised twenty years before from seed and never grafted, producing a fruit half peach and half nectarine; subsequently it bore a perfect peach.

To sum up the foregoing facts: we have excellent evidence of peach-stones producing nectarine-trees, and of nectarine-stones producing peach-trees,—of the same tree bearing peaches and nectarines,—of peach-trees suddenly producing by bud-variation nectarines (such nectarines reproducing nectarines by seed), as well as fruit in part nectarine and in part peach,—and lastly of one nectarine-tree first bearing half-and-half fruit, and subsequently true peaches. As the peach came into existence before the nectarine, it might have been expected from the law of reversion that

---

[43] 'Transact. Hort. Soc.,' vol. i. p. 103.

[44] Loudon's 'Gardener's Mag.,' 1826, vol. i. p. 471.

[45] Ibid., 1828, p. 53.

[46] Ibid., 1830, p. 597.

[47] 'Gardener's Chronicle,' 1841, p. 617.

[48] 'Gardener's Chronicle,' 1844, p. 589.

[49] 'Phytologist,' vol. iv. p. 299.

[50] 'Gardener's Chron.,' 1856, p. 531.

[51] Godron, 'De l'Espèce,' tom. ii. p. 97.

[52] 'Gardener's Chron.,' 1856, p. 531.

nectarines would give birth by bud-variation or by seed to peaches, oftener than peaches to nectarines; but this is by no means the case.

Two explanations have been suggested to account for these conversions. First, that the parent-trees have been in every case hybrids[53] between the peach and nectarine, and have reverted by bud-variation or by seed to one of their pure parent-forms. This view in itself is not very improbable; for the Mountaineer peach, which was raised by Knight from the red nutmeg peach by pollen of the violette hâtive nectarine,[54] produces peaches, but these are said *sometimes* to partake of the smoothness and flavour of the nectarine. But let it be observed that in the previous list no less than six well-known varieties and several other unnamed varieties of the peach have once suddenly produced perfect nectarines by bud-variation; and it would be an extremely rash supposition that all these varieties of the peach, which have been cultivated for years in many districts, and which show not a vestige of a mixed parentage, are, nevertheless, hybrids. A second explanation is, that the fruit of the peach has been directly affected by the pollen of the nectarine: although this certainly is possible, it cannot here apply; for we have not a shadow of evidence that a branch which has borne fruit directly affected by foreign pollen is so profoundly modified as afterwards to produce buds which continue to yield fruit of the new and modified form. Now it is known that when a bud on a peach-tree has once borne a nectarine the same branch has in several instances gone on during successive years producing nectarines. The Carclew nectarine, on the other hand, first produced half-and-half fruit, and subsequently pure peaches. Hence we may confidently accept the common view that the nectarine is a variety of the peach, which may be produced either by bud-variation or from seed. In the following chapter many analogous cases of bud-variation will be given.

The varieties of the peach and nectarine run in parallel lines. In both classes the kinds differ from each other in the flesh of the fruit being white, red, or yellow; in being clingstones or freestones; in the flowers being large or small, with certain other characteristic differences; and in the leaves being serrated without glands, or crenated and furnished with globose or reniform glands.[55] We can hardly account for this parallelism by supposing that each variety of the nectarine is descended from a corresponding variety of the peach; for though our nectarines are certainly the descendants of several kinds of peaches, yet a large number are the descendants of other nectarines, and they vary so much when thus reproduced that we can scarcely admit the above explanation.

The varieties of the peach have largely increased in number since the Christian era, when from two to five varieties alone were known;[56] and the nectarine was unknown. At the present time, besides many varieties said to exist in China, Downing describes in the United States seventy-nine

---

[53] Alph. De Candolle, 'Géograph. Bot.,' p. 886.

[54] Thompson, in Loudon's 'Encyclop. of Gardening,' p. 911.

[55] 'Catalogue of Fruit in Garden of Hort. Soc.,' 1842, p. 105.

[56] Dr. A. Targioni-Tozzetti, 'Journal Hort. Soc.,' vol. ix. p. 167. Alph. De Candolle, 'Géograph. Bot.,' p. 885.

native and imported varieties of the peach; and a few years ago Lindley[57] enumerated one hundred and sixty-four varieties of the peach and nectarine grown in England. I have already indicated the chief points of difference between the several varieties. Nectarines, even when produced from distinct kinds of peaches, always possess their own peculiar flavour, and are smooth and small. Clingstone and freestone peaches, which differ in the ripe flesh either firmly adhering to the stone, or easily separating from it, also differ in the character of the stone itself; that of the freestones or melters being more deeply fissured, with the sides of the fissures smoother than in clingstones. In the various kinds, the flowers differ not only in size, but in the larger flowers the petals are differently shaped, more imbricated, generally red in the centre and pale towards the margin; whereas in the smaller flowers the margins of the petal are usually more darkly coloured. One variety has nearly white flowers. The leaves are more or less serrated, and are either destitute of glands, or have globose or reniform glands;[58] and some few peaches, such as the Brugnon, bear on the same tree both globular and kidney-shaped glands.[59] According to Robertson[60] the trees with glandular leaves are liable to blister, but not in any great degree to mildew; whilst the non-glandular trees are more subject to curl, to mildew, and to the attacks of aphides. The varieties differ in the period of their maturity, in the fruit keeping well, and in hardiness,—the latter circumstance being especially attended to in the United States. Certain varieties, such as the Bellegarde, stand forcing in hot-houses better than other varieties. The flat-peach of China is the most remarkable of all the varieties; it is so much depressed towards the summit, that the stone is here covered only by roughened skin and not by a fleshy layer.[61] Another Chinese variety, called the Honey-peach, is remarkable from the fruit terminating in a long sharp point; its leaves are glandless and widely dentate.[62] The Emperor of Russia peach is a third singular variety, having deeply and doubly serrated leaves; the fruit is deeply cleft with one-half projecting considerably beyond the other; it originated in America, and its seedlings inherit similar leaves.[63]

The peach has also produced in China a small class of trees valued for ornament, namely the double-flowered; of these five varieties are now known in England, varying from pure white, through rose, to intense crimson.[64] One of these varieties, called the camellia-flowered, bears flowers above 2¼ inches in diameter, whilst those of the fruit-bearing kinds do not at most exceed 1¼ inch in diameter. The flowers of the double-

---

[57] 'Transact. Hort. Soc.,' vol. v. p. 554.

[58] Loudon's 'Encyclop. of Gardening,' p. 907.

[59] M. Carrière, in 'Gard. Chron.,' 1865, p. 1154.

[60] 'Transact. Hort. Soc.,' vol. iii. p. 332. *See* also 'Gardener's Chronicle,' 1865, p. 271, to same effect. Also 'Journal of Horticulture,' Sept. 26th, 1865, p. 254.

[61] 'Transact. Hort. Soc.,' vol. iv. p. 512.

[62] 'Journal of Horticulture,' Sept. 8th, 1863, p. 188.

[63] 'Transact. Hort. Soc.,' vol. vi. p. 412.

[64] 'Gardener's Chronicle,' 1857, p. 216.

flowered peaches have the singular property[65] of frequently producing double or treble fruit. Finally, there is good reason to believe that the peach is an almond profoundly modified; but whatever its origin may have been, there can be no doubt that it has yielded during the last eighteen centuries many varieties, some of them strongly characterised, belonging both to the nectarine and peach form.

Apricot (*Prunus armeniaca*).—It is commonly admitted that this tree is descended from a single species, now found wild in the Caucasian region.[66] On this view the varieties deserve notice, because they illustrate differences supposed by some botanists to be of specific value in the almond and plum. The best monograph on the apricot is by Mr. Thompson,[67] who describes seventeen varieties. We have seen that peaches and nectarines vary in a strictly parallel manner; and in the apricot, which forms a closely allied genus, we again meet with variations analogous to those of the peach, as well as to those of the plum. The varieties differ considerably in the shape of their leaves, which are either serrated or crenated, sometimes with ear-like appendages at their bases, and sometimes with glands on the petioles. The flowers are generally alike, but are small in the Masculine. The fruit varies much in size, shape, and in having the suture little pronounced or absent; in the skin being smooth, or downy as in the orange-apricot; and in the flesh clinging to the stone, as in the last-mentioned kind, or in readily separating from it, as in the Turkey-apricot. In all these differences we see the closest analogy with the varieties of the peach and nectarine. In the stone we have more important differences, and these in the case of the plum have been esteemed of specific value: in some apricots the stone is almost spherical, in others much flattened, being either sharp in front or blunt at both ends, sometimes channelled along the back, or with a sharp ridge along both margins. In the Moorpark, and generally in the Hemskirke, the stone presents a singular character in being perforated, with a bundle of fibres passing through the perforation from end to end. The most constant and important character, according to Thompson, is whether the kernel is bitter or sweet; yet in this respect we have a graduated difference, for the kernel is very bitter in Shipley's apricot; in the Hemskirke less bitter than in some other kinds; slightly bitter in the Royal; and "sweet like a hazel-nut" in the Breda, Angoumois, and others. In the case of the almond, bitterness has been thought by some high authorities to indicate specific difference.

In N. America the Roman apricot endures "cold and unfavourable situations, where no other sort, except the Masculine, will succeed; and its blossoms bear quite a severe frost without injury."[68] According to Mr. Rivers[69] seedling apricots deviate but little from the character of

---

[65] 'Journal of Hort. Soc.,' vol. ii. p. 283.

[66] Alph. De Candolle, 'Géograph. Bot.,' p. 879.

[67] 'Transact. Hort. Soc.' (2nd series), vol. i. 1835, p. 56. *See* also 'Cat. of Fruit in Garden of Hort. Soc.,' 3rd edit. 1842.

[68] Downing, 'The Fruits of America,' 1845, p. 157; with respect to the Alberge apricot in France, *see* p. 153.

[69] 'Gardener's Chronicle,' 1863, p. 364.

their race: in France the Alberge is constantly reproduced from seed with but little variation. In Ladakh, according to Moorcroft,[70] ten varieties of the apricot, very different from each other, are cultivated, and all are raised from seed, excepting one, which is budded.

Plums (*Prunus insititia*).—Formerly the sloe, *P. spinosa*, was thought to be the parent of all our plums; but now this honour is very commonly accorded to *P. insititia* or the bullace, which is found wild in the Caucasus and N.-Western India, and is naturalised in England.[71] It is not at all improbable, in accordance with some observations made by Mr. Rivers,[72] that both these forms, which some botanists rank as a single species, may be the parents of our domesticated plums. Another supposed parent-form, the *P. domestica*, is said to be found wild in the region of the

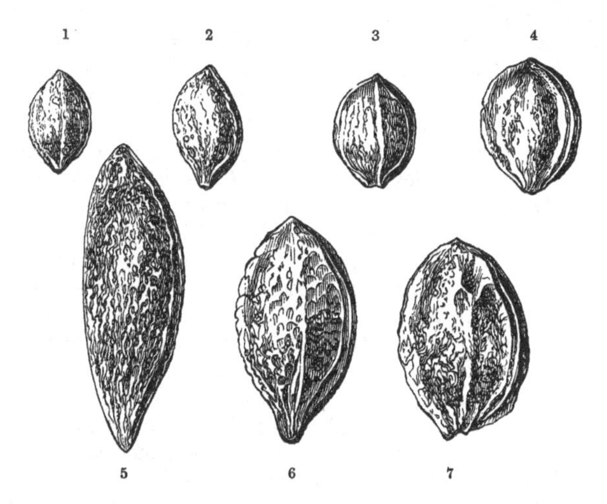

Fig. 43.—Plum Stones, of natural size, viewed laterally. 1. Bullace Plum. 2. Shropshire Damson. 3. Blue Gage. 4. Orleans. 5. Elvas. 6. Denyer's Victoria. 7. Diamond.

Caucasus. Godron remarks[73] that the cultivated varieties may be divided into two main groups, which he supposes to be descended from two aboriginal stocks; namely, those with oblong fruit and stones pointed at both ends, having narrow separate petals and upright branches; and those with rounded fruit, with stones blunt at both ends, with rounded petals and spreading branches. From what we know of the variability of the flowers in the peach and of the diversified manner of growth in our various fruit-trees, it is difficult to lay much weight on these latter

70 'Travels in the Himalayan Provinces,' vol. i. 1841, p. 295.

71 *See* an excellent discussion on this subject in Hewett C. Watson's 'Cybele Britannica,' vol. iv. p. 80.

72 'Gardener's Chronicle,' 1865, p. 27.

73 'De l'Espèce, tom. ii. p. 94. On the parentage of our plums, *see* also Alph. De Candolle, 'Géograph Bot.,' p. 878. Also Targioni-Tozzetti, 'Journal Hort. Soc.,' vol. ix. p. 164. Also Babington, 'Manual of Brit. Botany,' 1851, p. 87.

characters. With respect to the shape of the fruit, we have conclusive evidence that it is extremely variable: Downing[74] gives outlines of the plums of two seedlings, namely, the red and imperial gages, raised from the greengage; and the fruit of both is more elongated than that of the greengage. The latter has a very blunt broad stone, whereas the stone of the imperial gage is "oval and pointed at both ends." These trees also differ in their manner of growth: "the greengage is a very short-jointed, slow-growing tree, of spreading and rather dwarfish habit;" whilst its offspring, the imperial gage, "grows freely and rises rapidly, and has long dark shoots." The famous Washington plum bears a globular fruit, but its offspring, the emerald drop, is nearly as much elongated as the most elongated plum figured by Downing, namely, Manning's prune. I have made a small collection of the stones of twenty-five kinds, and they graduate in shape from the bluntest into the sharpest kinds. As characters derived from seeds are generally of high systematic importance, I have thought it worth while to give drawings of the most distinct kinds in my small collection; and they may be seen to differ in a surprising manner in size, outline, thickness, prominence of the ridges, and state of surface. It deserves notice that the shape of the stone is not always strictly correlated with that of the fruit: thus the Washington plum is spherical and depressed at the pole, with a somewhat elongated stone, whilst the fruit of the Goliath is more elongated, but the stone less so, than in the Washington. Again, Denyer's Victoria and Goliath bear fruit closely resembling each other, but their stones are widely different. On the other hand, the Harvest and Black Margate plums are very dissimilar, yet include closely similar stones.

The varieties of the plum are numerous, and differ greatly in size, shape, quality, and colour,—being bright yellow, green, almost white, blue, purple, or red. There are some curious varieties, such as the double or Siamese, and the Stoneless plum: in the latter the kernel lies in a roomy cavity surrounded only by the pulp. The climate of North America appears to be singularly favourable for the production of new and good varieties; Downing describes no less than forty, seven of which of first-rate quality have been recently introduced into England.[75] Varieties occasionally arise having an innate adaptation for certain soils, almost as strongly pronounced as with natural species growing on the most distinct geological formations; thus in America the imperial gage, differently from almost all other kinds, "is peculiarly fitted for *dry light* soils where many sorts drop their fruit," whereas on rich heavy soils the fruit is often insipid.[76] My father could never succeed in making the Wine-Sour yield even a moderate crop in a sandy orchard near Shrewsbury, whilst in some parts of the same county and in its native Yorkshire it bears abundantly: one of my rela-

---

[74] 'Fruits of America,' pp. 276, 278, 314, 284, 276, 310. Mr. Rivers raised ('Gard. Chron.,' 1863, p. 27) from the Prune-pêche, which bears large, round, red plums on stout robust shoots, a seedling which bears oval, smaller fruit on shoots that are so slender as to be almost pendulous.

[75] 'Gardener's Chronicle,' 1855, p. 726.

[76] Downing's 'Fruit Trees,' p. 278.

tions also repeatedly tried in vain to grow this variety in a sandy district in Staffordshire.

Mr. Rivers has given[77] a number of interesting facts, showing how truly many varieties can be propagated by seed. He sowed the stones of twenty bushels of the greengage for the sake of raising stocks, and closely observed the seedlings; "all had the smooth shoots, the prominent buds, and the glossy leaves of the greengage, but the greater number had smaller leaves and thorns." There are two kinds of damson, one the Shropshire with downy shoots, and the other the Kentish with smooth shoots, and these differ but slightly in any other respect: Mr. Rivers sowed some bushels of the Kentish damson, and all the seedlings had smooth shoots, but in some the fruit was oval, in others round or roundish, and in a few the fruit was small, and, except in being sweet, closely resembled that of the wild sloe. Mr. Rivers gives several other striking instances of inheritance: thus, he raised eighty thousand seedlings from the common German Quetsche plum, and "not one could be found varying in the least, in foliage or habit." Similar facts were observed with the Petite Mirabelle plum, yet this latter kind (as well as the Quetsche) is known to have yielded some well-established varieties; but, as Mr. Rivers remarks, they all belong to the same group with the Mirabelle.

Cherries (*Prunus cerasus*, *avium*, &c.).—Botanists believe that our cultivated cherries are descended from one, two, four, or even more wild stocks.[78] That there must be at least two parent-species we may infer from the sterility of twenty hybrids raised by Mr. Knight from the morello fertilized by pollen of the Elton cherry; for these hybrids produced in all only five cherries, and one alone of these contained a seed.[79] Mr. Thompson[80] has classified the varieties in an apparently natural method in two main groups by characters taken from the flowers, fruit, and leaves; but some varieties which stand widely separate in this classification are quite fertile when crossed; thus Knight's Early Black cherry is the product of a cross between two such kinds.

Mr. Knight states that seedling cherries are more variable than those of any other fruit-tree.[81] In the Catalogue of the Horticultural Society for 1842, eighty varieties are enumerated. Some varieties present singular characters: thus the flower of the Cluster cherry includes as many as twelve pistils, of which the majority abort; and they are said generally to produce from two to five or six cherries aggregated together and borne on a single peduncle. In the Ratafia cherry several flower-peduncles arise from a common peduncle, upwards of an inch in length. The fruit of Gascoigne's Heart has its apex produced into a globule or drop: that of the white

---

[77] 'Gardener's Chronicle,' 1863, p. 27. Sageret, in his 'Pomologie Phys.,' p. 346, enumerates five kinds which can be propagated in France by seed: *see* also Downing's 'Fruit Trees of America,' p. 305, 312, &c.

[78] Compare Alph. De Candolle, Géograph. Bot.,' p. 877; Bentham and Targioni-Tozzetti, in 'Hort. Journal,' vol. ix. p. 163; Godron, 'De l'Espèce,' tom. ii. p. 92.

[79] 'Transact. Hort. Soc.,' vol. v., 1824, p. 295.

[80] Ibid., second series, vol. i., 1835, p. 248.

[81] Ibid., vol. ii. p. 138.

Hungarian Gean has almost transparent flesh. The Flemish cherry is "a very odd-looking fruit," much flattened at the summit and base, with the latter deeply furrowed, and borne on a stout very short footstalk. In the Kentish cherry the stone adheres so firmly to the footstalk, that it can be drawn out of the flesh; and this renders the fruit well fitted for drying. The Tobacco-leaved cherry, according to Sageret and Thompson, produces gigantic leaves, more than a foot and sometimes even eighteen inches in length, and half a foot in breadth. The Weeping cherry, on the other hand, is valuable only as an ornament, and, according to Downing, is "a charming little tree with slender weeping branches, clothed with small almost myrtle-like foliage." There is also a peach-leaved variety.

Sageret describes a remarkable variety, *le griottier de la Toussaint*, which bears at the same time, even as late as September, flowers and fruit of all degrees of maturity. The fruit, which is of inferior quality, is borne on long, very thin footstalks. But the extraordinary statement is made that all the leaf-bearing shoots spring from old flower-buds. Lastly, there is an important physiological distinction between those kinds of cherries which bear fruit on young or on old wood; but Sageret positively asserts that a Bigarreau in his garden bore fruit on wood of both ages.[82]

Apple (*Pyrus malus*).—The one source of doubt felt by botanists with respect to the parentage of the apple is whether, besides *P. malus*, two or three other closely allied wild forms, namely, *P. acerba* and *præcox* or *paradisiaca*, do not deserve to be ranked as distinct species. The *P. præcox* is supposed by some authors [83] to be the parent of the dwarf paradise stock, which, owing to the fibrous roots not penetrating deeply into the ground, is so largely used for grafting; but the paradise stock, it is asserted,[84] cannot be propagated true by seed. The common wild crab varies considerably in England; but many of the varieties are believed to be escaped seedlings.[85] Every one knows the great difference in the manner of growth, in the foliage, flowers, and especially in the fruit, between the almost innumerable varieties of the apple. The pips or seeds (as I know by comparison) likewise differ considerably in shape, size, and colour. The fruit is adapted for eating or for cooking in different ways, and keeps for only a few weeks or for nearly two years. Some few kinds have the fruit covered with a powdery secretion, called bloom, like that on plums;

---

[82] These several statements are taken from the four following works, which may, I believe, be trusted. Thompson, in 'Hort. Transact.,' *see* above; Sageret's 'Pomologie Phys.,' 1830, pp. 358, 364, 367, 379; 'Catalogue of the Fruit in the Garden of Hort. Soc.,' 1842, pp. 57, 60; Downing, 'The Fruits of America,' 1845, pp. 189, 195, 200.

[83] Mr. Lowe states in his 'Flora of Madeira' (quoted in 'Gard. Chron.,' 1862, p. 215) that the *P. malus*, with its nearly sessile fruit, ranges farther south than the long-stalked *P. acerba*, which is entirely absent in Madeira, the Canaries, and apparently in Portugal. This fact supports the belief that these two forms deserve to be called species. But the characters separating them are of slight importance, and of a kind known to vary in other cultivated fruit-trees.

[84] *See* 'Journ. of Hort. Tour,' by Deputation of the Caledonian Hort. Soc., 1823, p. 459.

[85] H. C. Watson, 'Cybele Britannica,' vol. i. p. 334.

and "it is extremely remarkable that this occurs almost exclusively among varieties cultivated in Russia."[86] Another Russian apple, the white Astracan, possesses the singular property of becoming transparent, when ripe, like some sorts of crabs. The *api étoilé* has five prominent ridges, hence its name; the *api noir* is nearly black: the *twin cluster pippin* often bears fruit joined in pairs.[87] The trees of the several sorts differ greatly in their periods of leafing and flowering; in my orchard the *Court Pendu Plat* produces its leaves so late, that during several springs I have thought it dead. The Tiffin apple scarcely bears a leaf when in full bloom; the Cornish crab, on the other hand, bears so many leaves at this period that the flowers can hardly be seen.[88] In some kinds the fruit ripens in mid-summer; in others, late in the autumn. These several differences in leafing, flowering, and fruiting, are not at all necessarily correlated; for, as Andrew Knight has remarked,[89] no one can judge from the early flowering of a new seedling, or from the early shedding or change of colour of the leaves, whether it will mature its fruit early in the season.

The varieties differ greatly in constitution. It is notorious that our summers are not hot enough for the Newtown Pippin,[90] which is the glory of the orchards near New York; and so it is with several varieties which we have imported from the Continent. On the other hand, our Court of Wick succeeds well under the severe climate of Canada. The *Calville rouge de Micoud* occasionally bears two crops during the same year. The Burr Knot is covered with small excrescences, which emit roots so readily that a branch with blossom-buds may be stuck in the ground, and will root and bear a few fruit even during the first year.[91] Mr. Rivers has recently described[92] some seedlings valuable from their roots running near the surface. One of these seedlings was remarkable from its extremely dwarfed size, "forming itself into a bush only a few inches in height." Many varieties are particularly liable to canker in certain soils. But perhaps the strangest constitutional peculiarity is that the Winter Majetin is not attacked by the mealy bug or coccus; Lindley[93] states that in an orchard in Norfolk infested with these insects the Majetin was quite free, though the stock on which it was grafted was affected: Knight makes a similar statement with respect to a cider apple, and adds that he only once saw these insects just above the stock, but that three days afterwards they entirely disappeared; this apple, however, was raised from a cross between

---

[86] Loudon's 'Gardener's Mag.,' vol. vi., 1830, p. 83.

[87] *See* 'Catalogue of Fruit in Garden of Hort. Soc.,' 1842, and Downing's 'American Fruit Trees.'

[88] Loudon's 'Gardener's Magazine,' vol. iv., 1828, p. 112.

[89] 'The Culture of the Apple,' p. 43. Van Mons makes the same remark on the pear, 'Arbres Fruitiers,' tom. ii., 1836, p. 414.

[90] Lindley's 'Horticulture,' p. 116. *See* also Knight on the Apple-Tree, in 'Transact. of Hort. Soc.,' vol. vi. p. 229.

[91] 'Transact. Hort. Soc.,' vol. i., 1812, p. 120.

[92] 'Journal of Horticulture,' March 13th, 1866, p. 194.

[93] 'Transact. Hort. Soc.,' vol. iv. p. 68. For Knight's case, *see* vol. vi. p. 547. When the *coccus* first appeared in this country, it is said (vol. ii. p. 163) that it was more injurious to crab-stocks than to the apples grafted on them.

the Golden Harvey and the Siberian Crab; and the latter, I believe, is considered by some authors as specifically distinct.

The famous St. Valery apple must not be passed over; the flower has a double calyx with ten divisions, and fourteen styles surmounted by conspicuous oblique stigmas, but is destitute of stamens or corolla. The fruit is constricted round the middle, and is formed of five seed-cells, surmounted by nine other cells.[94] Not being provided with stamens, the tree requires artificial fertilisation; and the girls of St. Valery annually go to *"faire ses pommes,"* each marking her own fruit with a ribbon; and as different pollen is used, the fruit differs, and we here have an instance of the direct action of foreign pollen on the mother-plant. These monstrous apples include, as we have seen, fourteen seed-cells; the pigeon-apple,[95] on the other hand, has only four, instead of, as with all common apples, five cells; and this certainly is a remarkable difference.

In the catalogue of apples published in 1842 by the Horticultural Society, 897 varieties are enumerated; but the differences between most of them are of comparatively little interest, as they are not strictly inherited. No one can raise, for instance, from the seed of the Ribston Pippin, a tree of the same kind; and it is said that the "Sister Ribston Pippin" was a white, semi-transparent, sour-fleshed apple, or rather large crab.[96] Yet it is a mistake to suppose that with most varieties the characters are not to a certain extent inherited. In two lots of seedlings raised from two well-marked kinds, many worthless, crab-like seedlings will appear, but it is now known that the two lots not only usually differ from each other, but resemble to a certain extent their parents. We see this indeed in the several sub-groups of Russetts, Sweetings, Codlins, Pearmains, Reinettes, &c.,[97] which are all believed, and many are known, to be descended from other varieties bearing the same names.

Pears (Pyrus communis).—I need say little on this fruit, which varies much in the wild state, and to an extraordinary degree when cultivated, in its fruit, flowers, and foliage. One of the most celebrated botanists in Europe, M. Decaisne, has carefully studied the many varieties;[98] although he formerly believed that they were derived from more than one species, he is now convinced that all belong to one. He has arrived at this conclusion from finding in the several varieties a perfect gradation between the most extreme characters; so perfect is this gradation that he maintains it to be impossible to classify the varieties by any natural method. M. Decaisne raised many seedlings from four distinct kinds, and has carefully recorded the variations in each. Notwithstanding this extreme degree of

---

[94] 'Mém. de la Soc. Linn. de Paris,' tom. iii., 1825, p. 164; and Seringe, 'Bulletin Bot.,' 1830, p. 117.

[95] 'Gardener's Chronicle,' 1849, p. 24.

[96] R. Thompson, in 'Gardener's Chron.,' 1850, p. 788.

[97] Sageret, 'Pomologie Physiologique,' 1830, p. 263. Downing's 'Fruit Trees,' pp. 130, 134, 139, &c. Loudon's 'Gardener's Mag.,' vol. viii. p. 317. Alexis Jordan, 'De l'Origine des diverses Variétés,' in 'Mém. de l'Acad. Imp. de Lyon,' tom. ii., 1852, pp. 95, 114. 'Gardener's Chronicle,' 1850, pp. 774, 788.

[98] 'Comptes Rendus,' July 6th, 1863.

variability, it is now positively known that many kinds reproduce by seed the leading characters of their race.[99]

Strawberries (*Fragaria*).—This fruit is remarkable on account of the number of species which have been cultivated, and from their rapid improvement within the last fifty or sixty years. Let any one compare the fruit of one of the largest varieties exhibited at our Shows with that of the wild wood strawberry, or, which will be a fairer comparison, with the somewhat larger fruit of the wild American Virginian Strawberry, and he will see what prodigies horticulture has effected.[100] The number of varieties has likewise increased in a surprisingly rapid manner. Only three kinds were known in France, in 1746, where this fruit was early cultivated. In 1766 five species had been introduced, the same which are now cultivated, but only five varieties of *Fragaria vesca*, with some sub-varieties, had been produced. At the present day the varieties of the several species are almost innumerable. The species consist of, firstly, the wood or Alpine cultivated strawberries, descended from *F. vesca*, a native of Europe and of North America. There are eight wild European varieties, as ranked by Duchesne, of *F. vesca*, but several of these are considered species by some botanists. Secondly, the green strawberries, descended from the European *F. collina*, and little cultivated in England. Thirdly, the Hautbois, from the European *F. elatior*. Fourthly, the Scarlets, descended from *F. Virginiana*, a native of the whole breadth of North America. Fifthly, the Chili, descended from *F. Chiloensis*, an inhabitant of the west coast of the temperate parts both of North and South America. Lastly, the Pines or Carolinas (including the old Blacks), which have been ranked by most authors under the name of *F. grandiflora* as a distinct species, said to inhabit Surinam; but this is a manifest error. This form is considered by the highest authority, M. Gay, to be merely a strongly marked race of *F. Chiloensis*.[101] These five or six forms have been ranked by most botanists as specifically distinct; but this may be doubted, for Andrew Knight,[102] who raised no less than 400 crossed strawberries, asserts that the *F. Virginiana, Chiloensis*, and *grandiflora* "may be made to breed together indiscriminately," and he found, in accordance with the principle of analogous variation, "that similar varieties could be obtained from the seeds of any one of them."

Since Knight's time there is abundant and additional evidence[103] of the extent to which the American forms spontaneously cross. We owe

---

[99] 'Gardener's Chronicle,' 1856, p. 804; 1857, p. 820; 1862, p. 1195.

[100] Most of the largest cultivated strawberries are the descendants of *F. grandiflora* or *Chiloensis*, and I have seen no account of these forms in their wild state. Methuen's Scarlet (Downing, 'Fruits,' p. 527) has "immense fruit of the largest size," and belongs to the section descended from *F. Virginiana;* and the fruit of this species, as I hear from Prof. A. Gray, is only a little larger than that of *F. vesca*, or our common wood strawberry.

[101] 'Le Fraisier,' par le Comte L. de Lambertye, 1864, p. 50.

[102] 'Transact. Hort. Soc.,' vol. iii. 1820, p. 207.

[103] *See* an account by Prof. Decaisne, and by others in 'Gardener's Chronicle,' 1862, p. 335, and 1858, p. 172; and Mr. Barnet's paper in 'Hort. Soc. Transact.,' vol. vi., 1826, p. 170.

indeed to such crosses most of our choicest existing varieties. Knight did not succeed in crossing the European wood-strawberry with the American Scarlet or with the Hautbois. Mr. Williams, of Pitmaston, however, succeeded; but the hybrid offspring from the Hautbois, though fruiting well, never produced seed, with the exception of a single one, which reproduced the parent hybrid form.[104] Major R. Trevor Clarke informs me that he crossed two members of the Pine class (Myatt's B. Queen and Keen's Seedling), with the wood and hautbois, and that in each case he raised only a single seedling; one of these fruited, but was almost barren. Mr. W. Smith, of York, has raised similar hybrids with equally poor success.[105] We thus see[106] that the European and American species can with some difficulty be crossed; but it is improbable that hybrids sufficiently fertile to be worth cultivation will ever be thus produced. This fact is surprising, as these forms structurally are not widely distinct, and are sometimes connected in the districts where they grow wild, as I hear from Professor Asa Gray, by puzzling intermediate forms.

The energetic culture of the strawberry is of recent date, and the cultivated varieties can in most cases still be classed under some one of the above five native stocks. As the American strawberries cross so freely and spontaneously, we can hardly doubt that they will ultimately become inextricably confused. We find, indeed, that horticulturists at present disagree under which class to rank some few of the varieties; and a writer in the 'Bon Jardinier' of 1840 remarks that formerly it was possible to class all of them under some one species, but that now this is quite impossible with the American forms, the new English varieties having completely filled up the gaps between them.[107] The blending together of two or more aboriginal forms, which there is every reason to believe has occurred with some of our anciently cultivated productions, we now see actually occurring with our strawberries.

The cultivated species offer some variations worth notice. The Black Prince, a seedling from Keen's Imperial (this latter being a seedling of a very white strawberry, the white Carolina), is remarkable from "its peculiar dark and polished surface, and from presenting an appearance entirely unlike that of any other kind."[108] Although the fruit in the different varieties differs so greatly in form, size, colour, and quality, the so-called seed (which corresponds with the whole fruit in the plum), with the exception of being more or less deeply embedded in the pulp, is, according to De Jonghe,[109] absolutely the same in all; and this no doubt may be accounted for by the seed being of no value, and consequently not having been subjected to selection. The strawberry is properly three-leaved, but in 1761 Duchesne raised a single-leaved variety of the European wood-

---

[104] 'Transact. Hort. Soc.,' vol. v., 1824, p. 294.

[105] 'Journal of Horticulture,' Dec. 30th, 1862, p. 779. *See* also Mr. Prince to the same effect, idem, 1863, p. 418.

[106] For additional evidence *see* 'Journal of Horticulture,' Dec. 9th, 1862, p. 721.

[107] 'Le Fraisier,' par le Comte L. de Lambertye, pp. 221, 230.

[108] 'Transact. Hort. Soc.,' vol. vi. p. 200.

[109] 'Gardener's Chron.,' 1858, p. 173.

strawberry, which Linnæus doubtfully raised to the rank of a species. Seedlings of this variety, like those of most varieties not fixed by long-continued selection, often revert to the ordinary form, or present intermediate states.[110] A variety raised by Mr. Myatt,[111] apparently belonging to one of the American forms, presents a variation of an opposite nature, for it has five leaves; Godron and Lambertye also mention a five-leaved variety of *F. collina*.

The Red Bush Alpine strawberry (one of the *F. vesca* section) does not produce stolons or runners, and this remarkable deviation of structure is reproduced truly by seed. Another sub-variety, the White Bush Alpine, is similarly characterised, but when propagated by seed it often degenerates and produces plants with runners.[112] A strawberry of the American Pine section is also said to make but few runners.[113]

Much has been written on the sexes of strawberries; the true Hautbois properly bears the male and female organs on separate plants,[114] and was consequently named by Duchesne *dioica*; but it frequently produces hermaphrodites; and Lindley,[115] by propagating such plants by runners, at the same time destroying the males, soon raised a self-prolific stock. The other species often show a tendency towards an imperfect separation of the sexes, as I have noticed with plants forced in a hot-house. Several English varieties, which in this country are free from any such tendency, when cultivated in rich soils under the climate of North America [116] commonly produce plants with separate sexes. Thus a whole acre of Keen's Seedlings in the United States has been observed to be almost sterile from the absence of male flowers; but the more general rule is, that the male plants overrun the females. Some members of the Cincinnati Horticultural Society, especially appointed to investigate this subject, report that "few varieties have the flowers perfect in both sexual organs," &c. The most successful cultivators in Ohio, plant for every seven rows of "pistillata," or female plants, one row of hermaphrodites, which afford pollen for both kinds; but the hermaphrodites, owing to their expenditure in the production of pollen, bear less fruit than the female plants.

The varieties differ in constitution. Some of our best English kinds, such as Keen's Seedlings, are too tender for certain parts of North America, where other English and many American varieties succeed perfectly. That splendid fruit, the British Queen, can be cultivated but in few places either in England or France; but this apparently depends more on the nature of the soil than on the climate: a famous gardener says that "no mortal could grow the British Queen at Shrubland Park unless the whole nature of the soil was altered."[117] La Constantina is one of the

[110] Godron 'De l'Espèce,' tom. i. p. 161.

[111] 'Gardener's Chron.,' 1851, p. 440.

[112] F. Gloede, in 'Gardener's Chron.,' 1862, p. 1053.

[113] Downing's 'Fruits,' p. 532.

[114] Barnet, in 'Hort. Transact.,' vol. vi. p. 210.

[115] 'Gardener's Chron.,' 1847, p. 539.

[116] For the several statements with respect to the American strawberries, *see* Downing, 'Fruits,' p. 524; 'Gardener's Chronicle,' 1843, p. 188; 1847, p. 539; 1861, p. 717.

[117] Mr. D. Beaton, in 'Cottage Gardener,' 1860, p. 86. *See* also 'Cottage Gardener,' 1855, p. 88. and many other authorities. For the Continent, *see* F. Gloede, in 'Gardener's Chronicle,' 1862, p. 1053.

hardiest kinds, and can withstand Russian winters, but is easily burnt by the sun, so that it will not succeed in certain soils either in England or the United States.[118] The Filbert Pine Strawberry "requires more water than any other variety; and if the plants once suffer from drought, they will do little or no good afterwards."[119] Cuthill's Black Prince Strawberry evinces a singular tendency to mildew: no less than six cases have been recorded of this variety suffering severely, whilst other varieties growing close by, and treated in exactly the same manner, were not at all infested by this fungus.[120] The time of maturity differs much in the different varieties; some belonging to the wood or alpine section produce a succession of crops throughout the summer.

Gooseberry (*Ribes grossularia*).—No one, I believe, has hitherto doubted that all the cultivated kinds are sprung from the wild plant bearing this name, which is common in Central and Northern Europe; therefore it will be desirable briefly to specify all the points, though not very important, which have varied. If it be admitted that these differences are due to culture, authors perhaps will not be so ready to assume the existence of a large number of unknown wild parent-stocks for our other cultivated plants. The gooseberry is not alluded to by writers of the classical period. Turner mentions it in 1573, and Parkinson, in 1629, specifies eight varieties; the Catalogue of the Horticultural Society for 1842 gives 149 varieties, and the lists of the Lancashire nurserymen are said to include above 300 names.[121] In the 'Gooseberry Grower's Register for 1862' I find that 243 distinct varieties have at various periods won prizes; so that a vast number must have been exhibited. No doubt the difference between many of the varieties is very small; but Mr. Thompson in classifying the fruit for the Horticultural Society found less confusion in the nomenclature of the gooseberry than of any other fruit, and he attributes this "to the great interest which the prize-growers have taken in detecting sorts with wrong names," and this shows that all the kinds, numerous as they are, can be recognised with certainty.

The bushes differ in their manner of growth, being erect, or spreading, or pendulous. The periods of leafing and flowering differ both absolutely and relatively to each other; thus the Whitesmith produces early flowers, which from not being protected by the foliage, as it is believed, continually fail to produce fruit.[122] The leaves vary in size, tint, and in depth of lobes; they are smooth, downy, or hairy on the upper surface. The branches are more or less downy or spinose; "the Hedgehog has probably derived its name from the singular bristly condition of its shoots and fruit." The branches of the wild gooseberry, I may remark, are smooth, with the exception of thorns at the bases of the buds. The thorns themselves are either very small, few and single, or very large and triple; they are

118 Rev. W. F. Radclyffe, in 'Journal of Hort.,' March 14, 1865, p. 207.

119 Mr. H. Doubleday in 'Gardener's Chron.,' 1862, p. 1101.

120 'Gardener's Chronicle,' 1854, p. 254.

121 Loudon's 'Encyclop. of Gardening,' p. 930; and Alph. De Candolle, 'Géograph. Bot.,' p. 910.

122 Loudon's 'Gardener's Magazine,' vol. iv. 1828, p. 112.

sometimes reflexed and much dilated at their bases. In the different varieties the fruit varies in abundance, in the period of maturity, in hanging until shrivelled, and greatly in size, "some sorts having their fruit large during a very early period of growth, whilst others are small until nearly ripe." The fruit varies also much in colour, being red, yellow, green, and white—the pulp of one dark-red gooseberry being tinged with yellow; in flavour; in being smooth or downy,—few, however, of the Red gooseberries, whilst many of the so-called Whites, are downy; or in being so spinose that one kind is called Henderson's Porcupine. Two kinds acquire when mature a powdery bloom on their fruit. The fruit varies in the thickness and veining of the skin, and, lastly, in shape, being spherical, oblong, oval, or obovate.[123]

I cultivated fifty-four varieties, and, considering how greatly the fruit differs, it was curious how closely similar the flowers were in all these kinds. In only a few I detected a trace of difference in the size or colour of the corolla. The calyx differed in a rather greater degree, for in some kinds it was much redder than in others; and in one smooth white gooseberry it was unusually red. The calyx also differed in the basal part being smooth or woolly, or covered with glandular hairs. It deserves notice, as being contrary to what might have been expected from the law of correlation, that a smooth red gooseberry had a remarkably hairy calyx. The flowers of the Sportsman are furnished with very large coloured bracteæ; and this is the most singular deviation of structure which I have observed. These same flowers also varied much in the number of the petals, and occasionally in the number of the stamens and pistils; so that they were semi-monstrous in structure, yet they produced plenty of fruit. Mr. Thompson remarks that in the Pastime gooseberry "extra bracts are "often attached to the sides of the fruit."[124]

The most interesting point in the history of the gooseberry is the steady increase in the size of the fruit. Manchester is the metropolis of the fanciers, and prizes from five shillings to five or ten pounds are yearly given for the heaviest fruit. The 'Gooseberry Grower's Register' is published annually; the earliest known copy is dated 1786, but it is certain that meetings for the adjudication of prizes were held some years previously.[125] The 'Register' for 1845 gives an account of 171 Gooseberry Shows, held in different places during that year; and this fact shows on how large a scale the culture has been carried on. The fruit of the wild gooseberry is said[126] to weigh about a quarter of an ounce or 5 dwts., that is, 120 grains; about the year 1786 gooseberries were exhibited weighing 10 dwts., so that the weight was then doubled; in 1817 26 dwts. 17 grs. was attained; there was no advance till 1825, when 31 dwts. 16 grs. was reached; in

[123] The fullest account of the gooseberry is given by Mr. Thompson in 'Transact. Hort. Soc.,' vol. i., 2nd series, 1835, p. 218, from which most of the foregoing facts are given.

[124] 'Catalogue of Fruits of Hort. Soc. Garden,' 3rd edit. 1842.

[125] Mr. Clarkson, of Manchester, on the Culture of the Gooseberry, in Loudon's 'Gardener's Magazine,' vol. iv. 1828, p. 482.

[126] Downing's 'Fruits of America,' p. 213.

1830 "Teazer" weighed 32 dwts. 13 grs.; in 1841 "Wonderful" weighed 32 dwts. 16 grs.; in 1844 "London" weighed 35 dwts. 12 grs., and in the following year 36 dwts. 16 grs.; and in 1852 in Staffordshire the fruit of this same variety reached the astonishing weight of 37 dwts. 7 grs.,[127] or 895 grs.; that is, between seven and eight times the weight of the wild fruit. I find that a small apple, 6½ inches in circumference, has exactly this same weight. The "London" gooseberry (which in 1862 had altogether gained 343 prizes) has, up to the present year of 1864, never reached a greater weight than that attained in 1852. Perhaps the fruit of the gooseberry has now reached the greatest possible weight, unless in the course of time some quite new and distinct variety shall arise.

This gradual, and on the whole steady increase of weight from the latter part of the last century to the year 1852, is probably in large part due to improved methods of cultivation, for extreme care is now taken; the branches and roots are trained, composts are made, the soil is mulched, and only a few berries are left on each bush;[128] but the increase no doubt is in main part due to the continued selection of seedlings which have been found to be more and more capable of yielding such extraordinary fruit. Assuredly the "Highwayman" in 1817 could not have produced fruit like that of the "Roaring Lion" in 1825; nor could the "Roaring Lion," though it was grown by many persons in many places, gain the supreme triumph achieved in 1852 by the "London" Gooseberry.

Walnut (*Juglans regia*).—This tree and the common nut belong to a widely different order from the foregoing fruits, and are therefore here noticed. The walnut grows wild in the Caucasus and Himalaya, where Dr. Hooker[129] found the fruit of full size, but "as hard as a hickory-nut." In England the walnut presents considerable differences, in the shape and size of the fruit, in the thickness of the husk, and in the thinness of the shell; this latter quality has given rise to a variety called the thin-shelled, which is valuable, but suffers from the attacks of tom-tits.[130] The degree to which the kernel fills the shell varies much. In France there is a variety called the Grape or cluster-walnut, in which the nuts grow in "bunches of ten, fifteen, or even twenty together." There is another variety which bears on the same tree differently shaped leaves, like the heterophyllous hornbeam; this tree is also remarkable from having pendulous branches, and bearing elongated, large, thin-shelled nuts.[131] M. Cardan has minutely described[132] some singular physiological peculiarities in the June-leafing variety, which produces its leaves and flowers four or five weeks later, and retains its leaves and fruit in the autumn much longer, than the common varieties;

---

127 'Gardener's Chronicle,' 1844, p. 811, where a table is given; and 1845, p. 819. For the extreme weights gained, *see* 'Journal of Horticulture,' July 26, 1864, p. 61.

128 Mr. Saul, of Lancaster, in Loudon's 'Gardener's Mag.,' vol. iii. 1828, p. 421; and vol. x. 1834, p. 42.

129 'Himalayan Journals,' 1854, vol. ii. p. 334. Moorcroft ('Travels,' vol. ii. p. 146) describes four varieties cultivated in Kashmir.

130 'Gardener's Chronicle,' 1850, p. 723.

131 Paper translated in Loudon's 'Gardener's Mag.,' 1829, vol. v. p. 202.

132 Quoted in 'Gardener's Chronicle,' 1849, p. 101.

but in August is in exactly the same state with them. These constitutional peculiarities are strictly inherited. Lastly, walnut-trees, which are properly monoicous, sometimes entirely fail to produce male flowers.[133]

Nuts (*Corylus avellana*).—Most botanists rank all the varieties under the same species, the common wild nut.[134] The husk, or involucre, differs greatly, being extremely short in Barr's Spanish, and extremely long in filberts, in which it is contracted so as to prevent the nut falling out. This kind of husk also protects the nut from birds, for titmice (*Parus*) have been observed[135] to pass over filberts, and attack cobs and common nuts growing in the same orchard. In the purple-filbert the husk is purple, and in the frizzled-filbert it is curiously laciniated; in the red-filbert the pellicle of the kernel is red. The shell is thick in some varieties, but is thin in Cosford's-nut, and in one variety is of a bluish colour. The nut itself differs much in size and shape, being ovate and compressed in filberts, nearly round and of great size in cobs and Spanish nuts, oblong and longitudinally striated in Cosford's, and obtusely four-sided in the Downton Square nut.

Cucurbitaceous plants.—These plants have been for a long period the opprobrium of botanists; numerous varieties have been ranked as species, and, what happens more rarely, forms which now must be considered as species have been classed as varieties. Owing to the admirable experimental researches of a distinguished botanist, M. Naudin,[136] a flood of light has recently been thrown on this group of plants. M. Naudin, during many years, observed and experimented on above 1200 living specimens, collected from all quarters of the world. Six species are now recognised in the genus Cucurbita; but three alone have been cultivated and concern us, namely, *C. maxima* and *pepo*, which include all pumpkins, gourds, squashes, and vegetable marrow, and *C. moschata*, the water-melon. These three species are not known in a wild state; but Asa Gray[137] gives good reason for believing that some pumpkins are natives of N. America.

These three species are closely allied, and have the same general habit, but their innumerable varieties can always be distinguished, according to Naudin, by certain almost fixed characters; and what is still more important, when crossed they yield no seed, or only sterile seed; whilst the varieties spontaneously intercross with the utmost freedom. Naudin insists strongly (p. 15), that, though these three species have varied greatly in many characters, yet it has been in so closely an analogous manner that the varieties can be arranged in almost parallel series, as we have seen with the forms of wheat, with the two main races of the peach, and in other cases. Though some of the varieties are inconstant in character, yet others, when grown separately under uniform conditions of life, are, as Naudin repeatedly (pp. 6, 16, 35) urges, "douées d'une stabilité

[133] 'Gardener's Chronicle,' 1847, pp. 541 and 558.

[134] The following details are taken from the Catalogue of Fruits, 1842, in Garden of Hort. Soc., p. 103; and from Loudon's 'Encyclop. of Gardening,' p. 943.

[135] 'Gardener's Chron.,' 1860, p. 956.

[136] 'Annales des Sc. Nat. Bot.,' 4th series, vol. vi. 1856, p. 5.

[137] 'American Journ. of Science,' 2nd ser. vol. xxiv. 1857, p. 442.

presque comparable à celle des espèces les mieux caractérisées." One variety, l'Orangin (pp. 43, 63), has such prepotency in transmitting its character that when crossed with other varieties a vast majority of the seedlings come true. Naudin, referring (p. 47) to *C. pepo*, says that its races "ne diffèrent des espèces véritables qu'en ce qu'elles peuvent s'allier les unes aux autres par voie d'hybridité, sans que leur descendance perde la faculté de se perpétuer." If we were to trust to external differences alone, and give up the test of sterility, a multitude of species would have to be formed out of the varieties of these three species of Cucurbita. Many naturalists at the present day lay far too little stress, in my opinion, on the test of sterility; yet it is not improbable that distinct species of plants after a long course of cultivation and variation may have their mutual sterility eliminated, as we have every reason to believe has occurred with domesticated animals. Nor, in the case of plants under cultivation, should we be justified in assuming that varieties never acquire a slight degree of mutual sterility, as we shall more fully see in a future chapter when certain facts are given on the high authority of Gärtner and Kölreuter.[138]

The forms of *C. pepo* are classed by Naudin under seven sections, each including subordinate varieties. He considers this plant as probably the most variable in the world. The fruit of one variety (pp. 33, 46) exceeds in volume that of another by more than two thousand fold! When the fruit is of very large size, the number produced is few (p. 45); when of small size, many are produced. No less astonishing (p. 33) is the variation in the shape of the fruit; the typical form apparently is egg-like, but this becomes either drawn out into a cylinder, or shortened into a flat disc. We have also an almost infinite diversity in the colour and state of surface of the fruit, in the hardness both of the shell and of the flesh, and in the taste of the flesh, which is either extremely sweet, farinaceous, or slightly bitter. The seeds also differ in a slight degree in shape, and wonderfully in size (p. 34), namely, from six or seven to more than twenty-five millimètres in length.

In the varieties which grow upright or do not run and climb, the tendrils, though useless (p. 31), are either present or are represented by various semi-monstrous organs, or are quite absent. The tendrils are even absent in some running varieties in which the stems are much elongated. It is a singular fact that (p. 31), in all the varieties with dwarfed stems, the leaves closely resemble each other in shape.

Those naturalists who believe in the immutability of species often maintain that, even in the most variable forms, the characters which they consider of specific value are unchangeable. To give an example from a conscientious writer,[139] who, relying on the labours of M. Naudin and

---

[138] Gärtner, 'Bastarderzeugung,' 1849, s. 87, and s. 169 with respect to Maize; on Verbascum, idem, ss. 92 and 181; also his 'Kenntniss der Befruchtung,' s. 137. With respect to Nicotiana, *see* Kölreuter, 'Zweite Forts.,' 1764, s. 53; though this is a somewhat different case.

[139] 'De l'Espèce,' par M. Godron, tom. ii. p. 64.

referring to the species of Cucurbita, says, "au milieu de toutes les variations du fruit, les tiges, les feuilles, les calices, les corolles, les étamines restent invariables dans chacune d'elles." Yet M. Naudin in describing *Cucurbita pepo* (p. 30) says, "Ici, d'ailleurs, ce ne sont pas seulement les fruits qui varient, c'est aussi le feuillage et tout le port de la plante. Néanmoins, je crois qu'on la distinguera toujours facilement des deux autres espèces, si l'on veut ne pas perdre de vue les caractères différentiels que je m'efforce de faire ressortir. Ces caractères sont quelquefois peu marqués: il arrive même que plusieurs d'entre eux s'effacent presque entièrement, mais il en reste toujours quelques-uns qui remettent l'observateur sur la voie." Now let it be noted what a difference, with regard to the immutability of the so-called specific characters, this paragraph produces on the mind, from that above quoted from M. Godron.

I will add another remark: naturalists continually assert that no important organ varies; but in saying this they unconsciously argue in a vicious circle; for if an organ, let it be what it may, is highly variable, it is regarded as unimportant, and under a systematic point of view this is quite correct. But as long as constancy is thus taken as the criterion of importance, it will indeed be long before an important organ can be shown to be inconstant. The enlarged form of the stigmas, and their sessile position on the summit of the ovary, must be considered as important characters, and were used by Gasparini to separate certain pumpkins as a *distinct genus;* but Naudin says (p. 20) these parts have no constancy, and in the flowers of the Turban varieties of *C. maxima* they sometimes resume their ordinary structure. Again, in *C. maxima*, the carpels (p. 19) which form the Turban project even as much as two-thirds of their length out of the receptacle, and this latter part is thus reduced to a sort of platform; but this remarkable structure occurs only in certain varieties, and graduates into the common form in which the carpels are almost entirely enveloped within the receptacle. In *C. moschata* the ovarium (p. 50) varies greatly in shape, being oval, nearly spherical, or cylindrical, more or less swollen in the upper part, or constricted round the middle, and either straight or curved. When the ovarium is short and oval the interior structure does not differ from that of *C. maxima* and *pepo*, but when it is elongated the carpels occupy only the terminal and swollen portion. I may add that in one variety of the cucumber (*Cucumis sativus*) the fruit regularly contains five carpels instead of three.[140] I presume that it will not be disputed that we here have instances of great variability in organs of the highest physiological importance, and with most plants of the highest classificatory importance.

Sageret[141] and Naudin found that the cucumber (*C. sativus*) could not be crossed with any other species of the genus; therefore no doubt it is specifically distinct from the melon. This will appear to most persons a superfluous statement; yet we hear from Naudin[142] that there is a race

---

[140] Naudin, in 'Annal. des Sci. Nat.,' 4th ser. Bot. tom. xi. 1859, p. 28.

[141] 'Mémoire sur les Cucurbitacées,' 1826, pp. 6, 24.

[142] 'Flore des Serres,' Oct. 1861, quoted in 'Gardener's Chronicle,' 1861, p. 1135. I have also consulted and taken some facts from M. Naudin's

of melons, in which the fruit is so like that of the cucumber, "both externally and internally, that it is hardly possible to distinguish the one from the other except by the leaves." The varieties of the melon seem to be endless, for Naudin after six years' study has not come to the end of them: he divides them into ten sections, including numerous sub-varieties which all intercross with perfect ease.[143] Of the forms considered by Naudin to be varieties, botanists have made thirty distinct species! "and they had not the slightest acquaintance with the multitude of new forms which have appeared since their time." Nor is the creation of so many species at all surprising when we consider how strictly their characters are transmitted by seed, and how wonderfully they differ in appearance: "Mira est quidem foliorum et habitus diversitas, sed multo magis fructuum," says Naudin. The fruit is the valuable part, and this, in accordance with the common rule, is the most modified part. Some melons are only as large as small plums, others weigh as much as sixty-six pounds. One variety has a scarlet fruit! Another is not more than an inch in diameter, but sometimes more than a yard in length, "twisting about in all directions like a serpent." It is a singular fact that in this latter variety many parts of the plant, namely, the stems, the footstalks of the female flowers, the middle lobe of the leaves, and especially the ovarium, as well as the mature fruit, all show a strong tendency to become elongated. Several varieties of the melon are interesting from assuming the characteristic features of distinct species and even of distinct though allied genera: thus the serpent-melon has some resemblance to the fruit of *Trichosanthes anguina*; we have seen that other varieties closely resemble cucumbers; some Egyptian varieties have their seeds attached to a portion of the pulp, and this is characteristic of certain wild forms. Lastly, a variety of melon from Algiers is remarkable from announcing its maturity by "a spontaneous and almost sudden dislocation," when deep cracks suddenly appear, and the fruit falls to pieces; and this occurs with the wild *C. momordica*. Finally, M. Naudin well remarks that this "extraordinary production of races and varieties by a single species, and their permanence when not interfered with by crossing, are phenomena well calculated to cause reflection."

USEFUL AND ORNAMENTAL TREES.

TREES deserve a passing notice on account of the numerous varieties which they present, differing in their precocity, in their manner of growth, foliage, and bark. Thus of the common ash (*Fraxinus excelsior*) the catalogue of Messrs. Lawson of Edinburgh includes twenty-one varieties, some of which differ much in their bark; there is a yellow, a streaked reddish-white, a purple, a wart-barked and a fungous-barked variety.[144] Of hollies no less than eighty-four varieties are grown alongside each other in Mr.

Memoir on Cucumis in 'Annal. des Sc. Nat.,' 4th series, Bot. tom. xi. 1859, p. 5.

[143] *See* also Sageret's 'Mémoire,' p. 7.

[144] Loudon's 'Arboretum et Fruticetum,' vol. ii. p. 1217.

Paul's nursery.[145] In the case of trees, all the recorded varieties, as far as I can find out, have been suddenly produced by one single act of variation. The length of time required to raise many generations, and the little value set on the fanciful varieties, explains how it is that successive modifications have not been accumulated by selection; hence, also it follows that we do not here meet with sub-varieties subordinate to varieties, and these again subordinate to higher groups. On the Continent, however, where the forests are more carefully attended to than in England, Alph. De Candolle[146] says that there is not a forester who does not search for seeds from that variety which he esteems the most valuable.

Our useful trees have seldom been exposed to any great change of conditions; they have not been richly manured, and the English kinds grow under their proper climate. Yet in examining extensive beds of seedlings in nursery-gardens considerable differences may be generally observed in them; and whilst touring in England I have been surprised at the amount of difference in the appearance of the same species in our hedgerows and woods. But as plants vary so much in a truly wild state, it would be difficult for even a skilful botanist to pronounce whether, as I believe to be the case, hedgerow trees vary more than those growing in a primeval forest. Trees when planted by man in woods or hedges do not grow where they would naturally be able to hold their place against a host of competitors, and are therefore exposed to conditions not strictly natural: even this slight change would probably suffice to cause seedlings raised from such trees to be variable. Whether or not our half-wild English trees, as a general rule, are more variable than trees growing in their native forests, there can hardly be a doubt that they have yielded a greater number of strongly-marked and singular variations of structure.

In manner of growth, we have weeping or pendulous varieties of the willow, ash, elm, oak, and yew, and other trees; and this weeping habit is sometimes inherited, though in a singularly capricious manner. In the Lombardy poplar, and in certain fastigate or pyramidal varieties of thorns, junipers, oaks, &c., we have an opposite kind of growth. The Hessian oak,[147] which is famous from its fastigate habit and size, bears hardly any resemblance in general appearance to a common oak; "its acorns are not sure to produce plants of the same habit; some, however, turn out the same as the parent-tree." Another fastigate oak is said to have been found wild in the Pyrenees, and this is a surprising circumstance; it generally comes so true by seed, that De Candolle considered it as specifically distinct.[148] The fastigate Juniper (*J. suecica*) likewise transmits its character by seed.[149] Dr. Falconer informs me that in the Botanic Gardens at Calcutta the great heat causes apple-trees to become fastigate; and we

---

[145] 'Gardener's Chronicle,' 1866, p. 1096.

[146] 'Géograph. Bot.,' p. 1096.

[147] 'Gardener's Chron.,' 1842, p. 36.

[148] Loudon's 'Arboretum et Fruticetum,' vol. iii. p. 1731.

[149] Ibid., vol. iv. p. 2489.

thus see the same result following from the effects of climate and from an innate spontaneous tendency.[150]

In foliage we have variegated leaves which are often inherited; dark purple or red leaves, as in the hazel, barberry, and beech, the colour in these two latter trees being sometimes strongly and sometimes weakly inherited;[151] deeply-cut leaves; and leaves covered with prickles, as in the variety of the holly well called *ferox*, which is said to reproduce itself by seed.[152] In fact, nearly all the peculiar varieties evince a tendency, more or less strongly marked, to reproduce themselves by seed.[153] This is to a certain extent the case, according to Bosc,[154] with three varieties of the elm, namely, the broad-leafed, lime-leafed, and twisted elm, in which latter the fibres of the wood are twisted. Even with the heterophyllous hornbeam (*Carpinus betulus*), which bears on each twig leaves of two shapes, "several plants raised from seed all retained the same peculiarity."[155] I will add only one other remarkable case of variation in foliage, namely, the occurrence of two sub-varieties of the ash with simple instead of pinnated leaves, and which generally transmit their character by seed.[156] The occurrence, in trees belonging to widely different orders, of weeping and fastigate varieties, and of trees bearing deeply cut, variegated, and purple leaves, shows that these deviations of structure must result from some very general physiological laws.

Differences in general appearance and foliage, not more strongly marked than those above indicated, have led good observers to rank as distinct species certain forms which are now known to be mere varieties. Thus a plane-tree long cultivated in England was considered by almost every one as a North American species; but is now ascertained by old records, as I am informed by Dr. Hooker, to be a variety. So again the *Thuja pendula* or *filiformis* was ranked by such good observers as Lambert, Wallich, and others as a true species; but it is now known that the original plants, five in number, suddenly appeared in a bed of seedlings, raised at Mr. Loddige's nursery, from *T. orientalis*; and Dr. Hooker has adduced excellent evidence that at Turin seeds of *T. pendula* have reproduced the parent-form, *T. orientalis*.[157]

Every one must have noticed how certain individual trees regularly put forth and shed their leaves earlier or later than others of the same species. There is a famous horse-chesnut in the Tuileries which is named from

---

150 Godron ('De l'Espèce,' tom. ii. p. 91) describes four varieties of Robinia remarkable from their manner of growth.

151 'Journal of a Horticultural Tour, by Caledonian Hort. Soc.,' 1823, p. 107. Alph. De Candolle, 'Géograph. Bot.,' p. 1083. Verlot, 'Sur la Production des Variétés,' 1865, p. 55, for the Barberry.

152 Loudon's 'Arboretum et Fruticetum,' vol. ii. p. 508.

153 Verlot, 'Des Variétés,' 1865, p. 92.

154 Loudon's 'Arboretum et Fruticetum,' vol. iii. p. 1376.

155 'Gardener's Chronicle,' 1841, p. 687.

156 Godron, 'De l'Espèce,' tom. ii. p. 89. In Loudon's 'Gardener's Mag.,' vol. xii. 1836, p. 371, a variegated bushy ash is described and figured, as having simple leaves; it originated in Ireland.

157 'Gardener's Chron.,' 1861, p. 575.

leafing so much earlier than the others. There is also an oak near Edinburgh which retains its leaves to a very late period. These differences have been attributed by some authors to the nature of the soil in which the trees grow; but Archbishop Whately grafted an early thorn on a late one, and *vice versâ*, and both grafts kept to their proper periods, which differed by about a fortnight, as if they still grew on their own stocks.[158] There is a Cornish variety of the elm which is almost an evergreen, and is so tender that the shoots are often killed by the frost; and the varieties of the Turkish oak (*Q. cerris*) may be arranged as deciduous, sub-evergreen, and evergreen.[159]

Scotch Fir (*Pinus sylvestris*).—I allude to this tree as it bears on the question of the greater variability of our hedgerow trees compared with those under strictly natural conditions. A well-informed writer[160] states that the Scotch fir presents few varieties in its native Scotch forests; but that it "varies much in figure and foliage, and in the size, shape, and colour of its "cones, when several generations have been produced away from its native "locality." There is little doubt that the highland and lowland varieties differ in the value of their timber, and that they can be propagated truly by seed; thus justifying Loudon's remark, that "a variety is often of as much importance as a species, and sometimes far more so."[161] I may mention one rather important point in which this tree occasionally varies; in the classification of the Coniferæ, sections are founded on whether two, three, or five leaves are included in the same sheath; the Scotch fir has properly only two leaves thus enclosed, but specimens have been observed with groups of three leaves in a sheath.[162] Besides these differences in the semi-cultivated Scotch fir, there are in several parts of Europe natural or geographical races, which have been ranked by some authors as distinct species.[163] Loudon[164] considers *P. pumilio*, with its several sub-varieties, as *Mughus*, *nana*, &c., which differ much when planted in different soils and only come "tolerably true from seed," as alpine varieties of the Scotch fir; if this were proved to be the case, it would be an interesting fact as showing that dwarfing from long exposure to a severe climate is to a certain extent inherited.

The *Hawthorn* (*Cratægus oxycantha*) has varied much. Besides endless slighter variations in the form of the leaves, and in the size, hardness, fleshiness, and shape of the berries, Loudon[165] enumerates twenty-nine well-marked varieties. Besides those cultivated for their pretty flowers, there are others with golden-yellow, black, and whitish berries; others

---

[158] Quoted from Royal Irish Academy in 'Gardener's Chron.,' 1841, p. 767.

[159] Loudon's 'Arboretum et Fruticetum:' for Elm, *see* vol. iii. p. 1376; for Oak, p. 1846.

[160] 'Gardener's Chronicle,' 1849, p. 822.

[161] 'Arboretum et Fruticetum,' vol. iv. p. 2150.

[162] 'Gardener's Chron.,' 1852, p. 693.

[163] *See* 'Beiträge zur Kentniss Europäischer Pinus-arten von Dr. Christ: Flora, 1864.' He shows that in the Ober-Engadin *P. sylvestris* and *montana* are connected by intermediate links.

[164] 'Arboretum et Fruticetum,' vol. iv. pp. 2159 and 2189.

[165] Ibid., vol. ii. p. 830; Loudon's 'Gardener's Magazine,' vol. vi. 1830, p. 714.

with woolly berries, and others with recurved thorns. Loudon truly remarks that the chief reason why the hawthorn has yielded more varieties than most other trees, is that curious nurserymen select any remarkable variety out of the immense beds of seedlings which are annually raised for making hedges. The flowers of the hawthorn usually include from one to three pistils; but in two varieties, named *Monogyna* and *Sibirica*, there is only a single pistil; and d'Asso states that the common thorn in Spain is constantly in this state.[166] There is also a variety which is apetalous, or has its petals reduced to mere rudiments. The famous Glastonbury thorn flowers and leafs towards the end of December, at which time it bears berries produced from an earlier crop of flowers.[167] It is worth notice that several varieties of the hawthorn, as well as of the lime and juniper, are very distinct in their foliage and habit whilst young, but in the course of thirty or forty years become extremely like each other;[168] thus reminding us of the well-known fact that the deodar, the cedar of Lebanon, and that of the Atlas, are distinguished with the greatest ease whilst young, but with difficulty when old.

FLOWERS.

I SHALL not for several reasons treat the variability of plants which are cultivated for their flowers alone at any great length. Many of our favourite kinds in their present state are the descendants of two or more species crossed and commingled together, and this circumstance alone would render it difficult to detect the differences due to variation. For instance, our Roses, Petunias, Calceolarias, Fuchsias, Verbenas, Gladioli, Pelargoniums, &c., certainly have had a multiple origin. A botanist well acquainted with the parent-forms would probably detect some curious structural differences in their crossed and cultivated descendant; and he would certainly observe many new and remarkable constitutional peculiarities. I will give a few instances, all relating to the Pelargonium, and taken chiefly from Mr. Beck,[169] a famous cultivator of this plant: some varieties require more water than others; some are "very impatient of the knife if too greedily used in making cuttings;" some, when potted, scarcely "show a root at the outside of the ball of the earth;" one variety requires a certain amount of confinement in the pot to make it throw up a flower-stem; some varieties bloom well at the commencement of the season, others at the close; one variety is known,[170] which will stand "even pine-apple top and bottom heat, without looking any more drawn than if it had stood in a common greenhouse; and Blanche Fleur seems as if made on purpose for growing in winter, like many bulbs, and to rest all summer." These odd constitutional peculiarities would fit a plant when growing in a state of nature for widely different circumstances and climates.

---

166 Loudon's 'Arboretum et Fruticetum,' vol. ii. p. 834.

167 Loudon's 'Gardener's Mag.,' vol. ix. 1833, p. 123.

168 Ibid., vol. xi. 1835, p. 503.

169 'Gardener's Chron.,' 1845, p. 623.

170 D. Beaton, in 'Cottage Gardener,' 1860, p. 377. *See* also Mr. Beck, on the habits of Queen Mab, in 'Gardener's Chronicle,' 1845, p. 226.

Flowers possess little interest under our present point of view, because they have been almost exclusively attended to and selected for their beautiful colours, size, perfect outline, and manner of growth. In these particulars hardly one long-cultivated flower can be named which has not varied greatly. What does a florist care for the shape and structure of the organs of fructification, unless, indeed, they add to the beauty of the flower? When this is the case, flowers become modified in important points; stamens and pistils may be converted into petals, and additional petals may be developed, as in all double flowers. The process of gradual selection by which flowers have been rendered more and more double, each step in the process of conversion being inherited, has been recorded in several instances. In the so-called double flowers of the Compositæ, the corollas of the central florets are greatly modified, and the modifications are likewise inherited. In the columbine (*Aquilegia vulgaris*) some of the stamens are converted into petals having the shape of nectaries, one neatly fitting into the other; but in one variety they are converted into simple petals.[171] In the hose and hose primulæ, the calyx becomes brightly coloured and enlarged so as to resemble a corolla; and Mr. W. Wooler informs me that this peculiarity is transmitted; for he crossed a common polyanthus with one having a coloured calyx,[172] and some of the seedlings inherited the coloured calyx during at least six generations. In the "hen-and-chicken" daisy the main flower is surrounded by a brood of small flowers developed from buds in the axils of the scales of the involucre. A wonderful poppy has been described, in which the stamens are converted into pistils; and so strictly was this peculiarity inherited that, out of 154 seedlings, one alone reverted to the ordinary and common type.[173] Of the cock's-comb (*Celosia cristata*), which is an annual, there are several races in which the flower-stem is wonderfully "fasciated" or compressed; and one has been exhibited[174] actually eighteen inches in breadth. Peloric races of *Gloxinia speciosa* and *Antirrhinum majus* can be propagated by seed, and they differ in a wonderful manner from the typical form both in structure and appearance.

A much more remarkable modification has been recorded by Sir William and Dr. Hooker[175] in *Begonia frigida*. This plant properly produces male and female flowers on the same fascicles; and in the female flowers the perianth is superior; but a plant at Kew produced, besides the ordinary flowers, others which graduated towards a perfect hermaphrodite structure; and in these flowers the perianth was inferior. To show the importance of this modification under a classificatory point of view, I may quote what Prof. Harvey says, namely, that had it "occurred in a state of nature, and had a botanist collected a plant with such flowers, he would not only have

---

[171] Moquin-Tandon, 'Eléments de Tératologie,' 1841, p. 213.

[172] *See* also 'Cottage Gardener,' 1860, p. 133.

[173] Quoted by Alph. de Candolle, 'Bibl.Univ.,' November, 1862, p. 58.

[174] Knight, 'Transact. Hort. Soc.,' vol. iv. p. 322.

[175] 'Botanical Magazine,' tab. 5160, fig. 4; Dr. Hooker, in 'Gardener's Chron.,' 1860, p. 190; Prof. Harvey, in 'Gardener's Chron.,' 1860, p. 145; Mr. Crocker, in 'Gardener's Chron.,' 1861, p. 1092.

placed it in a distinct genus from Begonia, but would probably have considered it as the type of a new natural order." This modification cannot in one sense be considered as a monstrosity, for analogous structures naturally occur in other orders, as with Saxifragas and Aristolochiaceæ. The interest of the case is largely added to by Mr. C. W. Crocker's observation that seedlings from the *normal* flowers produced plants which bore, in about the same proportion as the parent-plant, hermaphrodite flowers having inferior perianths. The hermaphrodite flowers fertilised with their own pollen were sterile.

If florists had attended to, selected, and propagated by seed other modifications of structure besides those which are beautiful, a host of curious varieties would certainly have been raised; and they would probably have transmitted their characters so truly that the cultivator would have felt aggrieved, as in the case of culinary vegetables, if his whole bed had not presented a uniform appearance. Florists have attended in some instances to the leaves of their plant, and have thus produced the most elegant and symmetrical patterns of white, red, and green, which, as in the case of the pelargonium, are sometimes strictly inherited.[176] Any one who will habitually examine highly-cultivated flowers in gardens and greenhouses will observe numerous deviations in structure; but most of these must be ranked as mere monstrosities, and are only so far interesting as showing how plastic the organisation becomes under high cultivation. From this point of view such works as Professor Moquin-Tandon's 'Tératologie' are highly instructive.

Roses.—These flowers offer an instance of a number of forms generally ranked as species, namely, *R. centifolia, gallica, alba, damascena, spinosissima, bracteata, Indica, semperflorens, moschata,* &c., which have largely varied and been intercrossed. The genus Rosa is a notoriously difficult one, and, though some of the above forms are admitted by all botanists to be distinct species, others are doubtful; thus, with respect to the British forms, Babington makes seventeen, and Bentham only five species. The hybrids from some of the most distinct forms—for instance, from *R. Indica*, fertilised by the pollen of *R. centifolia*—produce an abundance of seed; I state this on the authority of Mr. Rivers,[177] from whose work I have drawn most of the following statements. As almost all the aboriginal forms brought from different countries have been crossed and recrossed, it is no wonder that Targioni-Tozzetti, in speaking of the common roses of the Italian gardens, remarks that "the native country and precise form of the wild type of most of them are involved in much uncertainty."[178] Nevertheless Mr. Rivers in referring to *R. Indica* (p. 68) says that the descendants of each group may generally be recognised by a close observer. The same author often speaks of roses as having been a little hybridised; but

[176] Alph. de Candolle, 'Géograph. Bot.,' p. 1083; 'Gard. Chronicle,' 1861, p. 433. The inheritance of the white and golden zones in Pelargonium largely depends on the nature of the soil. *See* D. Beaton, in 'Journal of Horticulture,' 1861, p. 64.

[177] 'Rose Amateur's Guide,' T. Rivers, 1837, p. 21.

[178] 'Journal Hort. Soc.,' vol. ix. 1855, p. 182.

it is evident that in very many cases the differences due to variation and to hybridisation can now only be conjecturally distinguished.

The species have varied both by seed and by buds; such modified buds being often called by gardeners sports. In the following chapter I shall fully discuss this latter subject, and shall show that bud-variations can be propagated not only by grafting and budding, but often even by seed. Whenever a new rose appears with any peculiar character, however produced, if it yields seed, Mr. Rivers (p. 4) fully expects it to become the parent-type of a new family. The tendency to vary is so strong in some kinds, as in the Village Maid (Rivers, p. 16), that when grown in different soils it varies so much in colour that it has been thought to form several distinct kinds. Altogether the number of kinds is very great: thus M. Desportes, in his Catalogue for 1829, enumerates 2562 as cultivated in France; but no doubt a large proportion of these are merely nominal.

It would be useless to specify the many points of difference between the various kinds, but some constitutional peculiarities may be mentioned. Several French roses (Rivers, p. 12) will not succeed in England; and an excellent horticulturist[179] remarks, that "Even in the same garden you will find that a rose that will do nothing under a south wall will do well under a north one. That is the case with Paul Joseph here. It grows strongly and blooms beautifully close to a north wall. For three years seven plants have done nothing under a south wall." Many roses can be forced, "many are totally unfit for forcing, among which is General Jacqueminot."[180] From the effects of crossing and variation Mr. Rivers enthusiastically anticipates (p. 87) that the day will come when all our roses, even moss-roses, will have evergreen foliage, brilliant and fragrant flowers, and the habit of blooming from June till November. "A distant view this seems, but perseverance in gardening will yet achieve wonders," as assuredly it has already achieved wonders.

It may be worth while briefly to give the well-known history of one class of roses. In 1793 some wild Scotch roses (*R. spinosissima*) were transplanted into a garden;[181] and one of these bore flowers slightly tinged with red, from which a plant was raised with semi-monstrous flowers, also tinged with red; seedlings from this flower were semi-double, and by continued selection, in about nine or ten years, eight sub-varieties were raised. In the course of less than twenty years these double Scotch roses had so much increased in number and kind, that twenty-six well-marked varieties, classed in eight sections, were described by Mr. Sabine. In 1841[182] it is said that three hundred varieties could be procured in the nursery-gardens near Glasgow; and these are described as blush, crimson, purple, red, marbled, two-coloured, white, and yellow, and as differing much in the size and shape of the flower.

---

[179] The Rev. W. F. Radclyffe, in 'Journal of Horticulture,' March 14, 1865, p. 207.

[180] 'Gardener's Chronicle,' 1861, p. 46.

[181] Mr. Sabine, in 'Transact. Hort. Soc.,' vol. iv. p. 285.

[182] 'An Encyclop. of Plants,' by J. C. Loudon, 1841, p. 443.

Pansy or Heartsease (*Viola tricolor*, &c.).—The history of this flower seems to be pretty well known; it was grown in Evelyn's garden in 1687; but the varieties were not attended to till 1810-1812, when Lady Monke, together with Mr. Lee the well-known nurseryman, energetically commenced their culture; and in the course of a few years twenty varieties could be purchased.[183] At about the same period, namely in 1813 or 1814, Lord Gambier collected some wild plants, and his gardener, Mr. Thomson, cultivated them together with some common garden varieties, and soon effected a great improvement. The first great change was the conversion of the dark lines in the centre of the flower into a dark eye or centre, which at that period had never been seen, but is now considered one of the chief requisites of a first-rate flower. In 1835 a book entirely devoted to this flower was published, and four hundred named varieties were on sale. From these circumstances this plant seemed to me worth studying, more especially from the great contrast between the small, dull, elongated, irregular flowers of the wild pansy, and the beautiful, flat, symmetrical, circular, velvet-like flowers, more than two inches in diameter, magnificently and variously coloured, which are exhibited at our shows. But when I came to inquire more closely, I found that, though the varieties were so modern, yet that much confusion and doubt prevailed about their parentage. Florists believe that the varieties [184] are descended from several wild stocks, namely, *V. tricolor*, *lutea*, *grandiflora*, *amœna*, and *Altaica*, more or less intercrossed. And when I looked to botanical works to ascertain whether these forms ought to be ranked as species, I found equal doubt and confusion. *Viola Altaica* seems to be a distinct form, but what part it has played in the origin of our varieties I know not; it is said to have been crossed with *V. lutea*. *Viola amœna* [185] is now looked at by all botanists as a natural variety of *V. grandiflora*; and this and *V. sudetica* have been proved to be identical with *V. lutea*. The latter and *V. tricolor* (including its admitted variety *V. arvensis*) are ranked as distinct species by Babington; and likewise by M. Gay,[186] who has paid particular attention to the genus; but the specific distinction between *V. lutea* and *tricolor* is chiefly grounded on the one being strictly and the other not strictly perennial, as well as on some other slight and unimportant differences in the form of the stem and stipules. Bentham unites these two forms; and a high authority on such matters, Mr. H. C. Watson,[187] says that, "while *V. tricolor* passes into *V. arvensis* on the one side, it approximates so much towards *V. lutea* and *V. Curtisii* on the other side, that a distinction becomes scarcely more easy between them."

---

[183] Loudon's 'Gardener's Magazine,' vol. xi. 1835, p. 427; also 'Journal of Horticulture,' April 14, 1863, p. 275.

[184] Loudon's 'Gardener's Magazine,' vol. viii. p. 575; vol. ix. p. 689.

[185] Sir J. E. Smith, 'English Flora,' vol. i. p. 306. H. C. Watson, 'Cybele Britannica,' vol. i. 1847, p. 181.

[186] Quoted from 'Annales des Sciences,' in the Companion to the 'Bot. Mag.,' vol. i. 1835, p. 159.

[187] 'Cybele Britannica,' vol. i. p. 173. *See* also Dr. Herbert on the changes of colour in transplanted specimens, and on the natural variations of V. grandiflora, in 'Transact. Hort. Soc.,' vol. iv. p. 19.

Hence, after having carefully compared numerous varieties, I gave up the attempt as too difficult for any one except a professed botanist. Most of the varieties present such inconstant characters, that when grown in poor soil, or when flowering out of their proper season, they produce differently coloured and much smaller flowers. Cultivators speak of this or that kind as being remarkably constant or true; but by this they do not mean, as in other cases, that the kind transmits its character by seed, but that the individual plant does not change much under culture. The principle of inheritance, however, does hold good to a certain extent even with the fleeting varieties of the Heartease, for to gain good sorts it is indispensable to sow the seed of good sorts. Nevertheless in every large seed-bed a few almost wild seedlings often reappear through reversion. On comparing the choicest varieties with the nearest allied wild forms, besides the difference in the size, outline, and colour of the flowers, the leaves are seen sometimes to differ in shape, as does the calyx occasionally in the length and breadth of the sepals. The differences in the form of the nectary more especially deserve notice; because characters derived from this organ have been much used in the discrimination of most of the species of Viola. In a large number of flowers compared in 1842 I found that in the greater number the nectary was straight; in others the extremity was a little turned upwards, or downwards, or inwards, so as to be completely hooked; in others, instead of being hooked, it was first turned rectangularly downwards, and then backwards and upwards; in others the extremity was considerably enlarged; and lastly, in some the basal part was depressed, becoming, as usual, laterally compressed towards the extremity. In a large number of flowers, on the other hand, examined by me in 1856 from a nursery-garden in a different part of England, the nectary hardly varied at all. Now M. Gay says that in certain districts, especially in Auvergne, the nectary of the wild *V. grandiflora* varies in the manner just described. Must we conclude from this that the cultivated varieties first mentioned were all descended from *V. grandiflora*, and that the second lot, though having the same general appearance, were descended from *V. tricolor*, of which the nectary, according to M. Gay, is subject to little variation? Or is it not more probable that both these wild forms would be found under other conditions to vary in the same manner and degree, thus showing that they ought not to be ranked as specifically distinct?

The *Dahlia* has been referred to by almost every author who has written on the variation of plants, because it is believed that all the varieties are descended from a single species, and because all have arisen since 1802 in France, and since 1804 in England.[188] Mr. Sabine remarks that "it seems as if some period of cultivation had been required before the fixed qualities of the native plant gave way and began to sport into those changes which now so delight us."[189] The flowers have been greatly modified in shape from a flat to a globular form. Anemone and ranun-

[188] Salisbury, in 'Transact. Hort. Soc.,' vol. i. 1812, pp. 84, 92. A semi-double variety was produced in Madrid in 1790.

[189] 'Transact. Hort. Soc.,' vol. iii. 1820, p. 225.

culus-like races,[190] which differ in the form and arrangement of the florets, have arisen; also dwarfed races, one of which is only eighteen inches in height. The seeds vary much in size. The petals are uniformly coloured or tipped or striped, and present an almost infinite diversity of tints. Seedlings of fourteen different colours[191] have been raised from the same plant; yet, as Mr. Sabine has remarked, "many of the seedlings follow their parents in colour." The period of flowering has been considerably hastened, and this has probably been effected by continued selection. Salisbury, writing 1808, says that they then flowered from September to November; in 1828 some new dwarf varieties began flowering in June;[192] and Mr. Grieve informs me that the dwarf purple Zelinda in his garden is in full bloom by the middle of June and sometimes even earlier. Slight constitutional differences have been observed between certain varieties: thus, some kinds succeed much better in one part of England than in another;[193] and it has been noticed that some varieties require much more moisture than others.[194]

Such flowers as the carnation, common tulip, and hyacinth, which are believed to be descended, each from a single wild form, present innumerable varieties, differing almost exclusively in the size, form, and colour of the flowers. These and some other anciently cultivated plants which have been long propagated by offsets, pipings, bulbs, &c., become so excessively variable, that almost each new plant raised from seed forms a new variety, "all of which to describe particularly," as old Gerarde wrote in 1597, "were to roll Sisyphus's stone, or to number the sands."

Hyacinth (*Hyacinthus orientalis*).—It may, however, be worth while to give a short account of this plant, which was introduced into England in 1596 from the Levant.[195] The petals of the original flower, says Mr. Paul, were narrow, wrinkled, pointed, and of a flimsy texture; now they are broad, smooth, solid, and rounded. The erectness, breadth, and length of the whole spike, and the size of the flowers, have all increased. The colours have been intensified and diversified. Gerarde, in 1597, enumerates four, and Parkinson, in 1629, eight varieties. Now the varieties are very numerous, and they were still more numerous a century ago. Mr. Paul remarks that "it is interesting to compare the Hyacinths of 1629 "with those of 1864, and to mark the improvement. Two hundred and "thirty-five years have elapsed since then, and this simple flower serves well "to illustrate the great fact that the original forms of nature do not remain "fixed and stationary, at least when brought under cultivation. While "looking at the extremes, we must not however forget that there are inter-"mediate stages which are for the most part lost to us. Nature will some-

---

[190] Loudon's 'Gardener's Mag.,' vol. vi. 1830, p. 77.

[191] Loudon's 'Encyclop. of Gardening,' p. 1035.

[192] 'Transact. Hort. Soc.,' vol. i. p. 91; and Loudon's 'Gardener's Mag.,' vol. iii. 1828, p. 179.

[193] Mr. Wildman, in 'Gardener's Chron.,' 1843, p. 87.

[194] 'Cottage Gardener,' April 8, 1856, p. 33.

[195] The best and fullest account of this plant which I have met with is by a famous horticulturist, Mr. Paul of Waltham, in the 'Gardener's Chronicle,' 1864, p. 342.

"times indulge herself with a leap, but as a rule her march is slow and "gradual." He adds that the cultivator should have "in his mind an "ideal of beauty, for the realisation of which he works with head and "hand." We thus see how clearly Mr. Paul, an eminently successful cultivator of this flower, appreciates the action of methodical selection.

In a curious and apparently trustworthy treatise, published at Amsterdam[196] in 1768, it is stated that nearly 2000 sorts were then known; but in 1864 Mr. Paul found only 700 in the largest garden at Haarlem. In this treatise it is said that not an instance is known of any one variety reproducing itself truly by seed: the white kinds, however, now[197] almost always yield white hyacinths, and the yellow kinds come nearly true. The hyacinth is remarkable from having given rise to varieties with bright blue, pink, and distinctly yellow flowers. These three primary colours do not occur in the varieties of any other species; nor do they often all occur even in the distinct species of the same genus. Although the several kinds of hyacinths differ but slightly from each other except in colour, yet each kind has its own individual character, which can be recognised by a highly educated eye; thus the writer of the Amsterdam treatise asserts (p. 43) that some experienced florists, such as the famous G. Voorholm, seldom failed in a collection of above twelve hundred sorts to recognise each variety by the bulb alone! This same writer mentions some few singular variations: for instance, the hyacinth commonly produces six leaves, but there is one kind (p. 35) which scarcely ever has more than three leaves; another never more than five; whilst others regularly produce either seven or eight leaves. A variety, called la Coriphée, invariably produces (p. 116) two flower-stems, united together and covered by one skin. The flower-stem in another kind (p. 128) comes out of the ground in a coloured sheath, before the appearance of the leaves, and is consequently liable to suffer from frost. Another variety always pushes a second flower-stem after the first has begun to develop itself. Lastly, white hyacinths with red, purple, or violet centres (p. 129) are the most liable to rot. Thus, the hyacinth, like so many previous plants, when long cultivated and closely watched, is found to offer many singular variations.

In the two last chapters I have given in some detail the range of variation, and the history, as far as known, of a considerable number of plants, which have been cultivated for various purposes. But some of the most variable plants, such as Kidney-beans, Capsicum, Millets, Sorghum, &c., have been passed over; for botanists are not agreed which kinds ought to rank as species and which as varieties; and the wild parent-species are unknown.[198] Many plants long cultivated in tropical

196 'Des Jacinthes, de leur Anatomie, Reproduction, et Culture,' Amsterdam, 1768.

197 Alph. de Candolle, 'Géograph. Bot.,' p. 1082.

198 Alph. de Candolle, 'Géograph. Bot.,' p. 983.

countries, such as the Banana, have produced numerous varieties; but as these have never been described with even moderate care, they also are here passed over. Nevertheless a sufficient, and perhaps more than sufficient, number of cases have been given, so that the reader may be enabled to judge for himself on the nature and extent of the variation which cultivated plants have undergone.

CHAPTER XI.

ON BUD-VARIATION, AND ON CERTAIN ANOMALOUS MODES OF REPRODUCTION AND VARIATION.

BUD-VARIATIONS IN THE PEACH, PLUM, CHERRY, VINE, GOOSEBERRY, CURRANT, AND BANANA, AS SHOWN BY THE MODIFIED FRUIT — IN FLOWERS : CAMELLIAS, AZALEAS, CHRYSANTHEMUMS, ROSES, ETC. — ON THE RUNNING OF THE COLOUR IN CARNATIONS— BUD-VARIATIONS IN LEAVES —VARIATIONS BY SUCKERS, TUBERS, AND BULBS — ON THE BREAKING OF TULIPS — BUD-VARIATIONS GRADUATE INTO CHANGES CONSEQUENT ON CHANGED CONDITIONS OF LIFE — CYTISUS ADAMI, ITS ORIGIN AND TRANSFORMATION — ON THE UNION OF TWO DIFFERENT EMBRYOS IN ONE SEED — THE TRIFACIAL ORANGE — ON REVERSION BY BUDS IN HYBRIDS AND MONGRELS — ON THE PRODUCTION OF MODIFIED BUDS BY THE GRAFTING OF ONE VARIETY OR SPECIES ON ANOTHER — ON THE DIRECT OR IMMEDIATE ACTION OF FOREIGN POLLEN ON THE MOTHER-PLANT — ON THE EFFECTS IN FEMALE ANIMALS OF A FIRST IMPREGNATION ON THE SUBSEQUENT OFFSPRING — CONCLUSION AND SUMMARY.

THIS chapter will be chiefly devoted to a subject in many respects important, namely, bud-variation. By this term I include all those sudden changes in structure or appearance which occasionally occur in full-grown plants in their flower-buds or leaf-buds. Gardeners call such changes "Sports;" but this, as previously remarked, is an ill-defined expression, as it has often been applied to strongly marked variations in seedling plants. The difference between seminal and bud reproduction is not so great as it at first appears; for each bud is in one sense a new and distinct individual; but such individuals are produced through the formation of various kinds of buds without the aid of any special apparatus, whilst fertile seeds are produced by the concourse of the two sexual elements. The modifications which arise through bud-variation can generally be propagated to any extent by grafting, budding, cuttings, bulbs, &c., and occasionally even by seed. Some few of our most beautiful and useful productions have arisen by bud-variation.

Bud-variations have as yet been observed only in the vegetable kingdom; but it is probable that if compound animals, such as corals, &c., had been subjected to a long course of domestication, they would have varied by buds; for they resemble plants in many respects. Thus any new or peculiar character presented by a compound animal is propagated by budding, as occurs with differently coloured Hydras, and as Mr. Gosse has shown to be the case with a singular variety of a true coral. Varieties of the Hydra have also been grafted on other varieties, and have retained their character.

I will in the first place give all the cases of bud-variations which I have been able to collect, and afterwards show their importance. These cases prove that those authors who, like Pallas, attribute all variability to the crossing either of distinct races, or of individuals belonging to the same race but somewhat different from each other, are in error; as are those authors who attribute all variability to the mere act of sexual union. Nor can we account in all cases for the appearance through bud-variation of new characters by the principle of reversion to long-lost characters. He who wishes to judge how far the conditions of life directly cause each particular variation ought to reflect well on the cases immediately to be given. I will commence with bud-variations, as exhibited in the fruit, and then pass on to flowers, and finally to leaves.

Peach (*Amygdalus Persica*).—In the last chapter I gave two cases of a peach-almond and double-flowered almond which suddenly produced fruit closely resembling true peaches. I have also recorded many cases of peach-trees producing buds, which, when developed into branches, have yielded nectarines. We have seen that no less than six named and several unnamed varieties of the peach have thus produced several varieties of nectarine. I have shown that it is highly improbable that all these peach-trees, some of which are old varieties, and have been propagated by the million, are hybrids from the peach and nectarine, and that it is opposed to all analogy to attribute the occasional production of nectarines on peach-trees to the direct action of pollen from some neighbouring nectarine-tree. Several of the cases are highly remarkable, because, firstly, the fruit thus produced has sometimes been in part a nectarine and in part a peach; secondly, because nectarines thus suddenly produced have reproduced themselves by seed; and thirdly, because nectarines are produced from peach-trees from seed as well as from buds. The seed of the nectarine, on the other hand, occasionally produces peaches; and we have seen in one instance that a nectarine-tree yielded peaches by bud-variation. As the peach is certainly the oldest or primary variety, the production of peaches from nectarines, either by seeds or buds, may perhaps be considered as a case of reversion. Certain trees have also been described as indifferently bearing peaches or nectarines, and this may be considered as bud-variation carried to an extreme degree.

The *grosse mignonne* peach at Montreuil produced "from a sporting branch" the *grosse mignonne tardive*, "a most excellent variety," which ripens its fruit a fortnight later than the parent tree, and is equally good.[1] This same peach has likewise produced by bud-variation the *early grosse mignonne*. Hunt's large tawny nectarine "originated from Hunt's small tawny nectarine, but not through seminal reproduction" [2]

Plums.—Mr. Knight states that a tree of the yellow magnum bonum plum, forty years old, which had always borne ordinary fruit, produced a branch which yielded red magnum bonums.[3] Mr. Rivers, of Sawbridgeworth, informs me (Jan. 1863) that a single tree out of 400 or 500 trees of the Early Prolific plum, which is a purple kind, descended from an old French variety bearing purple fruit, produced when about ten years old bright yellow plums; these differed in no respect except colour from those on the other trees, but were unlike any other known kind of yellow plum.[4]

Cherry (*Prunus cerasus*).—Mr. Knight has recorded (*idem*) the case of a branch of a May-Duke cherry, which, though certainly never grafted, always produced fruit, ripening later, and more oblong, than the fruit on the other branches. Another account has been given of two May-Duke cherry-trees in Scotland, with branches bearing oblong, and very fine fruit, which invariably ripened, as in Knight's case, a fortnight later than the other cherries.[5]

Grapes (*Vitis vinifera*).—The black or purple Frontignan in one case produced during two successive years (and no doubt permanently) spurs which bore white Frontignan grapes. In another case, on the same footstalk, the lower berries "were well-coloured black Frontignans; those next the stalk were white, with the exception of one black and one streaked berry;" and altogether there were fifteen black and twelve white berries on the same stalk. In another kind of grape black and amber-coloured berries were produced in the same cluster.[6] Count Odart describes a variety which often bears on the same stalk small round and large oblong berries; though the shape of the berry is generally a fixed character.[7] Here is another striking case given on the excellent authority of M. Carrière:[8] "a black Hamburgh grape (Frankenthal) was cut down, and produced three suckers; one of these was layered, and after a time produced much smaller berries, which always ripened at least a fortnight

[1] 'Gardener's Chron.,' 1854, p. 821.

[2] 'Lindley's Guide to Orchard,' as quoted in 'Gard. Chronicle,' 1852, p. 821. For the *Early mignonne peach*, *see* 'Gardener's Chron.,' 1864, p. 1251.

[3] 'Transact. Hort. Soc.,' vol. ii. p. 160.

[4] *See* also 'Gardener's Chron.,' 1863, p. 27.

[5] 'Gard. Chron.,' 1852, p. 821.

[6] 'Gardener's Chron.,' 1852, p. 629; 1856, p. 648; 1864, p. 986. Other cases are given by Braun, 'Rejuvenescence,' in 'Ray Soc. Bot. Mem.,' 1853, p. 314.

[7] 'Ampélographie,' &c., 1849, p. 71.

[8] 'Gardener's Chronicle,' 1866, p. 970.

earlier than the others. Of the remaining two suckers, one produced every year fine grapes, whilst the other, although it set an abundance of fruit, matured only a few, and these of inferior quality.

Gooseberry (*Ribes grossularia*).—A remarkable case has been described by Dr. Lindley[9] of a bush which bore at the same time no less than four kinds of berries, namely, hairy and red,—smooth, small and red,—green,—and yellow tinged with buff; the two latter kinds had a different flavour from the red berries, and their seeds were coloured red. Three twigs on this bush grew close together; the first bore three yellow berries and one red; the second twig bore four yellow and one red; and the third four red and one yellow. Mr. Laxton also informs me that he has seen a Red Warrington gooseberry bearing both red and yellow fruit on the same branch.

Currant (*Ribes rubrum*).—A bush purchased as the Champagne, which is a variety that bears blush-coloured fruit intermediate between red and white, produced during fourteen years, on separate branches and mingled on the same branch, berries of the red, white, and champagne kinds.[10] The suspicion naturally arises that this variety may have originated from a cross between a red and white variety, and that the above transformation may be accounted for by reversion to both parent-forms; but from the foregoing complex case of the gooseberry this view is doubtful. In France, a branch of a red-currant bush, about ten years old, produced near the summit five white berries, and lower down, amongst the red berries, one berry half red and half white.[11] Alexander Braun[12] also has often seen branches bearing red berries on white currants.

Pear (*Pyrus communis*).—Dureau de la Malle states that the flowers on some trees of an ancient variety, the *doyenné galeux*, were destroyed by frost: other flowers appeared in July, which produced six pears; these exactly resembled in their skin and taste the fruit of a distinct variety, the *gros doyenné blanc*, but in shape were like the *bon-chrétien*: it was not ascertained whether this new variety could be propagated by budding or grafting. The same author grafted a *bon-chrétien* on a quince, and it produced, besides its proper fruit, an apparently new variety, of a peculiar form, with thick and rough skin.[13]

Apple (*Pyrus malus*).—In Canada, a tree of the variety called Pound Sweet, produced,[14] between two of its proper fruit, an apple which was well russetted, small in size, different in shape, and with a short peduncle. As no russet apple grew anywhere near, this case apparently cannot be accounted for by the direct action of foreign pollen. I shall hereafter give

---

[9] 'Gardener's Chronicle,' 1855, pp. 597, 612.

[10] 'Gardener's Chron.,' 1842, p. 873; 1855, p. 646. In the 'Chronicle,' 1866, p. 876, Mr. P. Mackenzie states that the bush still continues to bear the three kinds of fruit, "although they have not been every year alike."

[11] 'Revue Horticole,' quoted in 'Gard. Chronicle,' 1844, p. 87.

[12] 'Rejuvenescence in Nature,' 'Bot. Memoirs Ray Soc.,' 1853, p. 314.

[13] 'Comptes Rendus,' tom. xli., 1855, p. 804. The second case is given on the authority of Gaudichaud, idem, tom. xxxiv., 1852, p. 748.

[14] This case is given in the 'Gard. Chronicle,' 1867, p. 403.

cases of apple-trees which regularly produce fruit of two kinds, or half-and-half fruit; these trees are generally supposed, and probably with truth, to be of crossed parentage, and that the fruit reverts to both parent-forms.

Banana (*Musa sapientium*).—Sir R. Schomburgk states that he saw in St. Domingo a raceme on the Fig Banana which bore towards the base 125 fruits of the proper kind; and these were succeeded, as is usual, higher up the raceme, by barren flowers, and these by 420 fruits, having a widely different appearance, and ripening earlier than the proper fruit. The abnormal fruit closely resembled, except in being smaller, that of the *Musa Chinensis* or *Cavendishii*, which has generally been ranked as a distinct species.[15]

FLOWERS.—Many cases have been recorded of a whole plant, or single branch, or bud, suddenly producing flowers different from the proper type in colour, form, size, doubleness, or other character. Half the flower, or a smaller segment, sometimes changes colour.

Camellia.—The myrtle-leaved species (*C. myrtifolia*), and two or three varieties of the common species, have been known to produce hexagonal and imperfectly quadrangular flowers; and the branches producing such flowers have been propagated by grafting.[16] The Pompone variety often bears "four distinguishable kinds of flowers,—the pure white and the "red-eyed, which appear promiscuously; the brindled pink and the rose-"coloured, which may be kept separate with tolerable certainty by "grafting from the branches that bear them." A branch, also, on an old tree of the rose-coloured variety has been seen to "revert to the pure "white colour, an occurrence less common than the departure from it."[17]

Cratægus oxycantha.—A dark pink hawthorn has been known to throw out a single tuft of pure white blossoms;[18] and Mr. A. Clapham, nursery-man, of Bradford, informs me that his father had a deep crimson thorn grafted on a white thorn, which, during several years, always bore, high above the graft, bunches of white, pink, and deep crimson flowers.

Azalea Indica is well known often to produce by buds new varieties. I have myself seen several cases. A plant of *Azalea Indica variegata* has been exhibited bearing a truss of flowers of *A. Ind. Gledstanesii* "as true as could possibly be produced, thus evidencing the origin of that fine variety." On another plant of *A. Ind. variegata* a perfect flower of *A. Ind. lateritia* was produced; so that both *Gledstanesii* and *lateritia* no doubt originally appeared as sporting branches of *A. Ind. variegata*.[19]

Cistus tricuspis.—A seedling of this plant, when some years old, produced, at Saharunpore,[20] some branches "which bore leaves and flowers widely different from the normal form." "The abnormal leaf is much less

---

15 'Journal of Proc. Linn. Soc.,' vol. ii. Botany, p. 131.

16 'Gard. Chronicle,' 1847, p. 207.

17 Herbert, 'Amaryllidaceæ,' 1838, p. 369.

18 'Gardener's Chronicle,' 1843, p. 391.

19 Exhibited at Hort. Soc., London. Report in 'Gardener's Chron.,' 1844, p. 337.

20 Mr. W. Bell, Bot. Soc. of Edinburgh, May, 1863.

divided, and not acuminated. The petals are considerably larger, and quite entire. There is also in the fresh state a conspicuous, large, oblong gland, full of a viscid secretion, on the back of each of the calycine segments."

Althœa rosea.—A double yellow Hollyock suddenly turned one year into a pure white single kind; subsequently a branch bearing the original double yellow flowers reappeared in the midst of the branches of the single white kind.[21]

Pelargonium.—These highly cultivated plants seem eminently liable to bud-variation. I will give only a few well-marked cases. Gärtner has seen[22] a plant of *P. zonale* with a branch having white-edged leaves, which remained constant for years, and bore flowers of a deeper red than usual. Generally speaking, such branches present little or no difference in their flowers: thus a writer[23] pinched off the leading shoot of a seedling *P. zonale,* and it threw out three branches, which differed in the size and colour of their leaves and stems; but on all three branches "the flowers were identical," except in being largest in the green-stemmed variety, and smallest in that with variegated foliage: these three varieties were subsequently propagated and distributed. Many branches, and some whole plants, of a variety called *compactum,* which bears orange-scarlet flowers, have been seen to produce pink flowers.[24] Hill's Hector, which is a pale red variety, produced a branch with lilac flowers, and some trusses with both red and lilac flowers. This apparently is a case of reversion, for Hill's Hector was a seedling from a lilac variety.[25] Of all Pelargoniums, Rollisson's Unique seems to be the most sportive; its origin is not positively known, but is believed to be from a cross. Mr. Salter, of Hammersmith, states[26] that he has himself known this purple variety to produce the lilac, the rose-crimson or *conspicuum,* and the red or *coccineum* varieties; the latter has also produced the *rose d'amour;* so that altogether four varieties have originated by bud variation from Rollisson's Unique. Mr. Salter remarks that these four varieties "may now be considered as fixed, although they " occasionally produce flowers of the original colour. This year *coccineum* " has pushed flowers of three different colours, red, rose, and lilac, upon " the same truss, and upon other trusses are flowers half red and half "lilac." Besides these four varieties, two other scarlet Uniques are known to exist, both of which occasionally produce lilac flowers identical with Rollisson's Unique;[27] but one at least of these did not arise through bud-variation, but is believed to be a seedling from Rollisson's Unique.[28] There are, also, in the trade[29] two other slightly different varieties, of unknown origin, of Rollisson's Unique: so that altogether we have a curiously complex case

---

[21] 'Revue Horticole,' quoted in 'Gard. Chron.,' 1845, p. 475.

[22] 'Bastarderzeugung,' 1849, s. 76.

[23] 'Journal of Horticulture,' 1861, p. 336.

[24] W. P. Ayres, in 'Gardener's Chron.,' 1842, p. 791.

[25] W. P. Ayres, idem.

[26] 'Gardener's Chron.,' 1861, p. 968.

[27] Idem, 1861, p. 945.

[28] W. Paul, in 'Gardener's Chron.,' 1861, p. 968.

[29] Idem, p. 945.

of variation both by buds and seeds.[30] An English wild plant, the *Geranium pratense*, when cultivated in a garden, has been seen to produce on the same plant both blue and white, and striped blue and white flowers.[31]

Chrysanthemum.—This plant frequently sports, both by its lateral branches and occasionally by suckers. A seedling raised by Mr. Salter has produced by bud-variation six distinct sorts, five different in colour and one in foliage, all of which are now fixed.[32] The varieties which were first introduced from China were so excessively variable, "that it was extremely difficult to tell which was the original colour of the variety, and which was the sport." The same plant would produce one year only buff-coloured, and next year only rose-coloured flowers; and then would change again, or produce at the same time flowers of both colours. These fluctuating varieties are now all lost, and, when a branch sports into a new variety, it can generally be propagated and kept true; but, as Mr. Salter remarks, "every sport should be thoroughly "tested in different soils before it can be really considered as fixed, as many "have been known to run back when planted in rich compost; but when "sufficient care and time are expended in proving, there will exist little "danger of subsequent disappointment." Mr. Salter informs me that with all the varieties the commonest kind of bud-variation is the production of yellow flowers, and, as this is the primordial colour, these cases may be attributed to reversion. Mr. Salter has given me a list of seven differently coloured chrysanthemums, which have all produced branches with yellow flowers; but three of them have also sported into other colours. With any change of colour in the flower, the foliage generally changes in a corresponding manner in lightness or darkness.

Another Compositous plant, namely, *Centauria cyanus*, when cultivated in a garden, not unfrequently produces on the same root flowers of four different colours, viz., blue, white, dark-purple, and particoloured.[33] The flowers of Anthemis also vary on the same plant.[34]

Roses.—Many varieties of the rose are known or are believed to have originated by bud-variation.[35] The common double moss-rose was imported into England from Italy about the year 1735.[36] Its origin is unknown, but from analogy it probably arose from the Provence rose (*R. centifolia*) by bud-variation; for branches of the common moss-rose have several times been known to produce Provence roses, wholly or partially destitute of moss: I have seen one such instance, and several others have been recorded.[37]

---

[30] For other cases of bud-variation in this same variety, see 'Gardener's Chron.,' 1861, pp. 578, 600, 925. For other distinct cases of bud-variation in the genus Pelargonium, *see* 'Cottage Gardener,' 1860, p. 194.

[31] Rev. W. T. Bree, in Loudon's 'Gard. Mag.,' vol. viii., 1832, p. 93.

[32] 'The Chrysanthemum, its History and Culture,' by J. Salter, 1865, p. 41, &c.

[33] Bree, in Loudon's 'Gard. Mag.,' vol. viii., 1832, p. 93.

[34] Bronn, 'Geschichte der Natur,' B. ii. s. 123.

[35] T. Rivers, 'Rose Amateur's Guide,' 1837, p. 4.

[36] Mr. Shailer, quoted in 'Gardener's Chron.,' 1848, p. 759.

[37] 'Transact. Hort. Soc.,' vol. iv., 1822, p. 137; 'Gard. Chron.,' 1842, p. 422.

Mr. Rivers also informs me that he raised two or three roses of the Provence class from seed of the old single moss-rose;[38] and this latter kind was produced in 1807 by bud-variation from the common moss-rose. The white moss-rose was also produced in 1788 by an offset from the common red moss-rose: it was at first pale blush-coloured, but became white by continued budding. On cutting down the shoots which had produced this white moss-rose, two weak shoots were thrown up, and buds from these yielded the beautiful striped moss-rose. The common moss-rose has yielded by bud-variation, besides the old single red moss-rose, the old scarlet semi-double moss-rose, and the sage-leaf moss-rose, which "has a delicate shell-like form, and is of a beautiful blush colour; it is now (1852) nearly extinct."[39] A white moss-rose has been seen to bear a flower half white and half pink.[40] Although several moss-roses have thus certainly arisen by bud-variation, the greater number probably owe their origin to seed of moss-roses. For Mr. Rivers informs me that his seedlings from the old single moss-rose almost always produced moss-roses; and the old single moss-rose was, as we have seen, the product by bud-variation of the double moss-rose originally imported from Italy. That the original moss-rose was the product of bud-variation is probable, from the facts above given and from the moss-rose de Meaux (also a var. of *R. centifolia*)[41] having appeared as a sporting branch on the common rose de Meaux.

Prof. Caspary has carefully described[42] the case of a six-year-old white moss-rose, which sent up several suckers, one of which was thorny, and produced red flowers, destitute of moss, exactly like those of the Provence rose (*R. centifolia*): another shoot bore both kinds of flowers and in addition longitudinally striped flowers. As this white moss-rose had been grafted on the Provence rose, Prof. Caspary attributes the above changes to the influence of the stock; but from the facts already given, and from others to be given, bud-variation, with reversion, is probably a sufficient explanation.

Many other instances could be added of roses varying by buds. The white Provence rose apparently thus originated.[43] The double and highly-coloured Belladonna rose has been known[44] to produce by suckers both semi-double and almost single white roses; whilst suckers from one of these semi-double white roses reverted to perfectly characterised Belladonnas. Varieties of the China rose propagated by cuttings in St. Domingo often revert after a year or two into the old China rose.[45] Many cases

[38] *See* also Loudon's 'Arboretum,' vol. ii. p. 780.

[39] All these statements on the origin of the several varieties of the moss-rose are given on the authority of Mr. Shailer, who, together with his father, was concerned in their original propagation, in 'Gard. Chron.,' 1852, p. 759.

[40] 'Gard. Chron.,' 1845, p. 564.

[41] 'Transact. Hort. Soc., vol. ii. p. 242.

[42] 'Schriften der Phys. Ökon. Gesell. zu Königsberg,' Feb. 3, 1865, s. 4. *See* also Dr. Caspary's paper in 'Transactions of the Hort. Congress of Amsterdam,' 1865.

[43] 'Gard. Chron.,' 1852, p. 759.

[44] 'Transact. Hort. Soc.,' vol. ii. p. 242.

[45] Sir R. Schomburgk, 'Proc. Linn. Soc. Bot.,' vol. ii. p. 132.

have been recorded of roses suddenly becoming striped or changing their character by segments: some plants of the Comtesse de Chabrillant, which is properly rose-coloured, were exhibited in 1862,[46] with crimson flakes on a rose ground. I have seen the Beauty of Billiard with a quarter and with half the flower almost white. The Austrian bramble (*R. lutea*) not rarely[47] produces branches with pure yellow flowers; and Prof. Henslow has seen exactly half the flower of a pure yellow, and I have seen narrow yellow streaks on a single petal, of which the rest was of the usual copper colour.

The following cases are highly remarkable. Mr. Rivers, as I am informed by him, possessed a new French rose with delicate smooth shoots, pale glaucous-green leaves, and semi-double pale flesh-coloured flowers striped with dark red; and on branches thus characterised there suddenly appeared, in more than one instance, the famous old rose called the Baronne Prevost, with its stout thorny shoots, and immense, uniformly and richly coloured, double flowers; so that in this case the shoots, leaves, and flowers, all at once changed their character by bud-variation. According to M. Verlot[48] a variety called *Rosa cannabifolia*, which has peculiarly shaped leaflets, and differs from every member of the family in the leaves being opposite instead of alternate, suddenly appeared on a plant of *R. alba* in the gardens of the Luxembourg. Lastly, "a running shoot" was observed by Mr. H. Curtis[49] on the old Aimée Vibert Noisette, and he budded it on Celine; thus a climbing Aimée Vibert was first produced and afterwards propagated.

Dianthus.—It is quite common with the Sweet William (*D. barbatus*) to see differently coloured flowers on the same root; and I have observed on the same truss four differently coloured and shaded flowers. Carnations and pinks (*D. caryophyllus*, &c.) occasionally vary by layers; and some kinds are so little certain in character that they are called by floriculturists "catch-flowers."[50] Mr. Dickson has ably discussed the "running" of particoloured or striped carnations, and says it cannot be accounted for by the compost in which they are grown: "layers from the same clean flower would come "part of them clean and part foul, even when subjected to precisely the "same treatment; and frequently one flower alone appears influenced by "the taint, the remainder coming perfectly clean."[51] This running of the parti-coloured flowers apparently is a case of reversion by buds to the original uniform tint of the species.

I will briefly mention some other cases of bud-variation to show how many plants belonging to many orders have varied in their flowers; numerous cases might be added. I have seen on a snap-dragon (*Antirrhinum majus*) white, pink, and striped flowers on the same plant, and branches with striped flowers on a red-coloured variety. On a double stock (*Matthiola incana*) I have seen a branch bearing single flowers; and

46 'Gard. Chron.,' 1862, p. 619.

47 Hopkirk's 'Flora Anomala,' p. 167.

48 'Sur la Production et la Fixation des Variétés,' 1865, p. 4.

49 'Journal of Horticulture,' March, 1865, p. 233.

50 'Gard. Chron.,' 1843, p. 135.

51 Ibid., 1842, p. 55.

on a dingy-purple, double variety of the wall-flower (*Cheiranthus cheiri*) a branch which had reverted to the ordinary copper colour. On other branches of the same plant, some flowers were exactly divided across the middle, one half being purple and the other coppery; but some of the smaller petals towards the centre of these same flowers were purple longitudinally streaked with coppery colour, or coppery streaked with purple. A Cyclamen [52] has been observed to bear white and pink flowers of two forms, the one resembling the Persicum strain, and the other the Coum strain. *Oenothera biennis* has been seen [53] bearing flowers of three different colours. The hybrid *Gladiolus colvillii* occasionally bears uniformly coloured flowers, and one case is recorded [54] of all the flowers on a plant thus changing colour. A Fuchsia has been seen [55] bearing two kinds of flowers. *Mirabilis jalapa* is eminently sportive, sometimes bearing on the same root pure red, yellow, and white flowers, and others striped with various combinations of these three colours.[56] The plants of the Mirabilis which bear such extraordinarily variable flowers, in most, probably in all cases, owe their origin, as shown by Prof. Lecoq, to crosses between differently-coloured varieties.

Leaves and Shoots.—Changes, through bud-variation, in fruits and flowers have hitherto been treated of, but incidentally some remarkable modifications in the leaves and shoots of the rose and Cistus, and in a lesser degree in the foliage of the Pelargonium and Chrysanthemum, have been noticed. I will now add a few more cases of variation in leaf-buds. Verlot [57] states that on *Aralia trifoliata,* which properly has leaves with three leaflets, branches bearing simple leaves of various forms frequently appear; these can be propagated by buds or grafting, and have given rise, as he states, to several nominal species.

With respect to trees, the history of but few of the many varieties with curious or ornamental foliage is known; but several probably have originated by bud-variation. Here is one case:—An old ash-tree (*Fraxinus excelsior*) in the grounds of Necton, as Mr. Mason states, "for many years has had one bough of a totally different character to the rest of the tree, or of any other ash-tree which I have seen; being short-jointed and densely covered with foliage." It was ascertained that this variety could be propagated by grafts.[58] The varieties of some trees with cut leaves, as the oak-leaved laburnum, the parsley-leaved vine, and especially the fern-leaved beech, are apt to revert by buds to the common form.[59] The fern-like leaves of the beech sometimes revert only partially, and the branches display here and there sprouts bearing common leaves, fern-like, and variously shaped leaves. Such cases differ but little from the so-called

---

[52] 'Gard. Chron.,' 1867, p. 235.

[53] Gärtner, 'Bastarderzeugung,' s. 305.

[54] Mr. D. Beaton, in 'Cottage Gardener,' 1860, p. 250.

[55] 'Gard. Chron.,' 1850, p. 536.

[56] Braun, 'Ray Soc. Bot. Mem.,' 1853, p. 315; Hopkirk's 'Flora Anomala,' p. 164; Lecoq, 'Géograph. Bot. de l'Europe,' tom. iii., 1854, p. 405; and 'De la Fécondation,' 1862, p. 303.

[57] 'Des Variétés,' 1865, p. 5.

[58] W. Mason, in 'Gard. Chron.,' 1843, p. 878.

[59] Alex. Braun, 'Ray Soc. Bot. Mem.,' 1853, p. 315; 'Gard. Chron.,' 1841, p. 329.

heterophyllous varieties, in which the tree habitually bears leaves of various forms; but it is probable that most heterophyllous trees have originated as seedlings. There is a sub-variety of the weeping willow with leaves rolled up into a spiral coil; and Mr. Masters states that a tree of this kind kept true in his garden for twenty-five years, and then threw out a single upright shoot bearing flat leaves.[60]

I have often noticed single twigs and branches on beech and other trees with their leaves fully expanded before those on the other branches had opened; and as there was nothing in their exposure or character to account for this difference, I presume that they had appeared as bud-variations, like the early and late fruit-maturing varieties of the peach and nectarine.

Cryptogamic plants are liable to bud-variation, for fronds on the same fern are often seen to display remarkable deviations of structure. Spores, which are of the nature of buds, taken from such abnormal fronds, reproduce, with remarkable fidelity, the same variety, after passing through the sexual stage.[61]

With respect to colour, leaves often become by bud-variation zoned, blotched, or spotted with white, yellow, and red; and this occasionally occurs even with plants in a state of nature. Variegation, however, appears still more frequently in plants produced from seed; even the cotyledons or seed-leaves being thus affected.[62] There have been endless disputes whether variegation should be considered as a disease. In a future chapter we shall see that it is much influenced, both in the case of seedlings and of mature plants, by the nature of the soil. Plants which have become variegated as seedlings, generally transmit their character by seed to a large proportion of their progeny; and Mr. Salter has given me a list of eight genera in which this occurred.[63] Sir F. Pollock has given me more precise information: he sowed seed from a variegated plant of *Ballota nigra* which was found growing wild, and thirty per cent. of the seedlings were variegated; seed from these latter being sown, sixty per cent. came up variegated. When branches become variegated by bud-variation, and the variety is attempted to be propagated by seed, the seedlings are rarely variegated; Mr. Salter found this to be the case with plants belonging to eleven genera, in which the greater number of the seedlings proved to be green-leaved; yet a few were slightly variegated, or were quite white, but none were worth keeping. Variegated plants, whether originally produced from seeds or buds, can generally be propagated by budding, grafting, &c.; but all are apt to revert by bud-variation to their ordinary foliage. This tendency, however, differs much in the varieties of even the same species; for instance, the golden-striped variety of *Euonymus Japonicus* "is very liable to run back to the green-leaved, while the silver-striped

---

60 Dr. M. T. Masters, 'Royal Institution Lecture,' March 16, 1860.

61 *See* Mr. W. K. Bridgman's curious paper in 'Annals and Mag. of Nat. Hist.,' December, 1861; also Mr. J. Scott, 'Bot. Soc. Edinburgh,' June 12, 1862.

62 'Journal of Horticulture,' 1861, p. 336: Verlot, 'Des Variétés,' p. 76.

63 *See* also Verlot, 'Des Variétés,' p. 74.

variety hardly ever changes."[64] I have seen a variety of the holly, with its leaves having a central yellow patch, which had everywhere partially reverted to the ordinary foliage, so that on the same small branch there were many twigs of both kinds. In the pelargonium, and in some other plants, variegation is generally accompanied by some degree of dwarfing, as is well exemplified in the "Dandy" pelargonium. When such dwarf varieties sport back by buds or suckers to the ordinary foliage, the dwarfed stature sometimes still remains.[65] It is remarkable that plants propagated from branches which have reverted from variegated to plain leaves[66] do not always (or never, as one observer asserts) perfectly resemble the original plain-leaved plant from which the variegated branch arose: it seems that a plant, in passing by bud-variation from plain leaves to variegated, and back again from variegated to plain, is generally in some degree affected so as to assume a slightly different aspect.

Bud-variation by Suckers, Tubers, and Bulbs.—All the cases hitherto given of bud-variation in fruits, flowers, leaves, and shoots, have been confined to buds on the stems or branches, with the exception of a few cases incidentally noticed of varying suckers in the rose, pelargonium, and chrysanthemum. I will now give a few instances of variation in subterranean buds, that is, by suckers, tubers, and bulbs; not that there is any essential difference between buds above and beneath the ground. Mr. Salter informs me that two variegated varieties of Phlox originated as suckers; but I should not have thought these worth mentioning, had not Mr. Salter found, after repeated trials, that he could not propagate them by "root-joints," whereas, the variegated *Tussilago farfara* can thus be safely propagated;[67] but this latter plant may have originated as a variegated seedling, which would account for its greater fixedness of character. The Barberry (*Berberis vulgaris*) offers an analogous case; there is a well-known variety with seedless fruit, which can be propagated by cuttings or layers; but suckers always revert to the common form, which produces fruit containing seeds.[68] My father repeatedly tried this experiment, and always with the same result.

Turning now to tubers: in the common Potato (*Solanum tuberosum*) a single bud or eye sometimes varies and produces a new variety; or, occasionally, and this is a much more remarkable circumstance, all the eyes in a tuber vary in the same manner and at the same time, so that the whole tuber assumes a new character. For instance, a single eye in a tuber of the

---

[64] 'Gard. Chron.,' 1844, p. 86.

[65] Ibid., 1861, p. 968.

[66] Ibid., 1861, p. 433. 'Cottage Gardener,' 1860, p. 2.

[67] M. Lemoine (quoted in 'Gard. Chron.,' 1867, p. 74) has lately observed that the Symphitum with variegated leaves cannot be propagated by division of the roots. He also found that out of 500 plants of a Phlox with striped flowers, which had been propagated by root-division, only seven or eight produced striped flowers. See also, on striped Pelargoniums, 'Gard. Chron.' 1867, p. 1000.

[68] Anderson's 'Recreations in Agriculture,' vol. v. p. 152.

old *Forty-fold potato*, which is a purple variety, was observed[69] to become white; this eye was cut out and planted separately, and the kind has since been largely propagated. *Kemp's Potato* is properly white, but a plant in Lancashire produced two tubers which were red, and two which were white; the red kind was propagated in the usual manner by eyes, and kept true to its new colour, and, being found a more productive variety, soon became widely known under the name of *Taylor's Forty-fold*.[70] The *Old Forty-fold* potato, as already stated, is a purple variety; but a plant long cultivated on the same ground produced, not as in the case above given a single white eye, but a whole white tuber, which has since been propagated and keeps true.[71] Several cases have been recorded of large portions of whole rows of potatoes slightly changing their character.[72]

Dahlias propagated by tubers under the hot climate of St. Domingo vary much; Sir R. Schomburgk gives the case of the "Butterfly variety," which the second year produced on the same plant "double and single flowers; "here white petals edged with maroon; there of a uniform deep maroon."[73] Mr. Bree also mentions a plant "which bore two different kinds of self-"coloured flowers, as well as a third kind which partook of both colours "beautifully intermixed."[74] Another case is described of a dahlia with purple flowers which bore a white flower streaked with purple.[75]

Considering how long and extensively many Bulbous plants have been cultivated, and how numerous are the varieties produced from seed, these plants have not varied so much by offsets,—that is, by the production of new bulbs,—as might have been expected. With the Hyacinth a case has been recorded of a blue variety which for three successive years gave offsets which produced white flowers with a red centre.[76] Another hyacinth has been described[77] as bearing on the same truss a perfectly pink and a perfectly blue flower.

Mr. John Scott informs me that in 1862 *Imatophyllum miniatum*, in the Botanic Gardens of Edinburgh, threw up a sucker which differed from the normal form, in the leaves being two-ranked instead of four-ranked. The leaves were also smaller, with the upper surface raised instead of being channelled.

In the propagation of *Tulips*, seedlings are raised, called *selfs* or *breeders*, which "consist of one plain colour on a white or yellow bottom. These, "being cultivated on a dry and rather poor soil, become broken or variegated "and produce new varieties. The time that elapses before they break "varies from one to twenty years or more, and sometimes this change "never takes place."[78] The various broken or variegated colours which give value to all tulips are due to bud-variation; for although the

69 'Gard. Chron.,' 1857, p. 662.

70 Ibid., 1841, p. 814.

71 Ibid., 1857, p. 613.

72 Ibid, 1857, p. 679. *See* also Phillips, 'Hist. of Vegetables,' vol. ii. p. 91, for other and similar accounts.

73 'Journal of Proc. Linn. Soc.,' vol. ii. Botany, p. 132.

74 Loudon's 'Gard. Mag.,' vol. viii., 1832, p. 94.

75 'Gard. Chron.,' 1850, p. 536; and 1842, p. 729.

76 'Des Jacinthes,' &c., Amsterdam, 1768, p. 122.

77 'Gard. Chron.,' 1845, p. 212.

78 Loudon's 'Encyclop. of Gardening,' p. 1024.

Bybloemens and some other kinds have been raised from several distinct breeders, yet all the Baguets are said to have come from a single breeder or seedling. This bud-variation, in accordance with the views of MM. Vilmorin and Verlot,[79] is probably an attempt to revert to that uniform colour which is natural to the species. A tulip, however, which has already become broken, when treated with too strong manure, is liable to flush or lose by a second act of reversion its variegated colours. Some kinds, as Imperatrix Florum, are much more liable than others to flushing; and Mr. Dickson maintains[80] that this can no more be accounted for than the variation of any other plant. He believes that English growers, from care in choosing seed from broken flowers instead of from plain flowers, have to a certain extent diminished the tendency in flowers already broken to flushing or secondary reversion.

During two consecutive years all the early flowers in a bed of *Tigridia conchiflora*[81] resembled those of the old *T. pavonia*; but the later flowers assumed their proper colour of fine yellow spotted with crimson. An apparently authentic account has been published[82] of two forms of Hemerocallis, which have been universally considered as distinct species, changing into each other; for the roots of the large-flowered tawny *H. fulva*, being divided and planted in a different soil and place, produced the small-flowered yellow *H. flava*, as well as some intermediate forms. It is doubtful whether such cases as these latter, as well as the "flushing" of broken tulips and the "running" of particoloured carnations,—that is, their more or less complete return to a uniform tint,—ought to be classed under bud-variation, or ought to be retained for the chapter in which I treat of the direct action of the conditions of life on organic beings. These cases, however, have this much in common with bud-variation, that the change is effected through buds and not through seminal reproduction. But, on the other hand, there is this difference—that in ordinary cases of bud-variation, one bud alone changes, whilst in the foregoing cases all the buds on the same plant were modified together; yet we have an intermediate case, for with the potato all the eyes in one tuber alone simultaneously changed their character.

I will conclude with a few allied cases, which may be ranked either under bud-variation, or under the direct action of the conditions of life. When the common Hepatica is transplanted from its native woods, the flowers change colour, even during the first year.[83] It is notorious that the improved varieties of the Heartsease (*Viola tricolor*) when transplanted often produce flowers widely different in size, form, and colour: for instance, I transplanted a large uniformly-coloured dark purple variety, whilst in full flower, and it then produced much smaller, more elongated flowers, with the lower petals yellow; these were succeeded by flowers marked with large purple spots, and ultimately, towards the end of the same summer, by the original large dark purple flowers. The slight changes which some

---

[79] 'Production des Variétés,' 1865, p. 63.

[80] 'Gard. Chron.,' 1841, p. 782; 1842, p. 55.

[81] 'Gard. Chron.,' 1849, p. 565.

[82] 'Transact. Linn. Soc.,' vol. ii. p. 354.

[83] Godron, 'De l'Espèce,' tom. ii. p. 84.

fruit-trees undergo from being grafted and regrafted on various stocks,[84] were considered by Andrew Knight[85] as closely allied to "sporting branches," or bud-variations. Again, we have the case of young fruit-trees changing their character as they grow old; seedling pears, for instance, lose with age their spines and improve in the flavour of their fruit. Weeping birch-trees, when grafted on the common variety, do not acquire a perfect pendulous habit until they grow old: on the other hand, I shall hereafter give the case of some weeping ashes which slowly and gradually assumed an upright habit of growth. All such changes, dependent on age, may be compared with the changes, alluded to in the last chapter, which many trees naturally undergo; as in the case of the Deodar and Cedar of Lebanon, which are unlike in youth and closely resemble each other in old age; and as with certain oaks, and with some varieties of the lime and hawthorn.[86]

Before giving a summary on Bud-variation I will discuss some singular and anomalous cases, which are more or less closely related to this same subject. I will begin with the famous case of Adam's laburnum or *Cytisus Adami*, a form or hybrid intermediate between two very distinct species, namely, *C. laburnum* and *purpureus*, the common and purple laburnum; but as this tree has often been described, I will be as brief as I can.

Throughout Europe, in different soils and under different climates, branches on this tree have repeatedly and suddenly reverted to both parent-species in their flowers and leaves. To behold mingled on the same tree tufts of dingy-red, bright yellow, and purple flowers, borne on branches having widely different leaves and manner of growth, is a surprising sight. The same raceme sometimes bears two kinds of flowers; and I have seen a single flower exactly divided into halves, one side being bright yellow and the other purple; so that one half of the standard-petal was yellow and of larger size, and the other half purple and smaller. In another flower the whole corolla was bright yellow, but exactly half the calyx was purple. In another, one of the dingy-red wing-petals had a bright yellow narrow stripe on it; and lastly, in another flower, one of the stamens, which had become slightly foliaceous, was half yellow and half purple; so that the tendency to segregation of character or reversion affects even single parts

[84] M. Carrière has lately described, in the 'Révue Horticole' (Dec. 1, 1866, p. 457), an extraordinary case. He twice inserted grafts of the *Aria vestita* on thorn-trees (*épines*) growing in pots; and the grafts, as they grew, produced shoots with bark, buds, leaves, petioles, petals, and flower-stalks all widely different from those of the Aria. The grafted shoots were also much hardier, and flowered earlier, than those on the ungrafted Aria.

[85] 'Transact. Hort. Soc.,' vol. ii. p. 160.

[86] For the cases of oaks *see* Alph. De Candolle in 'Bibl. Univers.,' Geneva, Nov. 1862; for limes, &c., Loudon's 'Gard. Mag.,' vol. xi., 1835, p. 503.

and organs.[87] The most remarkable fact about this tree is that in its intermediate state, even when growing near both parent-species, it is quite sterile; but when the flowers become pure yellow or pure purple they yield seed. I believe that the pods from the yellow flowers yield a full complement of seed; they certainly yield a large number. Two seedlings raised by Mr. Herbert from such seed[88] exhibited a purple tinge on the stalks of their flowers; but several seedlings raised by myself resembled in every character the common laburnum, with the exception that some of them had remarkably long racemes: these seedlings were perfectly fertile. That such purity of character and fertility should be suddenly reacquired from so hybridized and sterile a form is an astonishing phenomenon. The branches with purple flowers appear at first sight exactly to resemble those of *C. purpureus*; but on careful comparison I found that they differed from the pure species in the shoots being thicker, the leaves a little broader, and the flowers slightly shorter, with the corolla and calyx less brightly purple: the basal part of the standard-petal also plainly showed a trace of the yellow stain. So that the flowers, at least in this instance, had not perfectly recovered their true character; and in accordance with this, they were not perfectly fertile, for many of the pods contained no seed, some produced one, and very few contained as many as two seeds; whilst numerous pods on a tree of the pure *C. purpureus* in my garden contained three, four, and five fine seeds. The pollen, moreover, was very imperfect, a multitude of grains being small and shrivelled; and this is a singular fact; for, as we shall immediately see, the pollen-grains in the dingy-red and sterile flowers on the parent-tree, were, in external appearance, in a much better state, and included very few shrivelled grain. Although the pollen of the reverted purple flowers was in so poor a condition, the ovules were well-formed, and, when mature, germinated freely with me. Mr. Herbert also raised plants from seeds of the reverted purple flowers, and they differed *very little* from the usual state of *C. purpureus*; but this expression shows that they had not perfectly recovered their proper character.

Prof. Caspary has examined the ovules of the dingy-red and sterile flowers in several plants of *C. adami* on the Continent,[89] and finds them generally monstrous. In three plants examined by me in England, the ovules were likewise monstrous, the nucleus varying much in shape, and projecting irregularly beyond the proper coats. The pollen-grains, on the other hand, judging from their external appearance, were remarkably good, and readily protruded their tubes. By repeatedly counting, under the microscope, the proportional number of bad grains, Prof. Caspary ascertained that only 2·5 per cent. were bad, which is a less proportion than in the pollen of three pure species of Cytisus in their cultivated state, viz. *C. purpureus, laburnum*, and *alpinus*. Although the pollen of *C. adami* is thus in appearance good, it does not follow, according

[87] For analogous facts, *see* Braun, 'Rejuvenescence,' in 'Ray Soc. Bot. Mem.,' 1853, p. 320; and 'Gard. Chron.,' 1842, p. 397.

[88] 'Journal of Hort. Soc.,' vol. ii., 1847, p. 100.

[89] *See* 'Transact. of Hort. Congress of Amsterdam,' 1865; but I owe most of the following information to Prof. Caspary's letters.

to M. Naudin's observations[90] on Mirabilis, that it would be functionally effective. The fact of the ovules of *C. adami* being monstrous, and the pollen apparently sound, is all the more remarkable, because it is opposed to what usually occurs not only with most hybrids,[91] but with two hybrids in the same genus, namely in *C. purpureo-elongatus*, and *C. alpino-laburnum*. In both these hybrids, the ovules, as observed by Prof. Caspary and myself, were well-formed, whilst many of the pollen-grains were ill-formed; in the latter hybrid 20·3 per cent., and in the former no less than 84·8 per cent. of the grains were ascertained by Prof. Caspary to be bad. This unusual condition of the male and female reproductive elements in *C. adami* has been used by Prof. Caspary as an argument against this plant being considered as an ordinary hybrid produced from seed; but we should remember that with hybrids the ovules have not been examined nearly so frequently as the pollen, and they may be much oftener imperfect than is generally supposed. Dr. E. Bornet, of Antibes, informs me (through Mr. J. Traherne Moggridge) that with hybrid Cisti the ovarium is frequently deformed, the ovules being in some cases quite absent, and in other cases incapable of fertilisation.

Several theories have been propounded to account for the origin of *C. adami*, and for the transformations which it undergoes. These transformations have been attributed by some authors to simple bud-variation; but considering the wide difference between *C. laburnum* and *purpureus*, both of which are natural species, and considering the sterility of the intermediate form, this view may be summarily rejected. We shall presently see that, with hybrid plants, two different embryos may be developed within the same seed and cohere; and it has been supposed that *C. adami* might have thus originated. It is known that when a plant with variegated leaves is budded on a plain stock, the latter is sometimes affected, and it is believed by some that the laburnum has been thus affected. Thus Mr. Purser states[92] that a common laburnum-tree in his garden, into which three *grafts* of the *Cytisus purpureus* had been inserted, gradually assumed the character of *C. adami*; but more evidence and copious details would be requisite to make so extraordinary a statement credible.

Many authors maintain that *C. adami* is a hybrid produced in the common way by seed, and that it has reverted by buds to its two parent-forms. Negative results are of little value; but Reisseck, Caspary, and I myself, tried in vain to cross *C. laburnum* and *purpureus*; when I fertilised the former with pollen of the latter, I had the nearest approach to success, for pods were formed, but in sixteen days after the withering of the flowers they fell off. Nevertheless, the belief that *C. adami* is a spontaneously produced hybrid between these two species is strongly supported by the fact that hybrids between these species and two others have spontaneously

---

90 'Nouvelles Archives du Muséum,' tom. i. p. 143.

91 *See* on this head, Naudin, idem, p. 141.

92 The statement is believed by Dr. Lindley in 'Gard. Chron.,' 1857, pp. 382, 400.

arisen. In a bed of seedlings from *C. elongatus*, which grew near to *C. purpureus*, and was probably fertilised by it, through the agency of insects (for these, as I know by experiment, play an important part in the fertilisation of the laburnum), the sterile hybrid *C. purpureo-elongatus* appeared.[93] Thus, also, Waterer's laburnum, the *C. alpino-laburnum*,[94] spontaneously appeared, as I am informed by Mr. Waterer, in a bed of seedlings.

On the other hand, we have a clear and distinct account given by M. Adam, who raised the plant, to Poiteau,[95] showing that *C. adami* is not an ordinary hybrid. M. Adam inserted in the usual manner a shield of the bark of *C. purpureus* into a stock of *C. laburnum*; and the bud lay dormant, as often happens, for a year; the shield then produced many buds and shoots, one of which grew more upright and vigorous with larger leaves than the shoots of *C. purpureus*, and was consequently propagated. Now it deserves especial notice that these plants were sold by M. Adam, as a variety of *C. purpureus*, before they had flowered; and the account was published by Poiteau after the plants had flowered, but before they had exhibited their remarkable tendency to revert into the two parent-species. So that there was no conceivable motive for falsification, and it is difficult to see how there could have been any error. If we admit as true M. Adam's account, we must admit the extraordinary fact that two distinct species can unite by their cellular tissue, and subsequently produce a plant bearing leaves and sterile flowers intermediate in character between the scion and stock, and producing buds liable to reversion; in short, resembling in every important respect a hybrid formed in the ordinary way by seminal reproduction. Such plants, if really thus formed, might be called graft-hybrids.

I will now give all the facts which I have been able to collect illustrative of the above theories, not for the sake of merely throwing light on the origin of *C. adami*, but to show in how many extraordinary and complex methods one kind of plant may affect another, generally in connection with bud-variation. The supposition that either *C. laburnum* or *purpureus* produced by ordinary bud-variation the intermediate and the other form, may, as already remarked, be absolutely excluded, from the want of any evidence, from the great amount of change thus implied,

---

93 Braun, in 'Bot. Mem. Ray Soc.,' 1853, p. xxiii.

94 This hybrid has never been described. It is exactly intermediate in foliage, time of flowering, dark striæ at the base of the standard petal, hairiness of the ovarium, and in almost every other character, between *C. laburnum* and *alpinus*; but it approaches the former species more nearly in colour, and exceeds it in the length of the racemes. We have before seen that 20·3 per cent. of its pollen-grains are ill-formed and worthless. My plant, though growing not above thirty or forty yards from both parent-species, during some seasons yielded no good seeds; but in 1866 it was unusually fertile, and its long racemes produced from one to occasionally even four pods. Many of the pods contained no good seeds, but generally they contained a single apparently good seed, sometimes two, and in one case three seeds. Some of the seeds germinated.

95 'Annales de la Soc. de Hort. de Paris,' tom. vii., 1830, p. 93.

and from the sterility of the intermediate form. Nevertheless such cases as nectarines suddenly appearing on peach-trees, occasionally with the fruit half-and-half in nature,—moss-roses appearing on other roses, with the flowers divided into halves, or striped with different colours,—and other such cases, are closely analogous in the result produced, though not in origin, with the case of *C. adami*.

A distinguished botanist, Mr. G. H. Thwaites,[96] has recorded a remarkable case of a seed from *Fuchsia coccinea* fertilised by *F. fulgens*, which contained two embryos, and was "a true vegetable twin." The two plants produced from the two embryos were "extremely different in appearance and character," though both resembled other hybrids of the same parentage produced at the same time. These twin plants "were "closely coherent, below the two pairs of cotyledon-leaves, into a single "cylindrical stem, so that they had subsequently the appearance of being "branches on one trunk." Had the two united stems grown up to their full height, instead of dying, a curiously mixed hybrid would have been produced; but even if some of the buds had subsequently reverted to both parent-forms, the case, although more complex, would not have been strictly analogous with that of *C. adami*. On the other hand, a mongrel melon described by Sageret[97] perhaps did thus originate; for the two main branches, which arose from two cotyledon-buds, produced very different fruit,—on the one branch like that of the paternal variety, and on the other branch to a certain extent like that of the maternal variety, the melon of China.

The famous *bizzarria Orange* offers a strictly parallel case to that of *Cytisus adami*. The gardener who in 1644 in Florence raised this tree, declared that it was a seedling which had been grafted; and after the graft had perished, the stock sprouted and produced the bizzarria. Gallesio, who carefully examined several living specimens and compared them with the description given by the original describer P. Nato,[98] states that the tree produces at the same time leaves, flowers, and fruit, identical with the bitter orange and with the citron of Florence, and likewise compound fruit with the two kinds either blended together, both externally and internally, or segregated in various ways. This tree can be propagated by cuttings, and retains its diversified character. The so-called trifacial orange of Alexandria and Smyrna[99] resembles in its general nature the bizzarria, but differs from it in the *sweet* orange and citron being blended together in the same fruit, and separately produced on the same tree: nothing is known of its origin. In regard to the bizzarria, many authors believe that it is a graft-hybrid; Gallesio on the other hand thinks that it is an ordinary hybrid, with the habit of partially reverting

---

[96] 'Annals and Mag. of Nat. Hist.,' March, 1848.

[97] 'Pomologie Physiolog.,' 1830, p. 126.

[98] Gallesio, 'Gli Agrumi dei Giard. Bot. Agrar. di Firenze,' 1839, p. 11. In his 'Traité du Citrus,' 1811, p. 146, he speaks as if the compound fruit consisted in part of lemons, but this apparently was a mistake.

[99] 'Gard. Chron.,' 1855, p. 628. *See* also Prof. Caspary, in 'Transact. Hort. Congress of Amsterdam,' 1865.

by buds to the two parent-forms; and we have seen in the last chapter that the species in this genus often cross spontaneously.

Here is another analogous, but doubtful case. A writer in the 'Gardener's Chronicle'[100] states that an *Æsculus rubicunda* in his garden yearly produced on one of its branches "spikes of pale yellow flowers, smaller in size and somewhat similar in colour to those of *Æ. flava.*" If as the editor believes *Æsculus rubicunda* is a hybrid descended on one side from *Æ. flava*, we have a case of partial reversion to one of the parent-forms. If, as some botanists maintain, *Æ. rubicunda* is not a hybrid, but a natural species, the case is one of simple bud-variation.

The following facts show that hybrids produced from seed in the ordinary way, certainly sometimes revert by buds to their parent-forms. Hybrids between *Tropæolum minus* and *majus*[101] at first produced flowers intermediate in size, colour, and structure between their two parents; but later in the season some of these plants produced flowers in all respects like those of the mother-form, mingled with flowers still retaining the usual intermediate condition. A hybrid Cereus between *C. speciosissimus* and *phyllanthus*,[102] plants which are widely different in appearance, produced for the first three years angular, five-sided stems, and then some flat stems like those of *C. phyllanthus*. Kölreuter also gives cases of hybrid Lobelias and Verbascums, which at first produced flowers of one colour, and later in the season flowers of a different colour.[103] Naudin[104] raised forty hybrids from *Datura lævis* fertilised by *D. stramonium*; and three of these hybrids produced many capsules, of which a half, or quarter, or lesser segment was smooth and of small size like the capsule of the pure *D. lævis*, the remaining part being spinose and of larger size like the capsule of the pure *D. stramonium*: from one of these composite capsules, plants were raised which perfectly resembled both parent-forms.

Turning now to varieties. A *seedling* apple, conjectured to be of crossed parentage, has been described in France,[105] which bears fruit, with one half larger than the other, of a red colour, acid taste, and peculiar odour; the other side being greenish-yellow and very sweet: it is said scarcely ever to include perfectly developed seed. I suppose that this is not the same tree with that which Gaudichaud[106] exhibited before the French Institute, bearing on the same branch two distinct kinds of apples, one a *reinette rouge*, and the other like a *reinette canada jaunâtre*: this double-bearing variety can be propagated by grafts, and continues to produce both kinds; its origin is unknown. The Rev. J. D. La Touche sent me a coloured drawing of an apple which he brought from Canada, of which half, surrounding and including the whole of the calyx and the insertion of the

---

[100] 'Gard. Chron.,' 1851, p. 406.

[101] Gärtner, 'Bastarderzeugung,' s. 549. It is, however, doubtful whether these plants should be ranked as species or varieties.

[102] Gärtner, idem, s. 550.

[103] 'Journal de Physique,' tom. xxiii., 1783, p. 100. 'Act. Acad. St. Petersburgh,' 1781, part i. p. 249.

[104] 'Nouvelles Archives du Muséum,' tom. i. p. 49.

[105] L'Hermès, Jan. 14, 1837, quoted in Loudon's 'Gard. Mag.,' vol. xiii. p. 230.

[106] 'Comptes Rendus,' tom. xxxiv., 1852, p. 746.

footstalk, is green, the other half being brown and of the nature of the *pomme gris* apple, with the line of separation between the two halves exactly defined. The tree was a grafted one, and Mr. La Touche thinks that the branch which bore this curious apple sprung from the point of junction of the graft and stock: had this fact been ascertained, the case would probably have come into the small class of graft-hybrids presently to be given. But the branch may have sprung from the stock, which no doubt was a seedling.

Prof. H. Lecoq, who has made a great number of crosses between the differently coloured varieties of *Mirabilis jalapa*,[107] finds that in the seedlings the colours rarely combine, but form distinct stripes; or half the flower is of one colour and half of a different colour. Some varieties regularly bear flowers striped with yellow, white, and red; but plants of such varieties occasionally produce on the same root branches with uniformly coloured flowers of all three tints, and other branches with half-and-half coloured flowers and others with marbled flowers. Gallesio[108] crossed reciprocally white and red carnations, and the seedlings were striped; but some of the striped plants also bore entirely white and entirely red flowers. Some of these plants produced one year red flowers alone, and in the following year striped flowers; or conversely, some plants, after having borne for two or three years striped flowers, would revert and bear exclusively red flowers. It may be worth mentioning that I fertilised the *Purple Sweet-pea* (*Lathyrus odoratus*) with pollen from the light-coloured *Painted Lady:* seedlings raised from one and the same pod were not intermediate in character, but perfectly resembled both parents. Later in the summer, the plants which had at first borne flowers identical with those of the *Painted Lady*, produced flowers streaked and blotched with purple; showing in these darker marks a tendency to reversion to the mother-variety. Andrew Knight[109] fertilised two white grapes with pollen of the Aleppo grape, which is darkly variegated both in its leaves and fruit. The result was that the young seedlings were not at first variegated, but all became variegated during the succeeding summer; besides this, many produced on the same plant bunches of grapes which were all black, or all white, or lead-coloured striped with white, or white dotted with minute black stripes; and grapes of all these shades could frequently be found on the same footstalk.

In most of these cases of crossed varieties, and in some of the cases of crossed species, the colours proper to both parents appeared in the seedlings, as soon as they first flowered, in the form of stripes or larger segments, or as whole flowers or fruit of two kinds borne on the same plant; and in this case the appearance of the two colours cannot strictly be said to be due to reversion, but to some incapacity of fusion, leading to their

107 'Géograph. Bot. de l'Europe,' tom. iii., 1854, p. 405; and 'De la Fécondation,' 1862, p. 302.

108 'Traité du Citrus,' 1811, p. 45.

109 'Transact. Linn. Soc.,' vol. ix. p. 268.

segregation. When, however, the later flowers or fruit, produced during the same season or during a succeeding year or generation, become striped or half-in-half, &c., the segregation of the two colours is strictly a case of reversion by bud-variation. In a future chapter I shall show that, with animals of crossed parentage, the same individual has been known to change its character during growth, and to revert to one of its parents which it did not at first resemble. From the various facts now given there can be no doubt that the same individual plant, whether a hybrid or a mongrel, sometimes returns in its leaves, flowers, and fruit, either wholly or by segments, to both parent-forms, in the same manner as the *Cytisus adami*, and the *Bizzarria Orange*.

We will now consider the few facts which have been recorded in support of the belief that a variety when grafted or budded on another variety sometimes affects the whole stock, or at the point of junction gives rise to a bud, or graft-hybrid, which partakes of the characters of both stock and scion.

It is notorious that when the variegated Jessamine is budded on the common kind, the stock sometimes produces buds bearing variegated leaves: Mr. Rivers, as he informs me, has seen instances of this. The same thing occurs with the Oleander.[110] Mr. Rivers, on the authority of a trustworthy friend, states that some buds of a golden-variegated ash, which were inserted into common ashes, all died except one; but the ash-stocks were affected,[111] and produced, both above and below the points of insertion of the plates of bark bearing the dead buds, shoots which bore variegated leaves. Mr. J. Anderson Henry has communicated to me a nearly similar case: Mr. Brown, of Perth, observed many years ago, in a Highland glen, an ash-tree with yellow leaves; and buds taken from this tree were inserted into common ashes, which in consequence were affected, and produced the *Blotched Breadalbane Ash*. This variety has been propagated, and has preserved its character during the last fifty years. Weeping ashes, also, were budded on the affected stocks, and became similarly variegated. Many authors consider variegation as the result of disease; and on this view, which however is doubtful, for some variegated plants are perfectly healthy and vigorous, the foregoing cases may be looked at as the direct result of the inoculation of a disease. Variegation is much influenced, as we shall hereafter see, by the nature of the soil in which the

---

[110] Gärtner ('Bastarderzeugung,' s. 611) gives many references on this subject.

[111] A nearly similar account was given by Bradley, in 1724, in his 'Treatise on Husbandry,' vol. i. p. 199.

plants are grown; and it does not seem improbable that whatever change in the sap or tissues certain soils induce, whether or not called a disease, might spread from the inserted piece of bark to the stock. But a change of this kind cannot be considered to be of the nature of a graft-hybrid.

There is a variety of the hazel with dark-purple leaves, like those of the copper-beech: no one has attributed this colour to disease, and it apparently is only an exaggeration of a tint which may often be seen on the leaves of the common hazel. When this variety is grafted on the common hazel,[112] it sometimes colours, as has been asserted, the leaves below the graft; but I should add that Mr. Rivers, who has possessed hundreds of such grafted trees, has never seen an instance.

Gärtner[113] quotes two separate accounts of branches of dark and white-fruited vines which had been united in various ways, such as being split longitudinally, and then joined, &c.; and these branches produced distinct bunches of grapes of the two colours, and other bunches with grapes either striped or of an intermediate and new tint. Even the leaves in one case were variegated. These facts are the more remarkable because Andrew Knight never succeeded in raising variegated grapes by fertilising white kinds by pollen of dark kinds; though, as we have seen, he obtained seedlings with variegated fruit and leaves, by fertilising a white variety by the variegated dark Aleppo grape. Gärtner attributes the above-quoted cases merely to bud-variation; but it is a strange coincidence that the branches which had been grafted in a peculiar manner should alone have thus varied; and H. Adorne de Tscharner positively asserts that he produced the described result more than once, and could do so at will, by splitting and uniting the branches in the manner described by him.

I should not have quoted the following case had not the author of 'Des Jacinthes'[114] impressed me with the belief not only of his extensive knowledge, but of his truthfulness: he says that bulbs of blue and red hyacinths may be cut in two, and that they will grow together and throw up a united stem (and this I have myself seen), with flowers of the two colours on the opposite sides. But the remarkable point is, that flowers are sometimes produced with the two colours blended together, which makes the case closely analogous with that of the blended colours of the grapes on the united vine-branches.

Mr. R. Trail stated in 1867, before the Botanical Society of Edinburgh (and has since given me fuller information), that several years ago he cut about sixty blue and white potatoes into halves through the eyes or buds, and then carefully joined them, destroying at the same time the other eyes. Some of these united tubers produced white, and others blue tubers; and it is probable that in these cases the one half alone of the bud grew. Some, however, produced tubers partly white and partly blue; and the tubers from about four or five were regularly mottled with the two colours. In these latter cases we may conclude that a stem had been formed by

---

[112] Loudon's 'Arboretum,' vol. iv. p. 2595.

[113] 'Bastarderzeugung,' s. 619.

[114] Amsterdam, 1768, p. 124.

the union of the bisected buds; and as tubers are produced by the enlargement of subterranean branches arising from the main stem, their mottled colour apparently affords clear evidence of the intimate commingling of the two varieties. I have repeated these experiments on the potato and on the hyacinth on a large scale, but with no success.

The most reliable instance known to me of the formation of a graft-hybrid is one, recorded by Mr. Poynter,[115] who assures me, in a letter of the entire accuracy of the statement. *Rosa Devoniensis* had been budded some years previously on a white Banksian rose; and from the much enlarged point of junction, whence the Devoniensis and Banksian still continued to grow, a third branch issued, which was neither pure Banksian nor pure Devoniensis, but partook of the character of both; the flowers resembled, but were superior in character to those of the variety called *Lamarque* (one of the Noisettes), while the shoots were similar in their manner of growth to those of the Banksian rose, with the exception that the longer and more robust shoots were furnished with prickles. This rose was exhibited before the Floral Committee of the Horticultural Society of London. Dr. Lindley examined it, and concluded that it had certainly been produced by the mingling of *R. Banksiæ* with some rose like *R. Devoniensis*, "for while it was very greatly increased in vigour and in the size of all the parts, the leaves were half-way between a Banksian and Tea-scented rose." It appears that rose-growers were aware that the Banksian rose sometimes affects other roses. Had it not been for this latter statement, it might have been suspected that this new variety was simply due to bud-variation, and that it had occurred by a mere accident at the point of junction between the two old kinds.

To sum up the foregoing facts: the statement that *Cytisus adami* originated as a graft-hybrid is so precise that it can hardly be rejected, and, as we have just seen, some analogous facts render the statement to a certain extent probable. The peculiar, monstrous condition of the ovules, and the apparently sound condition of the pollen, favour the belief that it is not an ordinary or seminal hybrid. On the other hand, the fact that the same two species, viz. *C. laburnum* and *purpureus*, have spontaneously produced hybrids by seed, is a strong argument in support of the belief that *C. adami* originated in a similar manner. With respect to the extraordinary tendency which this tree exhibits to complete or partial reversion, we have seen that undoubted seminal hybrids and mongrels are similarly liable. On the whole, I am inclined to put trust in M. Adam's statement; and if it should ever be proved true, the same view would probably have

115 'Gard. Chron.,' 1860, p. 672, with a woodcut.

to be extended to the Bizzarria and Trifacial oranges and to the apples above described; but more evidence is requisite before the possibility of the production of graft-hybrids can be fully admitted. Although it is at present impossible to arrive at any certain conclusion with respect to the origin of these remarkable trees, the various facts above given appear to me to deserve attention under several points of view, more especially as showing that the power of reversion is inherent in Buds.

On the direct or immediate action of the Male Element on the Mother Form.—Another remarkable class of facts must be here considered, because they have been supposed to account for some cases of bud-variation: I refer to the direct action of the male element, not in the ordinary way on the ovules, but on certain parts of the female plant, or in the case of animals on the subsequent progeny of the female by a second male. I may premise that with plants the ovarium and the coats of the ovules are obviously parts of the female, and it could not have been anticipated that they would be affected by the pollen of a foreign variety or species, although the development of the embryo, within the embryonic sack, within the ovule, within the ovarium, of course depends on the male element.

Even as long ago as 1729 it was observed[116] that white and blue varieties of the Pea, when planted near each other, mutually crossed, no doubt through the agency of bees, and in the autumn blue and white peas were found within the same pods. Wiegmann made an exactly similar observation in the present century. The same result has followed several times when a variety with peas of one colour has been artificially crossed by a differently-coloured variety.[117] These statements led Gärtner, who was highly sceptical on the subject, carefully to try a long series of experiments: he selected the most constant varieties, and the result conclusively showed that the colour of the skin of the pea is modified when pollen of a differently coloured variety is used. This conclusion has since been confirmed by experiments made by the Rev. J. M. Berkeley.[118]

Mr. Laxton of Stamford, whilst making experiments on peas for the express purpose of ascertaining the influence of foreign pollen on the mother-plant, has recently[119] observed an important additional fact. He fertilised the Tall Sugar pea, which bears very thin green pods, becoming brownish-

[116] 'Philosophical Transact.,' vol. xliii., 1744-45, p. 525.

[117] Mr. Swayne, in 'Transact. Hort. Soc.,' vol. v. p. 234; and Gärtner, 'Bastarderzeugung,' 1849, s. 81 and 499.

[118] 'Gard. Chron.,' 1854, p. 404.

[119] Ibid., 1866, p. 900.

white when dry, with pollen of the Purple-podded pea, which, as its name expresses, has dark-purple pods with very thick skin, becoming pale reddish-purple when dry. Mr. Laxton has cultivated the tall sugar-pea during twenty years, and has never seen or heard of it producing a purple pod; nevertheless, a flower fertilised by pollen from the purple-pod yielded a pod clouded with purplish-red, which Mr. Laxton kindly gave to me. A space of about two inches in length towards the extremity of the pod, and a smaller space near the stalk, were thus coloured. On comparing the colour with that of the purple-pod, both pods having been first dried and then soaked in water, it was found to be identically the same; and in both the colour was confined to the cells lying immediately beneath the outer skin of the pod. The valves of the crossed pod were also decidedly thicker and stronger than those of the pods of the mother-plant, but this may have been an accidental circumstance, for I know not how far their thickness in the Tall Sugar-pea is a variable character.

The peas of the Tall Sugar-pea, when dry, are pale greenish-brown, thickly covered with dots of dark purple so minute as to be visible only through a lens, and Mr. Laxton has never seen or heard of this variety producing a purple pea; but in the crossed pod one of the peas was of a uniform beautiful violet-purple tint, and a second was irregularly clouded with pale purple. The colour lies in the outer of the two coats which surround the pea. As the peas of the purple-podded variety when dry are of a pale greenish-buff, it would at first appear that this remarkable change of colour in the peas in the crossed pod could not have been caused by the direct action of the pollen of the purple-pod: but when we bear in mind that this latter variety has purple flowers, purple marks on its stipules, and purple pods; and that the Tall sugar-pea likewise has purple flowers and stipules, and microscopically minute purple dots on the peas, we can hardly doubt that the tendency to the production of purple in both parents has in combination modified the colour of the peas in the crossed pod. After having examined these specimens, I crossed the same two varieties, and the peas in one pod, but not the pods themselves, were clouded and tinted with purplish-red in a much more conspicuous manner than the peas in the uncrossed pods produced at the same time by the same plants. I may notice as a caution that Mr. Laxton sent me various other crossed peas slightly, or even greatly, modified in colour; but the change in these cases was due, as had been suspected by Mr. Laxton, to the altered colour of the cotyledons, seen through the transparent coats of the peas; and as the cotyledons are parts of the embryo, these cases are not in any way remarkable.

Turning now to the genus Matthiola. The pollen of one kind of stock sometimes affects the colour of the seeds of another kind, used as the mother-plant. I give the following case the more readily, as Gärtner doubted similar statements with respect to the stock previously made by other observers. A well-known horticulturist, Major Trevor Clarke, informs me[120] that the seeds of the large red-flowered *biennial* stock

[120] *See* also a paper by this observer, read before the International Hort. and Bot. Congress of London, 1866.

(*Matthiola annua; Cocardeau* of the French) are light brown, and those of the purple branching Queen stock (*M. incana*) are violet-black; and he found that, when flowers of the red stock were fertilised by pollen from the purple stock, they yielded about fifty per cent. of *black* seeds. He sent me four pods from a red-flowered plant, two of which had been fertilised by their own pollen, and they included pale brown seed; and two which had been crossed by pollen from the purple kind, and they included seeds all deeply tinged with black. These latter seeds yielded purple-flowered plants like their father; whilst the pale brown seeds yielded normal red-flowered plants; and Major Clarke, by sowing similar seeds, has observed on a greater scale the same result. The evidence in this case of the direct action of the pollen of one species on the colour of the seeds of another species appears to me conclusive.

In the foregoing cases, with the exception of that of the purple-podded pea, the coats of the seeds alone have been affected in colour. We shall now see that the ovarium itself, whether forming a large fleshy fruit or a mere thin envelope, may be modified by foreign pollen, in colour, flavour, texture, size, and shape.

The most remarkable instance, because carefully recorded by highly competent authorities, is one of which I have seen an account in a letter written, in 1867, by M. Naudin to Dr. Hooker. M. Naudin states that he has seen fruit growing on *Chamærops humilis*, which had been fertilised by M. Denis with pollen from the Phœnix or date-palm. The fruit or drupe thus produced was twice as large as, and more elongated than, that proper to the Chamærops; so that it was intermediate in these respects, as well as in texture, between the fruit of the two parents. These hybridised seeds germinated, and produced young plants likewise intermediate in character. This case is the more remarkable as the Chamærops and Phœnix belong not only to distinct genera, but in the estimation of some botanists to distinct sections of the family.

Gallesio[121] fertilised the flowers of an orange with pollen from the lemon; and one fruit thus produced bore a longitudinal stripe of peel having the colour, flavour, and other characters of the lemon. Mr. Anderson[122] fertilised a green-fleshed melon with pollen from a scarlet-fleshed kind; in two of the fruits "a sensible change was perceptible; and four other fruits were somewhat altered both internally and externally." The seeds of the two first-mentioned fruits produced plants partaking of the good properties of both parents. In the United States, where Cucurbitaceæ are largely cultivated, it is the popular belief[123] that the fruit is thus directly affected by foreign pollen; and I have received a similar statement with respect to

---

[121] 'Traité du Citrus,' p. 40.

[122] 'Transact. Hort. Soc.,' vol. iv. p. 318. *See* also vol. v. p. 65.

[123] Prof. Asa Gray, 'Proc. Acad. Sc.,' Boston, vol. iv., 1860, p. 21.

the cucumber in England. It is known that grapes have been thus affected in colour, size, and shape: in France a pale-coloured grape had its juice tinted by the pollen of the dark-coloured Teinturier; in Germany a variety bore berries which were affected by the pollen of two adjoining kinds; some of the berries being only partially affected or mottled.[124] As long ago as 1751[125] it was observed that, when differently coloured varieties of maize grow near each other, they mutually affect each other's seeds, and this is now a popular belief in the United States. Dr. Savi[126] tried the experiment with care: he sowed yellow and black-seeded maize together, and on the same ear some of the seeds were yellow, some black, and some mottled,[127] the differently coloured seeds being arranged in rows or irregularly. Mr. Sabine states[128] that he has seen the form of the nearly globular seed-capsule of *Amaryllis vittata* altered by the application of the pollen of another species, of which the capsule has gibbous angles. Mr. J. Anderson Henry[129] crossed *Rhododendron Dalhousiæ* with the pollen of *R. Nuttallii*, which is one of the largest-flowered and noblest species of the genus. The largest pod produced by the former species, when fertilised with its own pollen, measured 1⅜ inch in length and 1½ in girth; whilst three of the pods which had been fertilised by pollen of *R. Nuttallii* measured 1⅝ inch in length and no less than 2 inches in girth. Here we see the effect of foreign pollen apparently confined to increasing the size of the ovarium; but we must be cautious in assuming, as the following case shows, that in this instance size has been directly transferred from the male parent to the capsule of the female plant. Mr. Henry fertilised *Arabis blepharophylla* with pollen of *A. Soyeri*, and the pods thus produced, of which he was so kind as to send me detailed measurements and sketches, were much larger in all their dimensions than those naturally produced by either the male or female parent-species. In a future chapter we shall see

---

[124] For the French case, *see* 'Proc. Hort. Soc.,' vol. i. new series, 1866, p. 50. For Germany, *see* M. Jack, quoted in Henfrey's 'Botanical Gazette,' vol. i. p. 277. A case in England has recently been alluded to by the Rev. J. M. Berkeley before the Hort. Soc. of London.'

[125] 'Philosophical Transactions,' vol. xlvii., 1751-52, p. 206.

[126] Gallesio, 'Teoria della Riproduzione,' 1816, p. 95.

[127] It may be worth while to call attention to the several means by which flowers and fruit become striped or mottled. Firstly, by the direct action of the pollen of another variety or species, as with the above-given cases of oranges and maize. Secondly, in crosses of the first generation, when the colours of the two parents do not readily unite, as in the cases of Mirabilis and Dianthus given a few pages back. Thirdly, in crossed plants of a subsequent generation, by reversion, through either bud or seminal generation. Fourthly, by reversion to a character not originally gained by a cross, but which had long been lost, as with white-flowered varieties, which we shall hereafter see often become striped with some other colour. Lastly, there are cases, as when peaches are produced with a half or quarter of the fruit like a nectarine, in which the change is apparently due to mere variation, through either bud or seminal generation.

[128] 'Transact. Hort. Soc.,' vol. v. p. 69.

[129] 'Journal of Horticulture,' Jan. 20, 1863, p. 46.

that the organs of vegetation in hybrid plants, independently of the character of either parent, are sometimes developed to a monstrous size; and the increased size of the pods in the foregoing cases may be an analogous fact.

No case of the direct action of the pollen of one variety on another is better authenticated or more remarkable than that of the common apple. The fruit here consists of the lower part of the calyx and of the upper part of the flower-peduncle[130] in a metamorphosed condition, so that the effect of the foreign pollen has extended even beyond the limits of the ovarium. Cases of apples thus affected were recorded by Bradley in the early part of the last century; and other cases are given in old volumes of the Philosophical Transactions;[131] in one of these a Russeting apple and an adjoining kind mutually affected each other's fruit; and in another case a smooth apple affected a rough-coated kind. Another instance has been given[132] of two very different apple-trees growing close to each other, which bore fruit resembling each other, but only on the adjoining branches. It is, however, almost superfluous to adduce these or other cases, after that of the St. Valery apple, which, from the abortion of the stamens, does not produce pollen, but, being annually fertilised by the girls of the neighbourhood with pollen of many kinds, bears fruit, "differing from each other in size, flavour, and colour, but resembling in character the hermaphrodite kinds by which they have been fertilised."[133]

I have now shown, on the authority of several excellent observers, in the case of plants belonging to widely different orders, that the pollen of one species or variety, when applied to a distinct form, occasionally causes the coats of the seeds and the ovarium or fruit, including even in one instance the calyx and upper part of the peduncle of the mother-plant, to become modified. Sometimes the whole of the ovarium or all the seeds are thus affected; sometimes only a certain number of the seeds, as in the case of the pea, or only a part of the ovarium, as with the striped orange, mottled grapes and maize, are thus affected. It must not be supposed that any direct or immediate effect invariably follows the use of foreign pollen: this is far from being the case; nor is it known on what conditions the result depends. Mr. Knight[134] expressly states that he has never seen

[130] *See* on this head the high authority of Prof. Decaisne, in a paper translated in 'Proc. Hort. Soc.,' vol. i. new series, 1866, p. 48.

[131] Vol. xliii., 1744-45, p. 525; vol. xlv., 1747-48, p. 602.

[132] 'Transact. Hort. Soc.,' vol. v. pp. 63 and 68. Puvis also has collected ('De la Dégénération,' 1837, p. 36) several other instances; but it is not in all cases possible to distinguish between the direct action of foreign pollen and bud-variations.

[133] T. de Clermont-Tonnerre, in 'Mém. de la Soc. Linn. de Paris,' tom. iii., 1825, p. 164.

[134] 'Transact. of Hort. Soc.,' vol. v. p. 68.

the fruit thus affected, though he has crossed thousands of apple and other fruit-trees. There is not the least reason to believe that a branch which has borne seed or fruit directly modified by foreign pollen is itself affected, so as subsequently to produce modified buds: such an occurrence, from the temporary connection of the flower with the stem, would be hardly possible. Hence but very few, if any, of the cases of sudden modifications in the fruit of trees, given in the early part of this chapter, can be accounted for by the action of foreign pollen; for such modified fruits have commonly been afterwards propagated by budding or grafting. It is also obvious that changes of colour in the flower which necessarily supervene long before it is ready for fertilisation, and changes in the shape or colour of the leaves, can have no relation to the action of foreign pollen: all such cases must be attributed to simple bud-variation.

The proofs of the action of foreign pollen on the mother-plant have been given in considerable detail, because this action, as we shall see in a future chapter, is of the highest theoretical importance, and because it is in itself a remarkable and apparently anomalous circumstance. That it is remarkable under a physiological point of view is clear, for the male element not only affects, in accordance with its proper function, the germ, but the surrounding tissues of the mother-plant. That the action is anomalous in appearance is true, but hardly so in reality, for apparently it plays the same part in the ordinary fertilisation of many flowers. Gärtner has shown,[135] by gradually increasing the number of pollen-grains until he succeeded in fertilising a Malva, that many grains are expended in the development, or, as he expresses it, in the satiation, of the pistil and ovarium. Again, when one plant is fertilised by a widely distinct species, it often happens that the ovarium is fully and quickly developed without any seeds being formed, or the coats of the seeds are developed without an embryo being formed within. Dr. Hildebrand also has lately shown in a valuable paper[136] that, with several Orchideæ, the action of the plant's own

[135] 'Beitrage zur Kenntniss der Befruchtung,' 1844, s. 347-351.

[136] 'Die Fruchtbildung der Orchideen, ein Beweis für die doppelte Wirkung des Pollen,' Botanische Zeitung, No. 44 et seq., Oct. 30, 1863; and 1865, s. 249.

pollen is necessary for the development of the ovarium, and that this development takes place not only long before the pollen-tubes have reached the ovules, but even before the placentæ and ovules have been formed; so that with these orchids the pollen apparently acts directly on the ovarium. On the other hand, we must not overrate the efficacy of pollen in this respect; for in the case of hybridised plants it might be argued that an embryo had been formed and had affected the surrounding tissues of the mother-plant before it perished at a very early age. Again, it is well known that with many plants the ovarium may be fully developed, though pollen be wholly excluded. And lastly, Mr. Smith, the late Curator at Kew (as I hear through Dr. Hooker), observed the singular fact with an orchid, the *Bonatea speciosa*, the development of the ovarium could be effected by mechanical irritation of the stigma. Nevertheless, from the number of the pollen-grains expended "in the satiation of the ovarium and pistil,"—from the generality of the formation of the ovarium and seed-coats in sterile hybridised plants,—and from Dr. Hildebrand's observations on orchids, we may admit that in most cases the swelling of the ovarium, and the formation of the seed-coats, are at least aided, if not wholly caused, by the direct action of the pollen, independently of the intervention of the fertilised germ. Therefore, in the previously-given cases we have only to add to our belief in the power of the plant's own pollen on the development of the ovarium and seed-coats, its further power, when applied to a distinct species or variety, of influencing the shape, size, colour, texture, &c., of these same parts.

Turning now to the animal kingdom. If we could imagine the same flower to yield seeds during successive years, then it would not be very surprising that a flower of which the ovarium had been modified by foreign pollen should next year produce, when self-fertilised, offspring modified by the previous male influence. Closely analogous cases have actually occurred with animals. In the case often quoted from Lord Morton,[137] a nearly purely-bred, Arabian, chesnut mare bore a hybrid to a quagga; she was subsequently sent to Sir Gore Ouseley, and produced

[137] 'Philos. Transact.,' 1821, p. 20.

two colts by a black Arabian horse. These colts were partially dun-coloured, and were striped on the legs more plainly than the real hybrid, or even than the quagga. One of the two colts had its neck and some other parts of its body plainly marked with stripes. Stripes on the body, not to mention those on the legs, and the dun-colour, are extremely rare,—I speak after having long attended to the subject,—with horses of all kinds in Europe, and are unknown in the case of Arabians. But what makes the case still more striking is that the hair of the mane in these colts resembled that of the quagga, being short, stiff, and upright. Hence there can be no doubt that the quagga affected the character of the offspring subsequently begot by the black Arabian horse. With respect to the varieties of our domesticated animals, many similar and well-authenticated facts have been published,[138] and others have been communicated to me, plainly showing the influence of the first male on the progeny subsequently borne by the mother to other males. It will suffice to give a single instance, recorded in the 'Philosophical Transactions,' in a paper following that by Lord Morton: Mr. Giles put a sow of Lord Western's black and white Essex breed to a wild boar of a deep chesnut colour; and the "pigs produced partook in appearance of both boar and sow, but in some the chesnut colour of the boar strongly prevailed." After the boar had long been dead, the sow was put to a boar of her own black and white breed,—a kind which is well known to breed very true and never to show any chesnut colour,—yet from this union the sow produced some young pigs which were plainly marked with the same chesnut tint as in the first litter. Similar cases have so frequently occurred, that careful breeders avoid putting a choice female to an inferior male on account of the injury to her subsequent progeny which may be expected to follow.

[138] Dr. Alex. Harvey on 'A remarkable Effect of Cross-breeding,' 1851. On the 'Physiology of Breeding,' by Mr. Reginald Orton, 1855. 'Intermarriage,' by Alex. Walker, 1837. 'L'Hérédité Naturelle,' by Dr. Prosper Lucas, tom. ii. p. 58. Mr. W. Sedgwick in 'British and Foreign Medico-Chirurgical Review,' 1863, July, p. 183. Bronn, in his 'Geschichte der Natur,' 1843, B. ii. s. 127, has collected several cases with respect to mares, sows, and dogs. Mr. W. C. L. Martin ('History of the Dog,' 1845, p. 104) says he can personally vouch for the influence of the male parent of the first litter on the subsequent litters by other fathers. A French poet, Jacques Savary, who wrote in 1665 on dogs, was aware of this singular fact.

Some physiologists have attempted to account for these remarkable results from a first impregnation by the close attachment and freely intercommunicating blood-vessels between the modified embryo and the mother. But it is a most improbable hypothesis that the mere blood of one individual should affect the reproductive organs of another individual in such a manner as to modify the subsequent offspring. The analogy from the direct action of foreign pollen on the ovarium and seed-coats of the mother-plant strongly supports the belief that the male element acts directly on the reproductive organs of the female, wonderful as is this action, and not through the intervention of the crossed embryo. With birds there is no such close connection between the embryo and mother as in the case of mammals: yet a careful observer, Dr. Chapuis, states[139] that with pigeons the influence of a first male sometimes makes itself perceived in the succeeding broods; but this statement, before it can be fully trusted, requires confirmation.

Conclusion and Summary of the Chapter.—The facts given in the latter half of this chapter are well worthy of consideration, as they show us in how many extraordinary modes one organic form may lead to the modification of another, and often without the intervention of seminal reproduction. There is ample evidence, as we have just seen, that the male element may either directly affect the structure of the female, or in the case of animals lead to the modification of her offspring. There is a considerable but insufficient body of evidence showing that the tissues of two plants may unite and form a bud having a blended character; or again, that buds inserted into a stock may affect all the buds subsequently produced by this stock. Two embryos, differing from each other and contained in the same seed, may cohere and form a single plant. Offspring from a cross between two species or varieties may in the first or in a succeeding generation revert in various degrees by bud-variation to their parent-forms; and this reversion or segregation of character may affect the whole flower, fruit, or leaf-bud, or only the half or smaller segment, or a single organ. In some cases this segregation of character apparently depends on some

[139] 'Le Pigeon Voyageur Belge,' 1865, p. 59.

incapacity of union rather than on reversion, for the flowers or fruit which are first produced display by segments the characters of both parents. In the *Cytisus adami* and the Bizzarria orange, whatever their origin may have been, the two parent species occur blended together under the form of a sterile hybrid, or reappear with their characters perfect and their reproductive organs effective; and these trees, retaining the same sportive character, can be propagated by buds. These various facts ought to be well considered by any one who wishes to embrace under a single point of view the various modes of reproduction by gemmation, division, and sexual union, the reparation of lost parts, variation, inheritance, reversion, and other such phenomena. In a chapter towards the close of the following volume I shall attempt to connect these facts together by a provisional hypothesis.

In the early half of this chapter I have given a long list of plants in which through bud-variation, that is, independently of reproduction by seed, the fruit has suddenly become modified in size, colour, flavour, hairiness, shape, and time of maturity; flowers have similarly changed in shape, colour, and doubleness, and greatly in the character of the calyx; young branches or shoots have changed in colour, in bearing spines, and in habit of growth, as in climbing and weeping; leaves have changed in colour, variegation, shape, period of unfolding, and in their arrangement on the axis. Buds of all kinds, whether produced on ordinary branches or on subterranean stems, whether simple or, as in tubers and bulbs, much modified and supplied with a stock of nutriment, are all liable to sudden variations of the same general nature.

In the list, many of the cases are certainly due to reversion to characters not acquired from a cross, but which were formerly present, and have been lost for a longer or shorter period of time;—as when a bud on a variegated plant produces plain leaves, or when variously-coloured flowers on the Chrysanthemum revert to the aboriginal yellow tint. Many other cases included in the list are probably due to the plants being of crossed parentage, and to the buds reverting to one of the two parent-forms. In illustration of the origin of *Cytisus adami*, several cases were given of partial or complete reversion, both with hybrid and mongrel plants; hence we may suspect that the strong tendency in the Chrysanthemum, for instance, to produce by bud-variation differently-coloured flowers, results from the varieties formerly having been intentionally or accidentally crossed; and that their descendants at the present day still occasionally revert by buds to the colours of the more persistent parent-varieties. This is almost certainly the case with Rollisson's Unique Pelargonium; and so it may be to a large extent with the bud-varieties of the Dahlia and with the "broken colours" of Tulips.

Many cases of bud-variation, however, cannot be attributed to reversion, but to spontaneous variability, such as so commonly occurs with cultivated plants when raised from seed. As a single variety of the Chrysanthemum has produced by buds six other varieties, and as one variety of the gooseberry has borne at the same time four distinct varieties of fruit, it is scarcely possible to believe that all these variations are reversions to former parents. We can hardly believe, as remarked in a previous chapter, that all the many peaches which have yielded nectarine-buds are of crossed parentage. Lastly, in such cases as that of the moss-rose with its peculiar calyx, and of the rose which bears opposite leaves, in that of the Imatophyllum, &c., there is no known natural species or seedling variety, from which the characters in question could have been derived by crossing. We must attribute all such cases to actual variability in the buds. The varieties which have thus arisen cannot be distinguished by any external character from seedlings; this is notoriously the case with the varieties of the Rose, Azalea, and many other plants. It deserves notice that all the plants which have yielded bud-variations have likewise varied greatly by seed.

These plants belong to so many orders that we may infer that almost every plant would be liable to bud-variation if placed under the proper exciting conditions. These conditions, as far as we can judge, mainly depend on long-continued and high cultivation; for almost all the plants in the foregoing lists are perennials, and have been largely propagated in many soils and under different climates, by cuttings, offsets, bulbs, tubers, and especially by budding or grafting. The instances of annuals varying by buds, or producing on the same plant differently coloured flowers, are comparatively rare: Hopkirk[140] has seen this with *Convolvulus tricolor;* and it is not rare with the Balsam and annual Delphinium. According to Sir R. Schomburgk, plants from the warmer temperate regions, when cultivated under the hot climate of St. Domingo, are eminently liable to bud-variation; but change of climate is by no means a necessary contingent, as we see with the gooseberry, currant, and some others. Plants living under their natural conditions are very rarely subject to bud-variation: variegated and coloured leaves have, however, been occasionally observed; and I have given an instance of the variation of buds on an ash-tree; but it is doubtful whether any tree planted in ornamental grounds can be considered as living under strictly natural conditions. Gärtner has seen white and dark-red flowers produced from the same root of the wild *Achillea millefolium;* and Prof. Caspary has seen *Viola lutea*, in a completely wild condition, bearing flowers of different colours and sizes.[141]

As wild plants are so rarely liable to bud-variation, whilst highly cultivated plants long propagated by artificial means have yielded by this form of reproduction many varieties, we are led through a series such as the following,—namely, all the eyes in the same tuber of the potato varying in the same manner,—all the fruit on a purple plum-tree suddenly becoming yellow,—all the fruit on a double-flowered almond suddenly becoming peach-like,—all the buds on grafted trees being in some very slight degree affected by the stock on which they have been worked,—all the flowers on a transplanted heartsease changing for a time in colour, size, and shape,—we are led through such facts to look at every case of bud-variation as the direct result of the particular conditions of life to which the plant has been exposed. But if we turn to the other end of the series, namely, to such cases as that of a peach-tree which, after having been cultivated by tens of thousands during many years in many countries, and after having annually produced thousands of buds, all of which have apparently been exposed to precisely the same conditions, yet at last suddenly produces a single bud with its whole character greatly transformed, we are driven to an opposite

140 'Flora Anomala,' p. 164.

141 'Schriften der Phys.-Ökon. Gesell. zu Königsberg,' Band vi., Feb. 3, 1865, s. 4.

conclusion. In such cases as the latter it would appear that the transformation stands in no *direct* relation to the conditions of life.

We have seen that varieties produced from seeds and from buds resemble each other so closely in general appearance, that they cannot possibly be distinguished. Just as certain species and groups of species, when propagated by seed, are more variable than other species or genera, so it is in the case of certain bud-varieties. Thus the Queen of England Chrysanthemum has produced by this latter process no less than six, and Rollisson's Unique Pelargonium four distinct varieties; moss-roses have also produced several other moss-roses. The Rosaceæ have varied by buds more than any other group of plants; but this may be in large part due to so many members having been long cultivated; but within this one group, the peach has often varied by buds, whilst the apple and pear, both grafted trees extensively cultivated, have afforded, as far as I can ascertain, extremely few instances of bud-variation.

The law of analogous variation holds good with varieties produced by buds, as with those produced from seed: more than one kind of rose has sported into a moss-rose; more than one kind of camellia has assumed an hexagonal form; and at least seven or eight varieties of the peach have produced nectarines.

The laws of inheritance seem to be nearly the same with seminal and bud-varieties. We know how commonly reversion comes into play with both, and it may affect the whole, or only segments, of a leaf, flower, or fruit. When the tendency to reversion affects many buds on the same tree, it becomes covered with different kinds of leaves, flowers, or fruit; but there is reason to believe that such fluctuating varieties have generally arisen from seed. It is well known that, out of a number of seedling varieties, some transmit their character much more truly by seed than others; so with bud-varieties some retain their character by successive buds more truly than others; of which instances have been given with two kinds of variegated Euonymus and with certain kinds of tulips. Notwithstanding the sudden production of bud-varieties, the characters thus acquired are sometimes capable of transmission by seminal reproduction: Mr. Rivers has found that moss-roses generally reproduce themselves by seed; and the mossy character has been transferred by crossing, from one species of rose to another. The Boston nectarine, which appeared as a bud-variation, produced by seed a closely allied nectarine. We have however seen, on the authority of Mr. Salter, that seed taken from a branch with leaves variegated through bud-variation, transmits this character very feebly; whilst many plants, which became variegated as seedlings, transmit variegation to a large proportion of their progeny.

Although I have been able to collect a good many cases of bud-variation, as shown in the previous lists, and might probably, by searching foreign horticultural works, have collected more cases, yet their total number is as nothing in comparison with that of seminal varieties. With seedlings raised from the more variable cultivated plants, the variations are almost infinitely numerous, but their differences are generally slight: only at long intervals of time a strongly marked modification appears. On the other hand, it is a singular and inexplicable fact that, when plants vary by buds, the variations, though they occur with comparative rarity, are often, or even generally, strongly pronounced. It struck me that this might perhaps be a delusion, and that slight changes often occurred in buds, but from being of no value were overlooked or not recorded. Accordingly I applied to two great authorities on this subject, namely, to Mr. Rivers with respect to fruit-trees, and to Mr. Salter with respect to flowers. Mr. Rivers is doubtful, but does not remember having noticed very slight variations in fruit-buds. Mr. Salter informs me that with flowers such do occur, but, if propagated, they generally lose their new character in the following year; yet he concurs with me that bud-variations usually at once assume a decided and permanent character. We can hardly doubt that this is the rule, when we reflect on such cases as that of the peach, which has been so carefully observed and of which such trifling seminal varieties have been propagated, yet this tree has repeatedly produced by bud-variation nectarines, and only twice (as far as I can learn) any other variety, namely, the Early and Late Grosse Mignonne peaches; and these differ from the parent-tree in hardly any character except the period of maturity.

To my surprise I hear from Mr. Salter that he brings the great principle of selection to bear on variegated plants propagated by buds, and has thus greatly improved and fixed several varieties. He informs me that at first a branch often produces variegated leaves on one side alone, and that the leaves are marked only with an irregular edging or with a few lines of white and yellow. To improve and fix such varieties, he finds it necessary to encourage the buds at the bases of the most distinctly marked leaves, and to propagate from them alone. By following with perseverance this plan during three or four successive seasons, a distinct and fixed variety can generally be secured.

Finally, the facts given in this chapter prove in how close and remarkable a manner the germ of a fertilised seed and the small cellular mass forming a bud resemble each other in function,—in their powers of inheritance with occasional reversion,—and in their capacity for variation of the same general nature, in obedience to the same laws. This resemblance, or rather identity, is rendered far more striking if the facts can be trusted which apparently render it probable that the cellular tissue of one species or variety, when budded or grafted on another, may give rise to a bud having an intermediate character. In this chapter we clearly see that variability is not necessarily contingent on sexual generation, though much more frequently its concomitant than on bud-reproduction. We see that bud-variability is not solely dependent on reversion or atavism to long-lost characters, or to those formerly acquired from a cross, but that it is often spontaneous. But when we ask ourselves what is the cause of any particular bud-variation, we are lost in doubt, being driven in some cases to look to the direct action of the external conditions of life as sufficient, and in other cases to feel a profound conviction that these have played a quite subordinate part, of not more importance than the nature of the spark which ignites a mass of combustible matter.

END OF VOL. I.

Printed in the United States
By Bookmasters